WOLF-RAYET AND HIGH-TEMPERATURE STARS

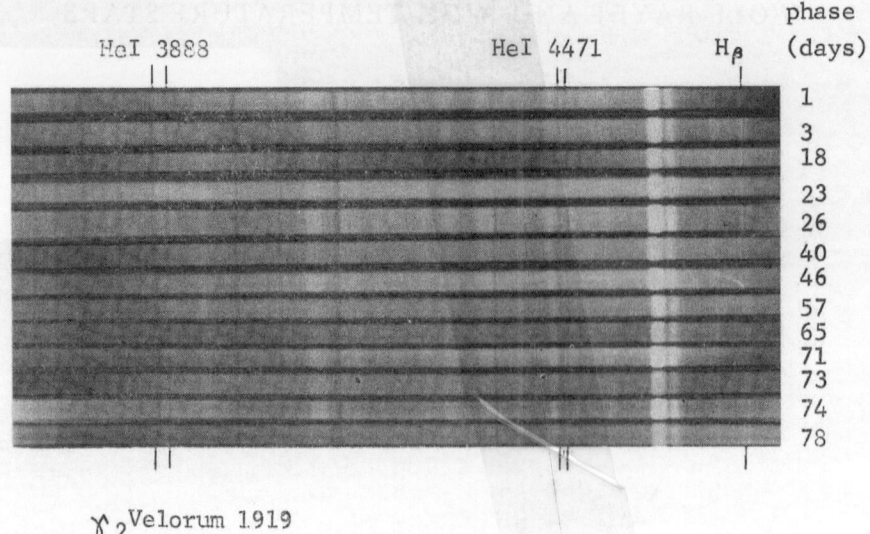

Some spectra of γ_2 Velorum taken by Perrine in 1919, showing the violet-shifted, variable absorption edge of the He I 3888 line.

INTERNATIONAL ASTRONOMICAL UNION
UNION ASTRONOMIQUE INTERNATIONALE

SYMPOSIUM No. 49

HELD IN BUENOS AIRES, ARGENTINA, AUGUST 9–14, 1971

WOLF-RAYET AND HIGH-TEMPERATURE STARS

EDITED BY

M. K. V. BAPPU

Indian Institute of Astrophysics, Kodaikanal, India

AND

J. SAHADE

Instituto de Astronomía y Física del Espacio, Buenos Aires, Argentina

D. REIDEL PUBLISHING COMPANY

DORDRECHT-HOLLAND / BOSTON-U.S.A.

1973

Published on behalf of
the International Astronomical Union
by
D. Reidel Publishing Company, Dordrecht, Holland

Solid and distributed in the U.S.A., Canada, and Mexico
by D. Reidel Publishing Company, Inc.
306 Dartmouth Street, Boston,
Mass. 02116, U.S.A.

Library of Congress Catalog Card Number 72–87470

ISBN-13:978-90-277-0361-3   e-ISBN-13:978-94-010-2511-9
DOI: 10.1007/978-94-010-2511-9

This volume is dedicated to

C. S. BEALS

B. EDLEN

CECILIA PAYNE-GAPOSCHKIN

P. SWINGS

For their contributions to our present degree of comprehension of
Wolf-Rayet Spectra

PREFACE

We have in this volume, compiled a connected account of the proceedings of the Symposium on Wolf-Rayet and High-Temperature Stars held at Buenos Aires. The Organizing Committee had assigned broad areas of topical interest to be reviewed by invited speakers. Each of these presentations was followed by lengthy discussions that were tape recorded and transcribed later. These discussions have been edited only to a limited extent. We have shortened them and rearranged them to bring about a greater coherence. We have, however, attempted to retain the tenor of the discussions, the flavour of impromptu remarks and the continuity of an argument. Much of the success of such a venture depends on the contributors to the discussions. To be able to make these thoughts available to a larger audience has been the task of those responsible for the elaborate tape recording of the proceedings. We thank those at the Instituto de Astronomía y Física del Espacio for the efficient way in which this responsibility has been discharged. Many at Buenos Aires and Kodaikanal have contributed efficient assistance to the preparation of this volume and we are deeply indebted for their help. In particular, two amongst these, Nora Martinez and A. M. Batcha have contributed overwhelmingly both to the organization of the symposium and the final preparation of the symposium volume.

Financial support for this symposium came from the International Astronomical Union and the Argentine National Research Council. The Faculty of Exact and Natural Sciences of the University of Buenos Aires also sponsored the Symposium. We are grateful to all these for the support they have provided us.

M. K. V. BAPPU

J. SAHADE

INTRODUCTION

It is with a feeling of very great pleasure that I, on behalf of the Organizing Committee, welcome all of you to the Symposium on Wolf-Rayet stars and high temperature
stars. Sponsored jointly by the International Astronomical Union and our local hosts
the National Research Council of Argentina, this occasion is the forty-ninth in the
series of an effort by the IAU in accordance with its role of fostering progress in
astronomical research. These symposia have aided active workers in a field to critic-
ally assess on such occasions, the current status of achievement in order to best orient
future efforts towards a maximum return. The measure of success has been varied in
degree, as one would naturally expect in the diversity of topics covered to date. But
if success as a parameter is measured by the yardstick of stimulus to many an individual, with no regard to national boundaries whatsoever, truly these occasions have
justified the faith placed in them by those who have conceived them.

We thank our hosts for their very gracious invitation to have this Symposium here
in Buenos Aires. It is never an easy task to examine the myriad of details that have to
be ensured for the practice of hospitality to have a successful impact on guests of
such diverse origins, tastes and requirements. In the few hours that we have been here
we have already begun to feel the results of their efforts in this direction. We are
confident that indeed our stay here will be memorable, invigorating and informative.
We are particularly grateful to Prof. Sahade and his team of collaborators who have
spent so much time and energy in looking into the various requirements that undoubtedly will ensure a successful symposium.

It would not be out of place for me to remark on the aptness of holding such a
meeting at such a southern location. The brightest objects of the species which forms
the principal theme of our discussions are in the Southern Hemisphere. Astronomy
and astronomical research activity in South America, by a series of fortunate circum-
stances, is experiencing an expansion at a rate and magnitude that has never been
witnessed anywhere before. Undoubtedly, many of the future developments in our
area of interest will be the result of efforts on these southern objects. Where, obviously,
should one generate that spark of enthusiasm, except where such a result is most
likely to originate?

In planning the details of this Symposium the Organizing Committee considered
the time opportune for taking a comprehensive stock of these objects and the limitations under which we operate currently. The Wolf-Rayet object is essentially one that
displays a phenomenon, when at a particular stage a distinct atmospheric condition
prevails that comes about for different objects with varying chemical compositions
from different causes along the diverse evolutionary paths. In talking of the Wolf-
Rayet stars our speakers will introduce the points of similarity as well as minor

discordances between objects as the WR stars and planetary nuclei, the Of stars and others that display characteristics that have a common factor. The dichotomy of spectral behaviour is one wherein we have as yet no clue as to the nature of the cause. The obvious non-equilibrium configuration of the atmosphere necessitates consideration of the mechanisms of excitation of the various levels and possible stratification effects that prevail. And unless we are clear in our minds about the details of the physical conditions prevalent in the atmosphere, we can hardly speculate on the causes which may be the origin of such behaviour.

To my mind, therefore, the target for this week of deliberation is to examine firstly our achievements in observation and inference in detail, with its limitations in precision and capability of evaluation, followed by detailed consideration of how we can fill in the lacunae in our information and ability to build up a picture of what constitutes a Wolf-Rayet star.

A little over a hundred years ago, Wolf and Rayet detected the spectacular appearance of the spectra of these objects located in Cygnus. A short interval later, at total eclipses of the Sun, spectroscopic detection of the solar prominences, solar chromosphere and the solar corona followed in rapid sequence. Four decades ago Meg Nad Saha speculated on the nature of the ultraviolet spectrum we would see if only we could by a new technology open a window of research in the electromagnetic spectrum, hitherto inaccessible. Less than a decade ago, we detected in the far ultraviolet spectra of early type supergiants violet-displaced absorption edges to emission features, of a magnitude that we had been accustomed to believe from the visual spectrum to exist only in a Wolf-Rayet star. These are the unifying factors, a common characteristic in extended atmospheres that we pick out in the Sun and the stars, massive, young and old. The problem of the Wolf-Rayet phenomenon is the problem of an extended atmosphere with its diverse sources of radiative and mechanical energy and kinematical and thermal characteristics. Treated thus with the magnitudes of the different features as variants, we have more than a ray of hope towards a successful solution. With this in mind I shall now request Dr. Thomas to commence our deliberations with some general comments on the problems of extended atmospheres.

M. K. V. BAPPU

Indian Institute of Astrophysics,
Kodaikanal, India

TABLE OF CONTENTS

LIST OF PARTICIPANTS

Albano, J., Instituto de Astronomía y Física del Espacio, Buenos Aires, Argentina.

Alcaino, G., Santiago, Chile.

Altizer, Robert J., Corralitos Observatory, New Mexico, U.S.A.

Azcarate, I., Instituto de Astronomía y Física del Espacio, Buenos Aires, Argentina.

Bappu, M. K. V., Indian Institute of Astrophysics, Kodaikanal-3, India.

Brandi, E., Observatorio Astronómico, Universidad Nacional de La Plata, Buenos Aires, Argentina.

Conti, P. S., Joint Institute for Laboratory Astrophysics, Colorado, U.S.A.

De Groot, M., European Southern Observatory, Santiago, Chile.

Duro, J. C., Instituto de Astronomía y Física del Espacio, Buenos Aires, Argentina.

Feinstein, A., Observatorio Astronómico, Universidad Nacional de La Plata, Buenos Aires, Argentina.

Frank, J., Instituto de Astronomía y Física del Espacio, Buenos Aires, Argentina.

Frank, M. C., Instituto de Astronomía y Física del Espacio, Buenos Aires, Argentina.

Gamba, Z., Departamento de Física, Universidad de Buenos Aires, Buenos Aires, Argentina.

Gerola, H., Departamento de Física, Universidad de Buenos Aires, Buenos Aires, Argentina.

Ghielmetti, H., Instituto de Astronomía y Física del Espacio, Buenos Aires, Argentina.

Gomez, A., Observatorio Astronómico, Universidad Nacional de La Plata, Buenos Aires, Argentina.

Goniadski, D., Departamento de Física, Universidad de Buenos Aires, Buenos Aires, Argentina.

Havlen, R. J., European Southern Observatory, Santiago, Chile.

Hernandez, A. M., Instituto de Astronomía y Física del Espacio, Buenos Aires, Argentina.

Hernandez, C. A., Observatorio Astronómico, Universidad Nacional de La Plata, Buenos Aires, Argentina.

Iglesias, E., Departamento de Física, Universidad de Buenos Aires, Buenos Aires, Argentina.

Johnson, H. M., Lockheed Missiles and Space Co., California, U.S.A.

Kuhi, L. V., Berkeley Astronomy Department, University of California, California, U.S.A.

Levato, H., Observatorio Astronómico, Universidad Nacional de La Plata, Buenos Aires, Argentina.

Lopez, L., Observatorio Astronómico, Universidad Nacional de La Plata, Buenos Aires, Argentina.

Lopez Garcia, F., Observatorio Astronómico, Universidad Nacional de La Plata, Buenos Aires, Argentina.

Lopez Garcia, Z., Observatorio Astronómico, Universidad Nacional de La Plata, Buenos Aires, Argentina.

Machado, M., Observatorio de Física Cósmica, Buenos Aires, Argentina.

Malaroda, S., Observatorio Astrónomico, Universidad Nacional de La Plata, Buenos Aires, Argentina.

Marraco, H., Observatorio Astronómico, Universidad Nacional de La Plata, Buenos Aires, Argentina.

Mendez, R., Instituto de Astronomía y Física del Espacio, Buenos Aires, Argentina.

Morton, D. C., Princeton University Observatory, New Jersey, U.S.A.

Muzzio, J. C., Observatorio Astronómico, Universidad Nacional de La Plata, Buenos Aires, Argentina.

Niemela, V., Instituto de Astronomía y Física del Espacio, Buenos Aires, Argentina.

Paczyński, B., Astronomical Observatory of Warsaw University, Warsaw, Poland.

Sahade, J., Instituto de Astronomía y Física del Espacio, Buenos Aires, Argentina.

Seggewiss, W., Universitäts-Sternwarte Bonn, 5568/Daun/Eifel, West Germany.

Smith, L. F., Université de Liége, Belgique.

Terlevich, R., Instituto de Astronomía y Física del Espacio, Buenos Aires, Argentina.

Thomas, R. N., Joint Institute for Laboratory Astrophysics, Colorado, U.S.A.

Underhill, A. B., Laboratory for Optical Astronomy, Goddard Space Flight Center, Maryland, U.S.A.

Van Blerkom, D., Department of Physics, University of Massachusetts, U.S.A.

Walborn, N. R., David Dunlap Observatory, Ontario, Canada.

Westerlund, B. E., European Southern Observatory, Santiago, Chile.

Wood, H. J., European Southern Observatory, Santiago, Chile.

SECTION I

CHAIRMAN: M. K. V. BAPPU

THE WOLF-RAYET STARS – THE GENERAL PROBLEMS
OF EXTENDED ATMOSPHERES
AND NON-CLASSICAL ATMOSPHERIC MODELS

RICHARD N. THOMAS

Joint Institute for Laboratory Astrophysics, Boulder, Colo., U.S.A.

1. Introduction

The Organizers of this Symposium asked that I make some introductory remarks on the general problems of extended atmospheres, the degree of common behaviour among stars having such a feature, and how the Wolf-Rayet stars may contribute to the progress of our inference in this area. In a Symposium entitled *Wolf-Rayet and High Temperature Stars*, not 'Stars With Extended Atmospheres', the implication of your charge to me is clear. You are implicitly assuming that the Wolf-Rayet phenomenon is primarily one of extended atmospheres. I can indeed give you a succinct summary of my outlook on these three points under such an implicit assumption, elaborating in more detail only to the extent as is necessary in order to be specific. And, as I am sure you had planned, because my outlook is considerably contrary to this implicit assumption, at least my remarks will spark controversy from the outset of this Symposium. I would like to emphasize that it is the summary of outlook to which I attach most weight and which I think is correct. The more specific details may change as we develop the mosaic of the summary, which has two main points. First, when you ask the physical picture of the general structure of a stellar atmosphere, you will eventually reach the conclusion that part of this general structure is an extended atmosphere for *all* stars. A focus upon some class, or classes, of them as exhibiting 'extended atmospheres' simply reflects the degree to which particular observations focus upon a particular part of the atmosphere as a function of the particular characteristics of the considered star. Second, in this context, I think that the implication that the Wolf-Rayet phenomenon is primarily one of extended atmospheres is incorrect. You are confusing a single system with a more deep-seated disease. My emphasis will lie on the disease as a whole, and the Wolf-Rayet stars as a guide to its diagnostics. I want to emphasize that this outlook on the general model of an atmosphere has developed jointly with K. B. Gebbie, J-C. Pecker, and F. Praderie; so most of what I say simply reflects this joint work.

First, I think it clear that the primary problem of extended atmospheres lies in the definition and understanding of what one means by an extended atmosphere, and how one interprets the observations which one thinks imply its existence. 'Extended' means relative to something, and of course that something is the classical model of a stellar atmosphere. When one interprets the observations, one again does so relative to the predictions of the classical model, to decide that it is an anomalous geometrical

M. K. V. Bappu and J. Sahade (eds.), Wolf-Rayet and High-Temperature Stars, 3–12. All Rights Reserved.

extent rather than an anomalous something else, which one needs to understand these observations. Now, the essential physical characteristic of the classical model lies in its representation of the atmosphere simply as the outer layer of the internal structure of the star. It is described by the same parameters as is the interior, and these parameters satisfy the same interrelations as in the interior. The earliest representation of the classical atmospheric model was that of a thin surface layer, homogeneous in the boundary values of these parameters. Later models permitted gradients in them, but the notion of a thin, surface layer, whose chief function is to give boundary values for the parameters describing the interior, persisted. An 'extended' atmosphere simply meant that observed atmospheric phenomena could not be represented by such a thin surface layer. At least, they could not be so represented if one retained the description of the atmosphere by the parameters, and inter relations among them, of the interior. Now, my own outlook rests on a preference for simply speaking of non-classical atmospheric models, dropping completely the notion of the atmosphere as a surface layer that is necessarily described by the parameters and interrelations of the interior. Rather, I suggest that we should view the atmosphere as a transition region, or a boundary region rather than layer, between the stellar interior and the rest of the Universe or, in practice, the interstellar medium. The parameters, and relations between them, required to describe such a transition or boundary region can differ very considerably from those characterizing either the interior of the star or the regions completely exterior to it. Different kinds of observations of a star may relate to quite different parts – or possibly different aspects of a given part – of this transition layer. For a particular star, the relation between type of observations and part of atmosphere may depend upon gross properties of the star – mass and chemical composition – and upon transfer characteristics of mass, momentum, and energy in this boundary-transition region, and upon the way in which these are described. In this sense, we would regard stars with unusual properties attributed to extended atmospheres simply as stars whose particular properties focused the observations on regions considerably removed from the thin, deepest-observed, layer usually discussed. But we must also be very careful that such unusual observational properties can indeed be uniquely attributed to atmospheric extent.

Secondly, I think we should rephrase the question asking degree of common behaviour of stars having extended atmospheres, if we adopt this view of the atmosphere as a transition region. Placing a star in the category of having an extended atmosphere corresponds to recognizing that the star has certain properties – either intrinsic or because of favourable geometric location – that permit us to observe parts of the general atmospheric structure that are generally not observed. But while such parts may be 'extended' relative to 'normally' observed regions, they may not be the same parts, for all stars. Nor may the reason why they are observed always be the same. Consequently, this question of possible common behavior embodies several questions. First, we ask what is an operationally-useful set of subdivisions of the atmosphere, viewed as a transition region. Second, we ask what is the variation in observing conditions, observational techniques, or peculiarities of the star that provide information on particular subdivisions. In particular, under what conditions will we observe 'extended' regions. Thirdly, we ask whether we can establish general categories of stars according to the relative importance of particular atmospheric subregions in interpreting their spectra. To answer these questions, we require specific details of the general atmospheric model, and an understanding of what physical and observational effects correspond to these atmospheric subdivisions. We proceed to consider these specific points in Section 2.

Thirdly, when we ask how the Wolf-Rayet stars may contribute to the progress of our inferences in this area, I interpret area to mean 'non-classical atmospheric models' rather than 'extended atmospheres'. To explain my assertion that the Wolf-Rayet phenomenon is not primarily one of extended atmospheres, and to suggest how study of Wolf-Rayet stars may extend our understanding, I return to a suggestion I made some years ago. One should use observations of the Sun and of Wolf-Rayet stars as a guide to developing general models of stellar atmospheres, because they represent extreme examples in the then-vague beginnings of attempts to develop non-classical atmospheric models. At that time, the idea of a non-classical model was intuitively associated with the idea of a mechanical energy flux producing a stellar chromosphere. The Sun represented an object with a chromosphere so small in its 'obvious' effect on the spectrum that it might be undetectable were the Sun not so close to us that the wealth of different kinds of observations permitted us to develop an understanding of just what observations were chromospheric indicators. We were able to distinguish between physical anomaly and observational uncertainty by requiring consistency between a large number and a large variety of observations. The Wolf-Rayet stars represented objects in which the chromospheric phenomenon was so well-developed that even at their large distances and with rudimentary theory the effects could not be confused. Or, so I claimed at that time. My suggestion then was that the solar atmosphere reflected almost wholly only the energy dissipation from a set of aerodynamic motions, while the Wolf-Rayet star reflected both energy dissipation *and* a strong momentum supply from such aerodynamic motions. Consequently, one should carefully study the similarities and differences in these two kinds of stars, as a guide to the general kinds of effects to be associated with such aerodynamic motions in stellar atmospheres. I particularly stress this suggestion here, in the light of the implicit charge that the Wolf-Rayet phenomenon be viewed as primarily one of extended atmospheres. For I remind you that the solar chromosphere phenomenon was long regarded as primarily one of an extended atmosphere, induced by a system of 'turbulence', whose only effect was to distend the atmosphere in the same way as would an enhanced thermal velocity, but which was not allowed to change either kinetic temperature or internal excitation state of the atoms. And this suggestion of mine rested on the then-current evolution in our thinking on the solar chromosphere *away* from regarding it as primarily reflecting an extended atmosphere. The chromosphere phenomenon was being re-interpreted as primarily reflecting a mechanical energy flux, which did indeed distend the atmosphere but also changed the excitation very considerably. Thus many of the observational anomalies were reconciled in terms of those excitation effects accompanying a mechanical energy dissipation. In the years since, our understanding of the solar chromosphere phenomenon has further evolved, so that we now understand it as the consortium of a number of effects. These are best described as those effects associated with the atmosphere viewed as a boundary-transition layer. So, in Section 2, I summarize this outlook and this model. Here, I would only again emphasize my belief in the utility of the Wolf-Rayet studies as a probe of the utility and sufficiency of this kind of nonclassical model, especially as a cooperative probe blended with solar studies, and with intermediate classes of stars once we identify them.

2. General Structure of a Stellar Atmosphere

I stress that this general model reflects the viewpoint that a stellar atmosphere is the transition region between the stellar interior and the interstellar medium. Therefore, it must be described in terms of those concepts, parameters, and relations we would apply to the free-boundary regions of a quasi-isolated gaseous ensemble, not to interior regions. Mainly, the difference lies in the quasi-homogenous conditions of the interior as opposed to the anisotropic, inhomogeneous situation characterizing a boundary-transition region. As a particularly-fitting example, recall the evolution of the boundary layer as developed in fluid mechanics, whose introduction changed enormously the understanding of the interaction between fluid flows and the solid boundaries containing the flow. The change was significant when one considered only the momentum interchange associated with drag problems. It became very much more so when one considered the energy and mass interchange associated with heating and ablation problems. Similarly, the simple radiative transfer problems in stellar atmospheres require careful examination in the boundary-transition regions as collisional control of source-functions gives way to radiative, because of the fall in particle concentration. But the mass, momentum, and energy transfer associated with various kinds of atmospheric instabilities in the transition regions assume even more importance in these transition regions, and their effect on the interpretation of observations.

I have stressed the example of the solar chromosphere phenomenon as illustrating an evolution from a focus on it as a wholly extended atmosphere effect to a more general one. A more contemporary description of the anomaly would focus on the symbiosis exhibited by the chromosphere – the simultaneous presence of effects which, under the classical atmospheric model, would signal the presence of several different kinds of atmospheres. The most timely example at the present Symposium would be Mrs. Gaposchkin's remark that the solar rocket spectrum would be classified WC6, lacking any other information. So we have the symbiosis of a G0 and WC6 star. The presence of the He lines, especially in the eclipse spectrum, suggest another stellar class. And the self-reversed Ca II and Mg II lines, together with the residual intensities of the Balmer lines, all suggest, on the classical model, a puzzling super-position of several atmospheres, ranging from 'cold' to 'hot' stars. So I suggest the primary starting point in analysing peculiar stellar spectra of the type we are here concerned with is the star as 'symbiotic' rather than as 'extended atmosphere'. Now, Katharine Geb-bie and I have recently presented (1971a) an extensive description of the evolution of the solar chromosphere anomaly, blending it with ideas on symbiotic stars, to reach the suggested general atmospheric structure. Pecker and I are trying to resolve some of the problems of infrared excesses in terms of choices of the source of the required additional energy; interior or contraction. And Françoise Praderie and I have been trying to develop a simple approach to obtaining the main features of the electron-temperature behaviour in such an atmosphere as a function of differing opacity sour-ces. And we are all trying to put all this together into a more coherent specific picture. So there is no point to duplicating in detail here what you will see elsewhere. But it is useful to summarize two things: the physical basis of the general model in terms of atmospheric subdivisions, and the pattern of these subdivisions.

2.1. Physical Basis for System of Atmospheric Subdivision

So long as all microscopic processes fixing populations of energy levels are either dominated by collisions, or dominated by radiative processes in detailed balance, and only radiative transport processes exist, the description by the classical atmosphere model suffices. So a deviation from that description begins either when unbalanced radiative processes become competitive with collisions, or when non-radiative transfer processes become important, or both. The former can occur independently of the existence of the latter, but not conversely; so it is convenient to consider the sequence of events as one moves outward in an atmosphere as though the *population* effects begin before *transfer* effects, although, they may in fact begin simultaneously.

By *population* effects, we mean a change in microscopic distribution functions accompanying a change in domination of distribution function from one microscopic process to another unaccompanied by any change in energy supply. Such occurs, for example, when radiative processes become comparable to collisional because of a decrease in particle concentration. It may also occur when one radiative process be-comes relatively more important because it departs from the condition of detailed balance associated with greater optical depth. In terms of the continuous spectrum, such population effects underlie the change from control of electron temperature by total energy density of radiation to control by its spectral distribution. It may further be affected by a change in the relative influence of different continua on this mean photoionization energy, induced because an increased unbalance in radiation processes in either continua or lines changes the populations of the absorbing levels of these continua. All these effects can produce an outward rise in T_e that is not caused by a change in radiation field or an addition of a mechanical energy supply. Thus the symbiotic effect of an outward increase in excitation level, for properly sensitive lines, may occur. Independently, this population effect can also produce the symbiotic appearance of high and low-excitation spectra in the same atmospheric region if, for example, the source-function for one line remains collision-controlled while that for another becomes photo-ionization-controlled. Clearly, what we call symbiotic affects can also, erroneously, be interpreted as 'extended atmosphere' effects if we infer the

existence of these from abnormally-low rates of outward decrease in excitation. This
was a false path in early solar studies.

By *transfer* effects, we mean changes in microscopic distribution functions, both
from their LTE form *and* from their classical outward direction of decrease, associated
with the onset of non-radiative energy, momentum, and mass transfer. With these
effects are associated all the usual questions of mechanical heating, an aerodynamical
momentum supply, mass ejection, etc. Because the *absorption* of radiation, not the
supply of radiation, *decreases* outward with decreasing particle concentration, while
mechanical effects arising from any kind of perturbation *increase* outward in amplitude, the likelihood is great that somewhere in this transition region representation of
the atmosphere, transfer effects set in. Once initiated, they introduce an importance
to upper atmospheric layers which the classical model did not permit. They can
enhance the symbiotic effects associated with population effects above the level permitted by the radiation field alone. They increase the range of possible kinds of observations, thus of the regions of this transition layer atmosphere that can be studied. In
this sense, they increase the probability that all stars can be regarded as having extended atmospheres, by increasing the range of atmospheric region that can be studied.
But it is not a priori clear that all the associated phenomena are best interpreted as
primarily 'extended atmospheric' effects. Such, for example, might be the temptation
if one ignored excitation effects and attributed emission lines always to the volumetric
effects of a greater emitting disk in the lines than in the continuum. He might infer an
erroneous mechanism of line formation, and an erroneous location and extent of the
region of origin of such lines.

Finally, I would stress that the increased importance of such mechanical effects, and
the consequent increased discrepancy between radiative and mechanical excitation
sources, increases the possibility of atmospheric instabilities. In turn, these instabilities
induce the onset of the horizontal inhomogeneous structure which becomes of increasing concern with the increased resolution of our observations.

Thus, we have population effects, transfer effects, and instabilities associated with
the presence of transfer effects as a guide to setting up atmospheric subdivisions within
this general picture of the atmosphere as a transition region.

2.2. Pattern of the System of Atmospheric Subdivisions

We define the bottom of the atmosphere as the deepest layer from which we receive
direct radiation in the most transparent part of the spectrum. Then, we ask whether,
at this bottom level, collisions dominate all distribution functions, and there is only
radiative energy transfer. Or, possibly this last is supplemented by a quasi-static convection. If so, we model this deepest atmospheric layer by the classical atmosphere,
and call the layer 'the classical photosphere'. Whether this condition holds cannot be
decided *a priori*, but only by computation of microscopic rate processes, using the
actual kinetic temperature and particle concentrations at this level, and by comparing
the empirical T_e – distribution with the theoretical. Explicit criteria can be established
for the existence of such a classical layer; in essence the region is restricted to densities

(gravities) exceeding a certain value, and T_e lying below a certain value. Gebbie and Thomas (1971b) give illustrative formulae for the case of hydrogen opacity.

Once we pass from such a classical photospheric region, we enter one where population effects must be considered: deepest in the atmosphere, in the continuum; higher in both lines and continuum. If *only* population effects occur, we call this the non-classical photospheric region. The most notable observational effect should be on the T_e value which, apart from line-blanketing effects, should differ in gradient from the classical one. Initially, it may show a steeper decrease; eventually, a slower, than the classical one (again, cf. Gebbie and Thomas (1971a), for illustration). The interplay between line and continuum effects, as a function of kind of line and continuum and their coupling when line and continuous opacity arise from the same ion, is yet in the development stage of understanding. If this region persists sufficiently high in the atmosphere that it encompasses the region of formation of some strong lines, the symbiotic effect reflecting the presence of lines of differing classes of source-functions may occur, and provide a very valuable diagnostic tool. (For an exposition of this diagnostics, under the term 'New Spectroscopy', cf. Thomas, 1965.)

When transfer effects occur, we enter what we call the 'outer atmosphere'. It is not necessarily an extended atmosphere in the conventional sense. When only energy transfer effects occur, we call the region a 'chromosphere'. The solar chromosphere is the prototype stellar chromosphere. If we note that according to the most recent thinking the chromosphere-corona transition occurs only some 1500 km above the level τ_5 (tangential) $= 1$, which itself is only some 500 km above the level τ_5 (radial) $= 1$, with a solar radius of 700000 km, we recognize that the solar chromosphere is hardly an 'extended atmosphere' in the usual sense. Diagnostic approaches stress the rise in excitation above that permitted by population effects alone.

When momentum transfer effects, in addition to energy transfer effects, occur, we call the region a 'corona'. In this sense, it is not clear that the solar corona satisfies everywhere this definition. Also in this coronal region, the likelihood is strong that instability associated with increased ratio of mechanical to radiative effects will occur, and, consequently, an inhomogeneous horizontal structure of the atmosphere. Diagnostic approaches to the region stress the difference in density gradient from that permitted by energy transfer alone. The horizontal inhomogeneities are also of importance, but these are only strongly favoured, not unique, to corona over chromosphere (and even photosphere, classical or non-classical).

When mass transfer effects, as well as momentum and energy, occur, we call the region an 'exosphere'. This term was originally introduced to describe the very outer region of the terrestrial atmosphere where mass exhaustion, under free-particle orbits, occurs. It seems a good term to borrow, pending a better suggestion. I would expect the outermost parts of the exosphere to become confused with the interstellar medium, even in the absence of expanding shells like the H II regions. In their presence, the confusion would become more so. And clearly, such confusion is desirable, in terms of our regarding the atmosphere as the transition region to the interstellar medium; *and the similar confusion at the base of the atmosphere, where the classical photo-*

sphere represents indeed the transition region from the interior and becomes confused with it.

3. Empirical Investigations Under the Suggested General Model for Atmospheric Structure

Note that the general model suggested in Section 2 is not obviously *a priori* required for all stellar atmospheres, in all its sub-divisions. The necessity for a non-classical as well as classical photosphere does seem to have been established by the analyses of the last 20 years, both from the standpoint of theoretical consistency and empirical investigations. Indeed, Mihalas (1970) refers to the new classical atmosphere; viz., the old one without the LTE assumption. The existence of chromospheres in many stars seems also well-established observationally. For any stellar types where some kind of aerodynamical instability seems required, the occurrence of a chromospheric region would also seem required theoretically. Examples are stars having hydrogen convection zones, rotating stars, pulsating stars, late-type stars where the interior convective zone extends into the atmosphere. In my own opinion, we will ultimately discover causes of mechanical instability in all stellar atmospheres, which is the reason I assert the universality of the chromosphere region. The question of whether coronas are universal is a more uncertain one; for momentum transport varies as v^2 while energy transport varies as v^3, making the former harder than the latter. I consider this problem open for investigation. Finally, the exosphere universality likewise remains open for investigation. The basic reason underlying Parker's suggestion of the stellar wind – boundary conditions a long distance from the star – would seem to suggest its universality. Whether its generation can be satisfied by a chromosphere, without corona, also remains open for question.

On this basis, I would suggest the following scheme of empirical investigations to map out the properties of stellar atmospheres. First, we should delineate those regions of the HR diagram where classical photospheres can exist. We have a guide from the considerations already suggested; minimal gravities and maximum T_e. Mapping classical versus nonclassical photospheres is essentially a problem of T_e-distributions compared with LTE predictions, plus those symbiotic effects associated with population effects only. But the delineation of the atmospheric regions successfully described by the non-classical photosphere represents the second goal. The third one lies in the mapping of stellar chromospheres. There is to be a Symposium, sponsored by IAU Commission 36 and hosted jointly by NASA Goddard and the Smithsonian Astrophysical Observatory, on this subject next February; so there is little point trying to anticipate its considerations here, except as they deal specifically with Wolf-Rayet and high temperature stars. But here, it seems to me that our largest problem lies in the next subdivision of the atmosphere, separating chromospheric, coronal, and exospheric effects.

I would suggest that the focal point of considerations on Wolf-Rayet and hot stars lies there; viz., in casting light on the question of when momentum and mass transfer effects become observationally significant. And here, I would agree to some extent with the direction of prejudice reflected in the emphasis on extended atmospheres in considering Wolf-Rayet and high temperature stars. A primary consideration in making even classical models of such stars lies in the question of the stability of atmospheric structure when the high temperature forces a consideration of radiation pressure, and so a considerable reduction in the effective gravity from the dynamical one. Not only do we have a greatly increased scale-height, on a dogmatic hydrostatic equilibrium approach; but we have the strong possibility of the initiation of instabilities, on a more general aerodynamic approach. I would only insist that one be very careful to take into account chromospheric effects in the analysis of observations, before restricting attention to coronal and exospheric ones.

In establishing the existence of coronas and exospheres, we are basically concerned with an empirical density distribution compared with a hydrostatic one under the T_e consistent with chromospheric effects. Therefore, the first step in the diagnostic approach is to establish the magnitude of chromospheric effects, and the resulting thermal structure. We essentially accomplish this by comparing the excitation level of the spectrum relative to that inferred from the continuous spectrum associated with the deepest observable layers. One approach is simply to compare line and continuous spectra. Another is to compare continuous emission at wavelengths corresponding to differing opacities. Clearly, interferometric studies of the type being performed by Hanbury Brown (1968, 1970) are a most valuable approach to this problem. Another approach is to mimic that found so valuable in solar studies: eclipse observations. If it is possible to establish gradients in excitation as a function of atmospheric height, we can duplicate the solar studies. Some indications exist that this might be possible in extended cool systems like 31 Cyg and ζ Aur. Kuhi's work on WR binaries is thus far inconclusive, or negative, in this respect.

Given some knowledge of chromospheric effects, the next step is to establish a measure of atmospheric density gradient. Again, eclipse studies are a direct approach. The proto-measures of Mrs Shapley and Kopal (1946) remain unique in this approach. The indications from them were of a distinctly coronal, or exospheric, component of the atmosphere, unless you will accept a kinetic temperature of some 10^7 K. But it is unsatisfactory to base much on one set of observations. Another approach would be a study of the spectrum for distinctly 'extended atmosphere' effects on the spectrum. The varieties of 'dilution' effects are examples. Others might be identified. Hopefully, this Symposium will produce suggestions along these lines. So, let us proceed to them.

References

Gebbie, K. B. and Thomas, R. N.: 1971a, in K. B. Gebbie (ed.), *Menzel Symposium on Solar Physics, Atomic Spectra, and Gaseous Nebulae*, National Bureau of Standards Special Publication 353, Washington, p. 84.

Gebbie, K. B. and Thomas, R. N.: 1971b, *Astrophys. J.* **168**, 461

Hanbury Brown, R.: 1968, in K. B. Gebbie and R. N. Thomas (eds.), *Wolf-Rayet Stars*, National Bureau of Standards Special Publication 307, Washington, p. 79.

Hanbury Brown, R., Davis, J., Herbison-Evans, D., and Allen, L. R.: 1970, *Monthly Notices Roy. Astron. Soc.* **148**, 103.

Kopal, Z. and Shapley, M.: 1946, *Astrophys. J.* **104**, 160.

Kuhi, L.: 1968, in K. B. Gebbie and R. N. Thomas (eds.), *Wolf-Rayet Stars*, National Bureau of Standards Special Publication 307, Washington, p. 103.

Mihalas, D.: 1970, *Stellar Atmospheres*, Freeman Publ., San Francisco.

Thomas, R. N.: 1965, *Non-Equilibrium Thermodynamics in the Presence of a Radiation Field*, Univ. of Colorado Press, Boulder, Colo.

DISCUSSION

Underhill: I would like to make just one comment. I go along pretty well with the final division that Dr Thomas has got to. This is indeed what we have to sort out. However, I think he is a little bit harsh on all the observing astronomers who have attempted to lay the ground-work for the study of Wolf-Rayet stars. He has renamed things and somewhat scathingly indicated that the word extended atmospheres was really a misnomer and he has called these atmospheres non-classical. I think that this just indicates that Thomas is a theoretician and the rest of us are observers and the history of astronomy, if you try to deduce it from the papers in the literature, is a study in misunderstandings between the theoreticians and the observers. The theoreticians start by being very simple minded and they necessarily remain simple minded because they cannot put mathematics and all the needed things into what the observers want. It is too complicated. The observers are generally very greatly impressed by the complexity of details they observe. They are usually rather inarticulate and they try to simplify these things a little bit for the benefit of their brothers doing theory and they describe it by talking about extended atmospheres. But all the time they know that things are not simple.

Van Blerkom: Are you suggesting that every one of those regions you have described is present to some extent in a Wolf-Rayet star?

Thomas: I want to be very specific in answering to Anne Underhill's question about each of the effects that we must introduce over the non-classical model and each of the atmospheric subdivisions to which they lead. It seems to me possible that in all stars you have to introduce each one of these effects. First, the population effects. Then, the three types of transfer effects. And those three types of transfer effects also give the possibility of an inhomogeneous structure of the outer part of the star. The chance that all stars show these, seems very high, but I agree this is simply prejudice; what I can actually prove is something else. Now of all the stars that we see, the Wolf-Rayet stars seem to me the most extreme and so the chance that they exhibit all of these effects seems to me almost inescapable. Whether all stars show all features remains to be seen. I hope that a focal point of the Symposium will be whether the WR stars can be shown to exhibit all these effects and all these atmospheric subdivisions. In answer to Anne Underhill, I would say that the structure is a logical and explicit one. If it is what you have always meant by 'extended atmosphere' but never bothered to spell out for some reason, I am delighted to cheerfully admit I am only an explicit recorder of your implicit thoughts.

Bappu: Would you also accept the inhomogeneity of distribution over the surface?

Thomas: Very much so. A lack of explicit emphasis on inhomogeneity of distribution and variability in time is something that Françoise Praderie criticized very heavily when I was discussing this with her before I came here. Also, I should have made a stronger point than I did for Anne Underhill's symposium on stellar chromospheres to be held next February in Goddard at which some of these points will be discussed. But on the inhomogeneities, you are absolutely right, Bappu. I have been very slow to get on the band wagon while other people have been emphasizing these. The only thing is, I think, we somehow have to ask how do they come about. It seems to me that the chief argument is, and again I would like to see it discussed here, that as I go out in the atmosphere, the radiation field present does not change but the amount of it absorbed does decrease. Whereas if you start with any kind of a mechanical disturbance in the outer atmosphere it will amplify as the density decreases rather than decrease. So that I have on the one hand the radiative effect decreasing and on the other hand the mechanical effects increasing, just the conditions necessary for introducing instabilities and inhomogeneities.

SECTION II

CHAIRMAN: P. CONTI

CLASSIFICATION AND DISTRIBUTION OF WR STARS
AND AN INTERPRETATION OF THE WN SEQUENCE

LINDSEY F. SMITH

*NASA Goddard Space Flight Centre, Greenbelt, Md., U.S.A.**

1. Introduction

In the three years since the Boulder Wolf-Rayet Conference, a great deal of observational material has been collected and processed. That conference was, of course, the stimulus for a great deal of the work, since it made clear where some of the large gaps in our observational knowledge lay. As a result the greatest progress has been in directions which were previously underexplored, and some subjects which were actively pursued before the last conference have been left almost untouched since. Thus, much of what was said in 1968 need not be reviewed again; I can confine my attention to topics which have been significantly changed by new observations. The new observations comprise, in particular, the systematic observation of spectra of many stars at moderately high dispersion, observation of far ultraviolet spectra, and discovery of some WR stars in M33 – the first outside of our own Galaxy and the Magellanic Clouds.

We appear to have achieved one major clarification: I hope to convince you, on almost purely observational grounds, that WN stars are helium stars of 8–14 solar masses with atmospheres that contain only a very low proportion of hydrogen, that the properties of the star are very sensitive to this small amount of hydrogen and that, consequently, this structure explains many of the general properties of the subclasses. However the variety of the subclasses observed is greater than can be explained by variation of the hydrogen abundance alone; which other parameters are vital, and how they operate is not clear at this time.

Data allowing similar specific statements about WC stars are not available. A few general deductions can be made and will be dealt with later in this Symposium.

2. The Atmospheric Abundances

The IAU classification (Beals, 1938) of WR stars recognized that the WR spectra are characterized by emission lines of He and N *or* of He, C and O. Thus, the spectra were classified WN or WC and it was noted that the range of excitation in the two sequences is similar. By omission, it was implicit that H lines are not apparent in the spectra.

There are two controversies that occur repeatedly in the literature:

(1) whether the lack of obvious H lines in the spectra indicates an under-abundance of H in the atmosphere of the stars, and

(2) whether there is a composition difference between stars in the two sequences.

---

* Present address: Max Planck Institut für Radioastronomie, 53 Bonn, Germany.

M. K. V. Bappu and J. Sahade (eds.), Wolf-Rayet and High-Temperature Stars, 15–41. All Rights Reserved.
Copyright © 1973 by the IAU.

2.1. H/He RATIOS

Determination of H/He ratios is made difficult by the near-coincidence of wavelengths of H I and He II lines. With the line widths characteristic of WR spectra, all the hydrogen lines are completely blended with helium lines. However, it has been realized for some time that an estimate of the H/He ratio could, in principle, be obtained from observations of the Pickering decrement. Since the hydrogen Balmer lines coincide with every *second* helium line, the presence of hydrogen in the spectrum would cause a 'bumpy' decrement to be observed.

In the past, it was not possible, due to lack of photometric observations of the continua of the stars and lack of knowledge of reddening corrections, to convert the measured equivalent widths to the fluxes required for the calculation of H/He ratios. Now, both continuum observations (Kuhi, 1966; Kuhi, 1968) and reddening corrections (Smith, 1968c; Smith and Kuhi, 1970) are available. Some knowledge of the conditions under which the lines are formed is also necessary. Fortunately, if the atmosphere is optically thin to the line radiation, the line strengths depend only on the numbers of electrons in the upper levels and on the transition probabilities. Due to the similarity of the He II and H I ions the transition probabilities and the Boltzmann factors for the relative populations of the levels are nearly equal. The full equations are given by Castor and Van Blerkom (1970) who find that, for level 14 of He II and level 7 of H I, the ratio of the transition probabilities is 0.94; a 6% correction is unimportant in the present context. The 'n' values considered here are large enough that the assumption that b_n is equal to 1 will not introduce an error greater than 10%. Thus, the number ratio of H II to He III equals the ratio of the fluxes of the contributions of hydrogen and helium to the Pickering lines.

Castor and Van Blerkom (*loc. cit.*) have found that the observed line strengths in the spectrum of HD 192163 (Smith and Kuhi, 1970) imply an atmosphere that is optically thin to the line radiation for Pickering lines with the quantum number, n, of the upper level greater than 10. Of the lines satisfying that criterion, λ 3968, upper level = 14, is least affected by blending with nitrogen lines. Using this line, Castor and Van Blerkom derived He III/H II > 2/1.

Figure 1 represents the line fluxes of the Pickering lines for 'single' WR stars from nearly all subclasses of the WN sequence plotted against the quantum number, n, of the upper level. Figure 2 presents similar data for binary stars. The data is from the (imminently to be published) *Atlas of Wolf-Rayet Line Profiles* of Kuhi and Smith (1973), and is based on 16 Å/mm and 32 Å/mm Lick coudé spectrograms. Equivalent widths of the emission lines have been converted to fluxes with the continuum observations of Kuhi (1966, 1968), corrected for a change in the estimated reddening (Smith and Kuhi, 1970). The error bars allow for observational uncertainty in the measurements and for blending with N III and He I lines.

Possible errors due to blending are indicated by extension of the error bars in the direction of reduced intensity. Estimates of the amount are based on degree of deformation of the profile of the helium line, and the strengths of other lines of N III and

He I in the spectrum. Blending is the most serious in the WN6 spectra. In lower excitation subclasses the lines are moderately well resolved; in higher excitation subclasses, the strengths of N III and He I lines are much less than in the WN6 spectra. It is believed that the error bars given represent realistic upper limits to the possible

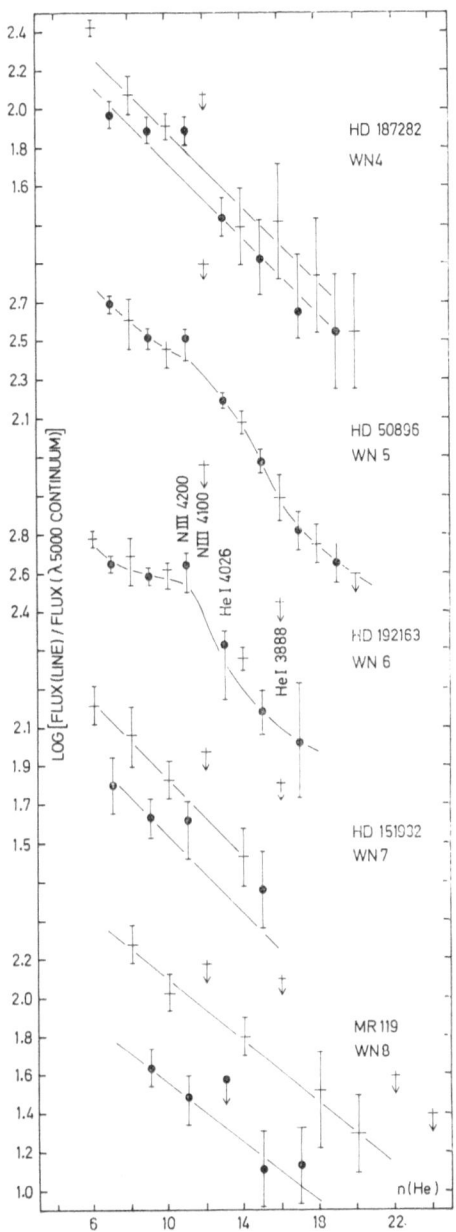

Fig. 1.   Graphs of the flux in the Pickering lines vs. the principle quantum number, n, of the upper level of the transition, for apparently single WN stars.

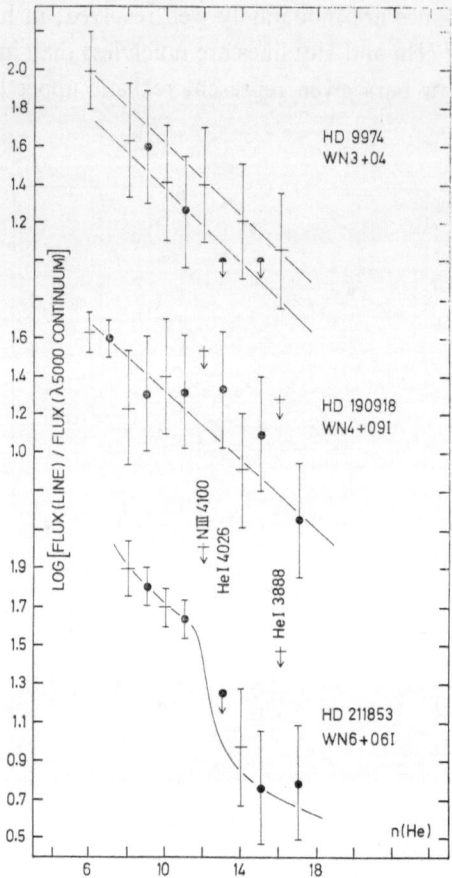

Fig. 2.   As for Figure 1, for binary WN stars.

errors; confirmation of this will, of course, only be possible when models for the atmosphere can be constructed and the relative strengths of the blended lines predicted. In spectra in subclasses other than WN5 and WN6 the H + He lines are consistently stronger than the pure He lines, and the ratio of line strengths appears to be independent of the strength of the lines. This strongly implies that, in these spectra, all lines of the series are optically thin.

In the graph for HD 192163 the H + He line that corresponds to $n = 14$, is relatively stronger than the two pure helium lines on either side, however, the stronger, low-n members of the series show no contribution from hydrogen to the even-n lines. This is most simply explained if the strong, low-n lines are optically thick so that the contribution due to hydrogen has little effect on the line strength. Since HeII lines in WN5 and WN6 spectra are about five times stronger than in other subclasses, optical thickness of the strongest lines is not surprising.

On the other hand, the fact that the HeI contribution to the $n = 16$ line is quite clearly visible indicates that the high-n members of the series are, even in this star, optically thin. (It should be noted that the same logic cannot be applied to the increased strength of the $n = 12$ line at $\lambda 4100$ since the N III and Si IV lines do not coincide closely in wavelength with the He II line, and the blend is very broad.) The rate of increase of the line intensities with decreasing n falls off for n less than 11. This is probably also due to optical thickness of the strong lines of the series, their intensities being reduced by self absorption. Thus the observations appear to be in accord with the calculations of Castor and Van Blerkom which suggest a transition from optically thick to optically thin at about $n = 10$.

The similarity of the graph for HD 50896 to that for HD 192163 suggests that, in this star also, the Pickering lines become optically thin for $n > 10$. (He I is essentially absent from this spectrum, thus the argument used above cannot be applied.)

The spectra of the binary stars appear to have similar properties to those of the single stars of the same subclass.

TABLE I

H/He abundances in WN atmospheres and their correlation with T_{eff}

HD	Sp. Type	$\Delta \log F$	N(H II)/ N(He III)	W(He I4471)/ W(He II4541)	$N(H)/N$(He)	T_{eff} (K)
50896	WN5	0.00 ± 0.02	0.00 ± 0.05	0.8/25	0.0 ± 0.05	53000
187282	WN4	0.15 ± 0.10	0.40 ± 0.30	0.0/5.7	0.4 ± 0.3	46000
192163	WN6	0.16 ± 0.06	0.45 ± 0.20	7.2/33	$\lesssim 0.4 \pm 0.2$	29000
151932	WN7	0.30 ± 0.15	1.00 ± 0.60	1.4/3.6	< 1.0	
MR 119	WN8	0.52 ± 0.10	2.30 ± 0.70	3.4/3.7	$\ll 2.3$	23000
9974	WN3 + 04	0.24 ± 0.10	0.75 ± 0.35	0.0/3.2	0.8 ± 0.4	
190918	WN4 + 09I	0.00 ± 0.05	0.00 ± 0.12	0.0/1.5	0.0 ± 0.1	
211853	WN6 + 0B	0.13 ± 0.10	0.35 ± 0.25	1.0/4.9	$\lesssim 0.3 \pm 0.2$	
COSMIC					10.0 ± 0.1	

Table I summarizes the numerical results. For subclasses other than WN5 and WN6, the mean difference, $\Delta \log F_\lambda$, was taken from the graphs, as indicated. For the WN6 stars the value depends only on the $n = 14$ line. The number ratio of H II/He III ranges from 0 to 2.3. To obtain values for the total H/He ratio we need to consider the possible presence of singly ionized helium. The relative strengths of He I and He II lines are indicated in Table I by the equivalent widths of He I 4471 and He II 4541. He I makes no contribution whatsoever to the visible spectra of WN3 and WN4 stars; thus, for these subclasses He II may be neglected. In WN5 and WN6 spectra He I lines are very weak, and we probably make little error by again neglecting He II. However, in WN7 and WN8 spectra He I lines are strong. In the WN8 atmospheres, in particular, most of the helium may be singly ionized. We have no method for determining the ionisation balance. In Table I, therefore, for the lower excitation classes, the values are given only as upper limits. Neutral hydrogen and helium should have low abundances out to regions of the atmosphere where densities are too low to greatly concern us and may probably be safely neglected.

We conclude that there is essentially no hydrogen in the atmospheres of WN5 stars, that there is a little hydrogen in WN4 and WN6 atmospheres, and that the H/He ratio may get as high as 1/1 in the atmospheres of WN8 and WN3 stars. The H/He ratios are an order of magnitude less than the cosmic abundance of 10/1 (Popper *et al.*, 1970). They are also much less than the value, 5/1, derived by Aller and Heap (1971) for HD 45166 using the same method, and the values between 4/1 and 14/1 derived by Oke (1954) for Of stars. Thus, severe underabundance of hydrogen in the atmosphere appears likely to be a necessary (though not a sufficient) condition for a WN star.

I emphasize that nothing has been assumed in the H/He derivations other than that the lines are optically thin and that the b_n are near one. It should be noted, in particular, that the temperature dependence of the strengths of the He II and H I lines are identical and thus the (unknown) temperature of the WR atmospheres does not enter the calculation. The assumption of optical thinness in is accord with the calculations of Castor and Van Blerkom (1970) and is substantiated by the observations as discussed above. The assumption that the $b_n = 1$ is not likely to be in error by more than 10%.

2.2. THE CNO ABUNDANCES

One can, in theory, use the same procedure to obtain number ratios of some ions of carbon, nitrogen and oxygen to helium. This possibility deserves further attention; at the present time it has not been attempted because it requires observation of lines due to transitions from high energy levels of the CNO ions and these lines are weak. The spectra are dominated by lines from lower levels of the ions and a theory for the relative strengths of these lines is not available at this time.

A further complication arises from the fact that C, N and O can each exist in many ionisation states, the relative populations of which are certainly variable through the atmosphere and are not known. Some of the ions do not have strong lines in the visual region of the spectrum, thus their presence or absence cannot be unequivocally determined from ground-based observations. However, at this time, OAO-II observations are available for HD 50896, an apparently-single WN5 star (Smith, 1972). Thus we now have observations of the spectrum from 1000–6000 Å, and within this spectral range all ions have some strong lines; we may, therefore, make a firm statement regarding the presence and relative strength of lines from all ions.

Figures 3 and 4 show spectra obtained by Bless; the line identifications are due to the present author. I present the raw data; information on the sensitivity functions of the satellite instruments are still tentative; however, their shapes are qualitatively similar to those of the continua that have been drawn on the diagrams. Figure 3 is a sum of two scans in the wavelength range 2000–4000 Å. The resolution is about 25 Å; data is taken every 21 Å. No simultaneous measurement of the background is available for spectra in this wavelength range; in Figure 3 the background is taken to be 10 counts, representing 5 counts per scan, in accordance with the estimate given by Savage (private communication). Figure 4 represents the sum of seven scans in the wavelength region 1000–2000 Å. Care was taken to line the spectra up as well as

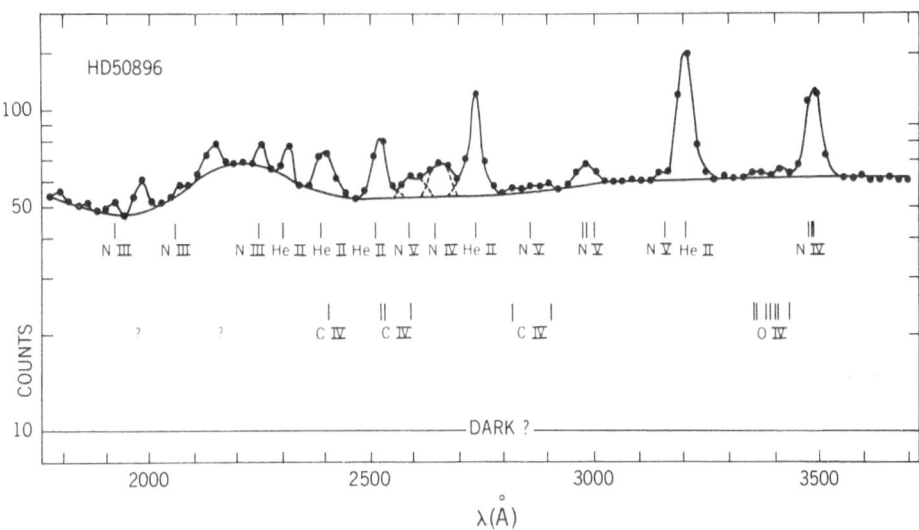

Fig. 3. The spectrum of HD 50896 in the wavelength range 2000–4000 Å obtained by Bless with OAO-II.

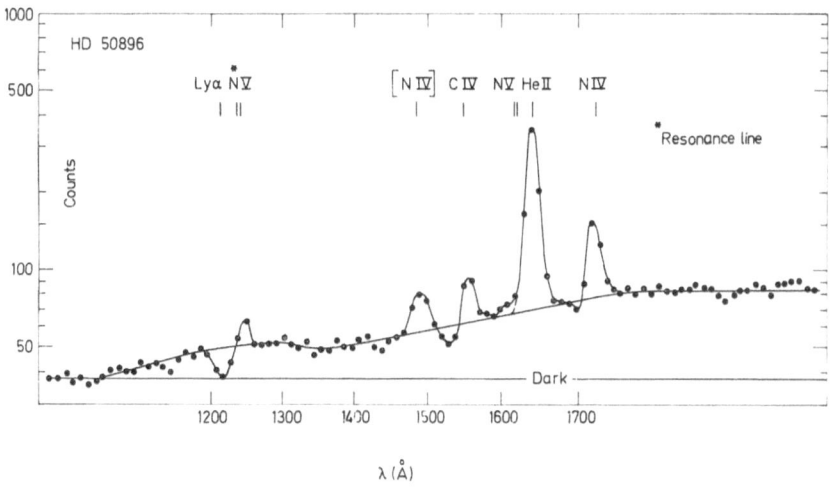

Fig. 4. The spectrum of HD 50896 in the wavelength range 1000–2000 Å obtained by Bless with OAO-II.

possible before adding them; however, data is taken every 10.5 Å and in some cases the spectra were clearly shifted by 5 Å. Thus, adding the scans produces some smearing; resolution on the integrated scan is probably about 15 Å. In this wavelength region the background may be estimated from the region shortward of 1000 Å where the system is insensitive to incoming radiation and only particle background and dark current are being measured.

LINDSEY F. SMITH

TABLE II
Equivalent widths in the spectrum of HD 50896

Lab Int (N)	C II	N III	O IV	Transition
10	Resonance	1751.7 ⎫	1338.6 ⎫	$2p^2\ {}^2P - 2p^3\ {}^2D^0$
9	lines	1747.8 ⎬ <2Å	1343.0 ⎬ <3Å	
6	λ 1335.7	1751.2 ⎭	1343.5 ⎭	
8	λ 1334.5	1184.5 ⎫ ?	923	$2p^2\ {}^2P - 2p\ {}^2P^0$
7	not	1183.0 ⎭		
10	observed	1885.2 <2Å	1068	$3d\ {}^2D - 4f\ {}^2F^0$
10		2064.0 ⎫	1164.3 ⎫	$3d\ {}^2F^0 - 4f\ {}^2G$
10		2063.5 ⎬ 1.5Å	1164.5 ⎭	
6		2068.2 ⎭		
10		4097.3 ⎫ ≤30Å	3063.5 ⎫ <1Å	$3s\ {}^2S - 3p\ {}^2P^0$
9		4103.4 ⎭ (He II, Si IV)	3071.7 ⎭	
10		4640.6 ⎫	3411.8 ⎫	$3p\ {}^2P^0 - 3\ {}^2D$
9		4634.2 ⎬ 7Å	3403.6 ⎪	
7		4641.9 ⎭	3413.7 ⎪	
7		4510.9 ⎫	3381.3 ⎬ 4Å	$3s'\ {}^4P^0 - 3p'\ {}^4D$
6		: ⎬ 3Å	: ⎪	
4		4534.6 ⎭	3409.8 ⎭	
10		4379.1 5Å		$4f\ {}^2F^0 - 5g\ {}^2G$
	C III	N IV	O V	
2	1909 ?(N III) <2Å	1486.5 35Å	1218.4 ?(Lα)	$2s^2\ {}^1S - 2p\ {}^3P^0$
20	2296.9 ?(He II) <3Å	1718.6 40Å	1371.3 <6Å	$2p\ {}^1P^0 - 2p^2\ {}^1D$
15	4647.4 ⎫ ?(N III)	3478.7 ⎫	2781.0 ⎫ ?(He II)	$3s\ {}^3S - 3p\ {}^3P^0$
14	4650.2 ⎬ <4Å	3483.0 ⎬ 91Å	2787.0 ⎬ <2Å	
13	4651.4 ⎭	3484.9 ⎭	2789.9 ⎭	
10	4067.9 ⎫ ?(N IV)	2645.6 ⎫	1643.7 ?(He II)	$4f\ {}^3F^0 - 5g\ {}^3G$
11	4068.9 ⎬ <12Å	2646.2 ⎬ 18Å		
12	4070.3 ⎭	2647.0 ⎭		
8	5695.9 <1Å	4057.8 23Å	3144.7 ?(N V) <3Å	$3p\ {}^1P^0 - 3d\ {}^1D$
	C IV	N V	O VI	
20	1548.2 ⎫ 20Å	1238.8 ⎫ >16Å	1031.9 ⎫	$2s\ {}^2S - 2p\ {}^2P^0$
19	1550.8 ⎭	1242.8 ⎭	1037.6 ⎭	
12	5801.5 ⎫ 28Å	4603.3 ⎫ 34Å	3811.4 ⎫ ?(He II)	$3s\ {}^2S - 3p\ {}^2P^0$
10	5812.1 ⎭	4619.1 ⎭	3834.2 ⎭ ≪1Å	
9	2524.2 ⎫ ?(He II)	1616.3 ⎫ ?(He II)		$4d\ {}^2D - 5f\ {}^2F^0$
12	2530.0 ⎭	1619.7 ⎭ ~4Å		$4f\ {}^2F^0 - 5g\ {}^2G$
6	4646.0 ⎫	2974.5 ⎫		$5d\ {}^2D - 6f\ {}^2F^0$
8	4658.3 ⎬ ?(He II)	2980.8 ⎬ 8Å		$5f\ {}^2F^0 - 6g\ {}^2G$
10		2981.3 ⎭		$5g\ {}^2G - 6h\ {}^2H^0$
9	7726.2 ᵃ yes	4944.6 6Å		$6h\ {}^2H^0 - 7i\ {}^2I$
6	2906.3 1.5Å	1860 1Å		$5g\ {}^2G - 7h\ {}^2H^0$
6	2404.4 ⎫ ?(He II)	1548 ⎫ ?(C IV)		$4p\ {}^2P^0 - 5d\ {}^2D$
	2405.1 ⎭	1549.3 ⎭		

ᵃ Kuhi (private communication) confirms presence of this line from scanner observations.
?( ) Possibly blended with –

Equivalent widths have been derived by numerical integration. These are presented in Table II which are arranged to show comparison between the isoelectronic ions of C, N and O. Strengths of lines in the visible region of the spectrum are taken from Kuhi and Smith (1971). The results may be summarized as follows:

(1) N IV and N V lines are strong. N III lines are also present but are weaker so that only lines in the visible (where the spectra are of much higher resolution) have been observed.

(2) Carbon appears only as C IV, and the lines are of comparable strength to lines due to equivalent transitions of N V, with which C IV is iso-electronic. C III lines are not observed, however the presence of weak lines cannot be excluded since most would be blended with lines of other ions.

(3) Oxygen occurs very weakly as O IV only. (O III is not included in the Table; however, its strong lines at λ 3759, λ 3774 and λ 3791 are not observed and an upper limit of 3 Å is estimated for their equivalent widths.)

The data emphasizes that, when an ion is present in a WR spectrum, the spectrum of that ion is well developed; selective excitation processes that generate only a few lines of a given ion are relatively unimportant.

TABLE III

Summary – relative equivalent widths in WN spectra
(ionisation potential)

	C	N	O
II	24.3	29.6	35.1
III	?47.9?	//47.4//	54.9
IV	///64.5///	////77.4////	/77.4/
V	392	///97.9///	113.8
		552	138.1
Cosmic Abundance	45	12	100

The relative strengths of spectra of different ions are summarised in Table III which gives the excitation potentials of each of the ions and indicates, by the number of slashes, the relative strengths of the observed spectra. C IV and O IV, which are the strongest ions of C and O, respectively, have ionization potentials nearest to that of N IV, the nitrogen ion showing the strongest spectrum. This situation is not surprising and does not require any specialized ionization processes. There is no evidence here for a peculiar distribution among the ionization states of O and C.

The general applicability of this argument is indicated by Table IV which is taken from Kuhi and Smith (1973) (and is included here in its entirety because of its general usefulness as a quick reference). The Table shows, in a qualitative manner, the relative strengths of the spectra of the various ions in the visual spectral regions of WN stars of all subclasses. In the present context, note that the C IV lines λ5801, λ5812 are present in nearly all subclasses but become very weak in the lowest excitation classes

TABLE IV
Qualitative description of WN spectra

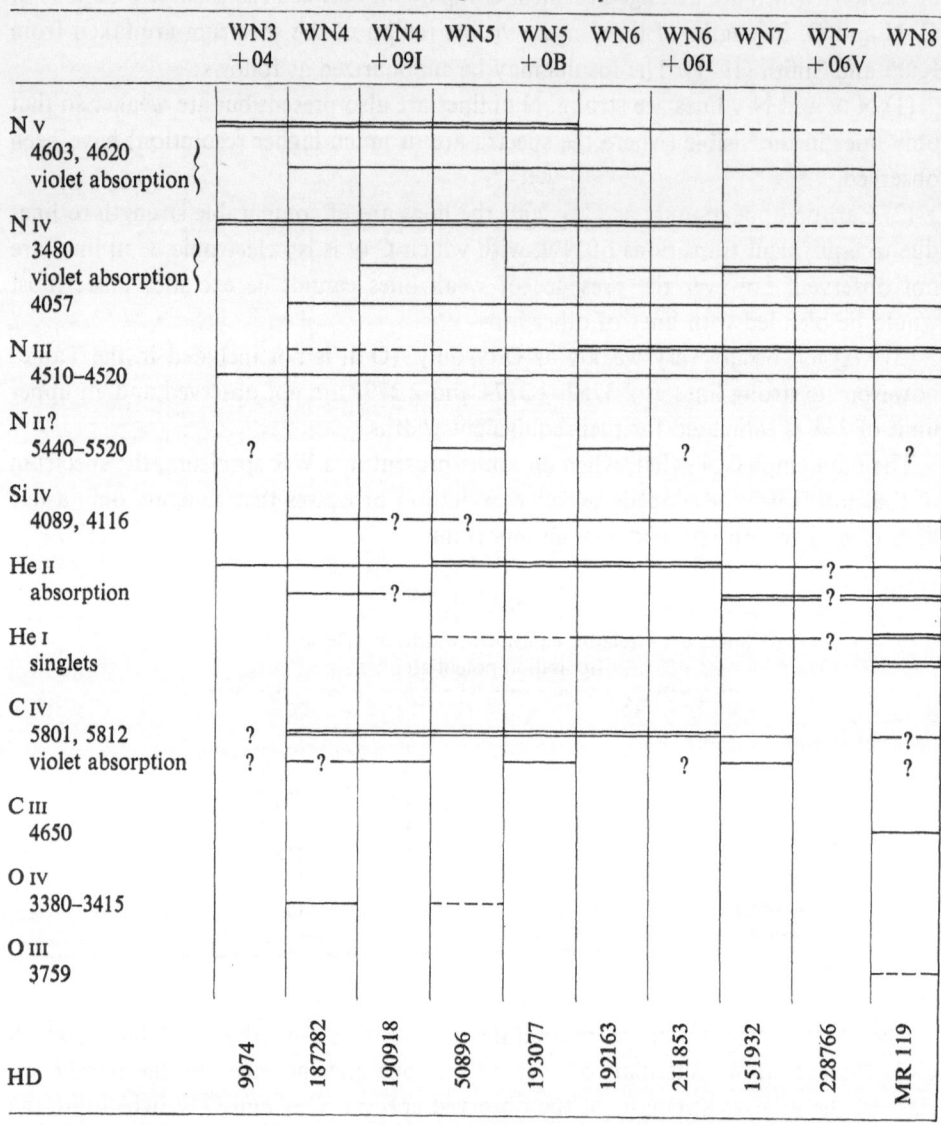

?   No red plate.
-?- Data inadequate to be sure – due to noise, blending, etc.
     Not visible.

Strong. ═════
Present. ─────
Weak. ‒ ‒ ‒ ‒

where N III becomes the dominant nitrogen ion; in the WN8 spectrum of MR 119, C III (whose ionization potential is approximately equal to that of N III) is observed also. Similarly, O IV appears in the highest excitation subclasses, and O III appears in the WN8 spectrum.

Thus, an explanation for the differences between the WN and WC stars in terms of different excitation or ionization mechanisms in the atmospheres is not supported by the observations. While it is possible that peculiarities in the WC atmospheres are alone responsible for the differences, I consider this unlikely. The data supports the hypothesis that there is an abundance difference between stars in the two sequences.

Which sequence, if either, has cosmic abundances ratios of He, C, N and O is an open question. In as much as the cosmic abundance ratio of H/He has been so drastically altered in the WN atmospheres, there seems no reason to expect that the other ratios should have retained their cosmic values. An overabundance of N with respect to C and O, and probably to He also, appears to be the obvious explanation for the line strengths summarised in Table III.

By analogy, low hydrogen abundance and high C and O abundances in WC atmospheres becomes a likely explanation for the very strong C and O lines in those spectra. It is unfortunate that H/He ratios and equivalent ultraviolet observations cannot be easily obtained for WC stars. Lines of the Pickering series are so badly blended with carbon lines that a meaningful H/He ratio cannot be derived. Gamma 2 Velorum is the only WC star that is bright enough to be observed in the ultraviolet with available equipment, and it is a binary in which the companion is the brighter; it is not clear to me which lines may be attributed to the WR star and which to the O star.

3. New Observations Relevant to the Classification System

Two new observations in the field of WR classifications should be noted at this time. First, Cowley and Hiltner (1969) have observed the star CPD-56°8032 and described its spectrum. While they do not give a print of the spectrogram it would appear, from the description, that we must classify this object WC10. The degree of excitation is lower than in WC9 stars and the lines are narrower, of the order of 2A! This is very narrow indeed for a WR star, but in as much as it appears to be a logical extension of the sequence, it should be provisionally accepted as such.

Of more basic importance to the subject of classification is the observation by R. Lynds of the spectra of two WR stars found by Wray and Corso (1972) in M33. These spectra show C III-IV $\lambda4650$ and C IV $\lambda5810$ as the strongest features, indicating that the classification should be WC5 or WC6. The weakness of other features would indicate that the objects are probably binary stars. However, the line widths are *very* much less than those observed in high excitation WC spectra of stars in the Galaxy and the Magellanic Clouds. This is a most startling observation! Line width is so uniformly and consistently correlated with excitation in WC stars that it is used as a classification criterion in the IAU system and in all modified systems proposed since (Hiltner and Schild, 1966; Smith ,1968a). Eventually it may be necessary to modify the classification system to incorporate this new dimension of possible spectral properties.

The absolute visual magnitudes of the two stars, $M_v = -4.9$ and -5.8, are to be compared with $M_v = -3.9$ to -5.1 for single WC5 stars in the Large Magellanic Cloud, and $M_v = -5.4$ to -8.1 for binaries (Smith, 1968b). The agreement is reasonable, although the fainter star is somewhat fainter than the LMC binaries.

4. Absolute Magnitudes of WR Stars

Absolute magnitudes of WR stars in the Galaxy have been reconsidered recently by Crampton (1971). He lists all WR stars which appear to be associated with nebulae. The distances of some of these nebulae may be estimated from spectroscopic parallaxes of O and B stars, or from kinematic methods. In general, the distances derived are less than those given by Smith (1968c) for the WR stars, although the derived reddening for the WR stars generally agrees well with those of the OB stars. The conclusion would appear to be that the luminosities of the WR stars are lower than estimated by Smith (1968b). Deriving luminosities for the galactic WR stars by subclass, Crampton arrives at the numbers in Table V; values derived by Smith are also given. Among the WN stars the largest discrepancy occurs between the two derived values for M_v for the WN6 stars. Smith's derivation depended mainly on three galactic stars; Crampton's depends on 7 stars and should definitely be taken in preference.

TABLE V

Absolute magnitudes of Wolf-Rayet stars

Class	Smith			Crampton		Adopted
	M_v	σ^a	n	M_v	n	
WN3	−4.5	0.1	2		1	−4.5
WN4	−3.9	0.3	5	−3.7	3	−3.9
WN5	−4.3	0.1	2		3	−4.3
WN6	−5.8		3	−4.8	7	−4.8
WN7	−6.8	1.0	4	−6.5	4	−6.8
WN8	−6.2	0.4	3			−6.2
WC5	−4.4	0.6	5⎱	−3.6	1	−4.4
WC6	−4.4		0⎰		2	−4.4
WC7	−4.4		2	−4.4	3	−4.4
WC8	−6.2		1	−5.4	1	−4.8
WC9	−6.2		0			−4.4

^{a} σ = standard deviation.

For the other subclasses, I would retain the values given by Smith since these depended on observations in the Large Magellanic Cloud and are free from the uncertainties inherent in distance determinations based on OB star luminosities or on kinematic models. The differences between Smith and Crampton's values are not large compared to the intrinsic range in the luminosities derived for the LMC stars. While a real difference between LMC and galactic WR stars is a possibility, the data does not warrant that conclusion yet. However, when classifications of the faint LMC WR stars (presently based mostly on photometric colours) have been checked with good spectrograms and when classifications including reliable luminosity classes are available for the OB stars on which the galactic distances depend, this problem should be reconsidered.

Among the WC stars, similar comments apply. Smith's derivation for M_v of WC8 stars depends only on the galactic star, γ_2 Velorum, a binary with components WC8 and O. Taking Smith's (1955) classification of the companion as O7 and assuming it had the luminosity of a main sequence star of that class, Smith believed the WC8 star to be the more luminous. Baschek and Scholz (1971) and Conti and Smith (1972) have derived the luminosity ratio of the two stars from comparisons of line strengths in the spectrum of γ_2 Vel with other (presumed single) O and WR stars. The conclusion is definitely that the O star (which appears to be a supergiant) is the more luminous. Baschek and Scholz estimate the luminosity of the WC8 star as between -3.6 and -4.5, but they use a distance modulus of 7.5 mag., rather than the larger figure, 8.3 mag., derived by Graham (1965) and independently substantiated by Brandt *et al.* (1971). Conti and Smith derive $M_v = -4.8 \pm 0.3$ with the larger modulus. This value is within the range 4.4 ± 0.6 mag. observed by Smith for the WC5 stars in the LMC. Thus there is no longer evidence for a difference in luminosity between the WC8 and WC5 stars. I have, therefore, adopted -4.4 as the M_v of WC9 stars also. Crampton derives a significantly lower value, -3.6 for the luminosities of WC5 and WC6 stars. As for the WN stars, this may indicate a difference between LMC and galactic WR stars (and would agaii ndicate a significant luminosity difference between WC8 and earlier subclasses). However, his value is based on only three stars and one should reserve judgement.

It is well to recall at this time that M_v refers to the narrow band system of Smith (1968b). This system effectively avoids emission lines in the spectra of WN stars, and magnitudes therefore refer to the continuum. This is not the case for WC stars, and while the effect of the emission lines is probably not greater than 0.1 or 0.2 mag., it has never been determined.

5. The Galactic Distribution

Revision of the absolute magnitudes revises the galactic distribution picture for the WR stars. The positions projected onto the galactic plane are shown in Figure 5. The diagram is a great deal tidier than the earlier version (Smith, 1968c); the big scatter of WN6 and WC9 stars over the galactic center region is removed. The H II regions whose distances are derived by Courtès *et al.* (1968) and by Georgelin and Georgelin (1970), the giant H II regions according to the diagram of Mezger (1970), and the outer edges of the OB star concentrations defined by Graham (1970) are also shown. The WR stars concentrate at approximately the same distances as do Courtès's H II regions, indicating that the distance scales of the two types of objects are in reasonable agreement. The WR stars also conveniently bridge the gap between the normal H II regions, which are, in general, only observed to distances of the order of 3 kpc, and the giant H II regions which are mostly at distances of 5 kpc or more.

The Carina arm is very clearly delineated and its connection to the Sagittarius arm in the north is re-inforced. The Sagittarius arm in the south appears to branch from the Carina-Sagittarius arm at a point between the Sun and the galactic centre. An inner arm at about 6 kpc from the galactic centre is weakly defined. The pitch angle

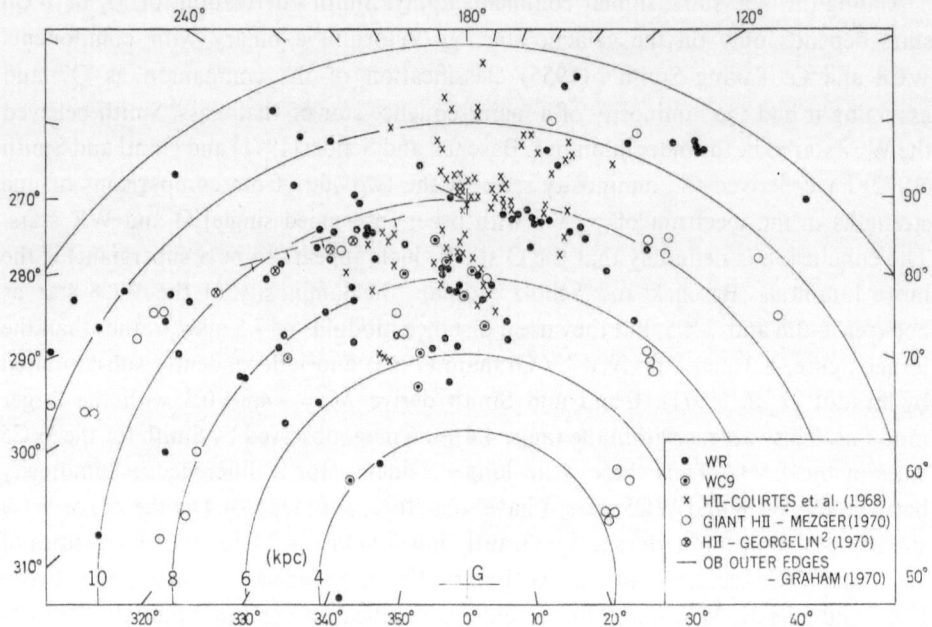

Fig. 5. The positions of WR stars projected onto the galactic plane, compared with position of
H II regions and the outer edge of the OB stars concentrations in Carina.

of about 20° for the Cygnus arm, advocated by Courtès *et al.* (1969) and many previous optical observers, is reinforced; this arm appears to stop about 1 kpc beyond
the Sun. The Cepheus arm appears to be rather irregular. The distances of the WR
stars agree moderately well with those of the H II regions in the nearer parts of the arm,
between $l^{II} = 115°$ and 140°; in these directions the arm lies about 11.5 kpc from the
galactic center. Between $l^{II} = 100°$ and 115°, where the arm is defined by WR stars alone,
its distance from the galactic center is about 13 kpc.

The Comparison to the H I distribution given by Kerr (1970) is shown in Figure 6. The
agreement is good for the distant parts of the Carina arm and the southern part of the
Sagittarius arm. However, the optically defined Cygnus arm and the nearer parts of
the Cepheus arm fall definitely closer to the galactic center than do the H I arms in
Kerr's diagram, and the connection between the Sun and Carina, shown in the H I
diagram, is definitely not substantiated by the optical objects. The lack of any optical
equivalent to the northern Sagittarius arm is largely due to heavy absorption. Some
WR stars are known in those directions but are not on the diagram due to lack of
photometric data.

The WC9 stars have been plotted with a different symbol to the other WR stars.
I argued at the 1968 meeting (Smith, 1968e) that the high visual luminosity ($M_v = -6.2$)
was necessary for these stars to explain their strong concentration towards the galactic
center direction. With the lower value now adopted for the visual absolute magnitude,
most of them fall in the section of spiral arm immediately interior to the Sun. There

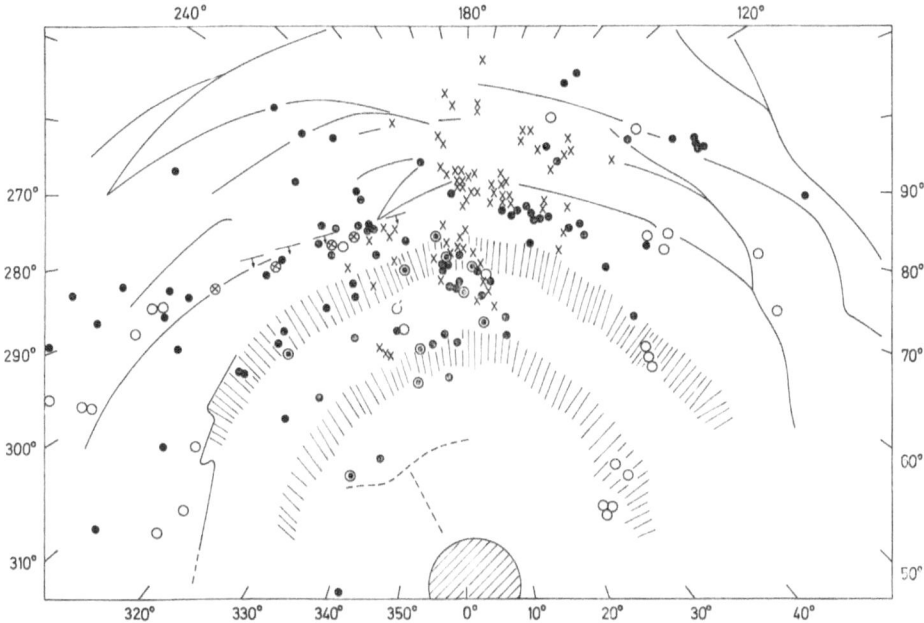

Fig. 6.   The positions of WR and H II regions shown in Figure 5 compared to the H I distribution
given by Kerr (1970).

is only one known in the rest of the southern Sagittarius arm and none in the Carina
arm. The distribution is without explanation at this time.

Apart from the WC9 stars, the discussion of galactic distribution of WR stars, of
the concentration of WC9, WN6 and WC7 stars to regions within 9kpc of the galactic
center, and the complete absence of subclasses WC6–9 and probably WN6 from the
Magellanic Clouds, remains as given in 1968.

6. The Distribution of WR Stars in M33

Wray and Corso (1972) have recently found, in M33, 25 objects which have strong
emission at wavelengths near $\lambda\,4670$ and do not have strong emission around $\lambda\,5007$.
They identify these as WR stars and have spectra of two of the objects (discussed in
Section 3) to verify that identification. The positions of the objects found are shown
in Figure 7 (Figure 1 of Wray and Corso). The circles in that figure represent the
boundaries of the regions observed. Clearly the stars fall within the spiral arms, justi-
fying again the assignment of WR stars to extreme population I and validating their
use as spiral tracers in our own Galaxy. Obviously, determination of the subclassifica-
tion of these stars is of great interest, in order to determine whether there is segregation
of some subclasses, as there appears to be in the Galaxy, or complete lack of some
subclasses, as occurs in the Magellanic Clouds.

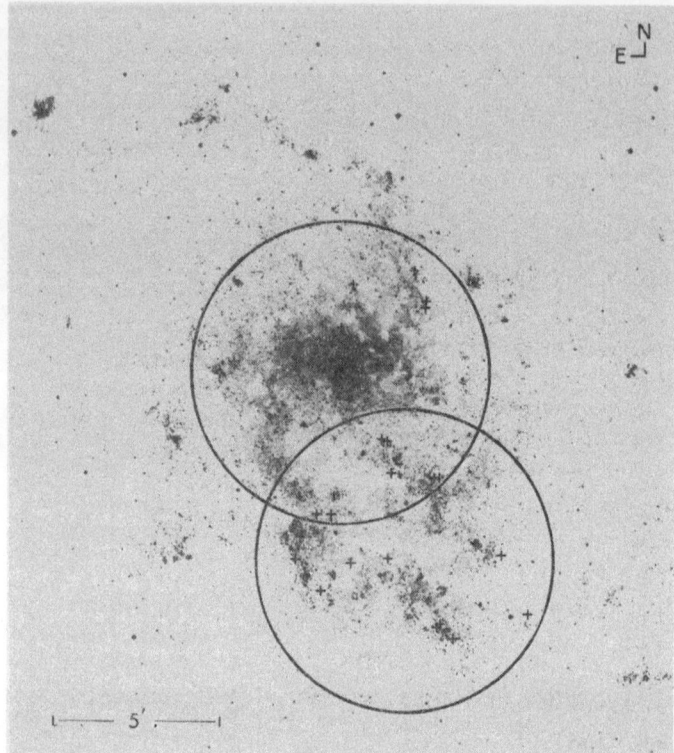

Fig. 7.   The positions of Wolf-Rayet stars in M33 identified by Wray and Corso (1971).

7. The Bolometric Luminosities, Radii and Compositions

Morton (1970) has derived temperature for a few WN stars. These are Zanstra temperatures based on the visual luminosities (in the continuum) of the WR stars and on radio observations of associated nebulae (Johnson and Hogg, 1965; Gebel, 1968; Smith and Bachelor, 1970). The method requires the assumption that the WR stars radiate like line-blanketed model atmospheres, a dangerous assumption. However, since the deviation of the true flux distributions from those of the models may be similar for all these stars, the relative temperatures may be approximately correct, even if there is a systematic error. Morton's results are given in row 2 of Table VI. In row 1 are the visual magnitudes as adopted in Section 4. In row 3 are bolometric corrections, as tabulated by Morton (1969), applicable to models at the derived temperatures. Row 4 gives the bolometric magnitudes. We have the surprising result that, despite a range of approximately 3 mag. in visual luminosity, all the WN stars appear to have bolometric magnitudes between -8.1 and -9.1. Thus, the total rate of energy production is approximately the same for all the stars. The different visual magnitudes appear to be due to different surface temperatures (causing differing portions of the energy to be observed in the visual region of the spectrum). Obviously, R^2T^4 must remain constant; thus the hottest stars are the smallest.

TABLE VI

Luminosities of WN stars

	WN5	WN4	WN6	WN7	WN8	Error	Range
M_v	−4.3	−3.9	−4.8	−6.8	−6.2	±0.5	3 mag.
T_{eff}[a]	54000°	46000°	40000°		23000°	±5000°?	
	52000		29000				
B.C.[b]	4.6	4.2	2.8, 3.7	2.3?	2.3	±0.4	
M_{bol}	−8.9	−8.1	−7.6, −8.5	−9.1	−8.5	±1.0	1 mag.

[a] Morton (1970).
[b] Morton (1969).

The derived temperatures are included in Table I, where the H/He ratios were derived; it is clear that there is a correlation between the temperatures and the H/He ratios in the sense that the stars with the lowest H/He ratios are the hottest. We deduce that the stars with the lowest H/He ratio are the smallest!

Bolometric magnitudes between −8 and −9 correspond to pure helium, helium-burning stars with masses between 8.5 and 14 $M_\odot$ or to pure hydrogen, hydrogen-burning stars with masses between 28 and 45 $M_\odot$ (see Table VII). Since WR stars are

TABLE VII

Comparative luminosities: helium-burning helium
stars to hydrogen-burning hydrogen stars

Log $L/L_\odot$	M_{bol}	M(He) [a] $(M_\odot)$	M(H) [b] $(M_\odot)$
4.9	−7.5	7	23
5.1	−8.0	8.5	28
5.3	−8.5	11.5	35
5.5	−9.0	14	45
5.7	−9.5	17	57

[a] Van der Borght and Meggitt (1963).
[b] Stothers (1963, 1965, 1966).

characteristically 11 $M_\odot$, not 30 $M_\odot$ (see Kuhi's discussion in this Symposium), the conclusion is that WN stars are not hydrogen burning. Pure helium, helium-burning stars are consistent with the observations; carbon cores and carbon burning are not excluded.

A small amount of hydrogen in the atmosphere causes a star to have a size much greater than that of a pure helium star. Thus, the deduction above, that stars with little or no hydrogen in their atmospheres are the smallest, is in complete accord with the theory.

It should be noted that this analysis has relied on data from binary and apparently-single stars, with the implicit assumption that these stars are intrinsically identical. In fact we do not know if the 'single' and binary stars have the same evolutionary history, so some caution may be in order. It can, however, by more careful use of the data, be shown that the masses and luminosities are almost certainly equal for the two types of object. The observations are as follows: the WR stars in binaries are observed to have masses of the order of 11 $M_\odot$. The H/He ratios are demonstrated to be the same for binary and 'single' stars of the same class. The luminosities derivations (Table VI) depend almost completely on 'single' stars.

Thus, the luminosities of the 'single' stars correspond to those of pure helium stars with masses equal to those derived for the binary stars. Since the very high helium abundance observed in the atmospheres are inconsistent with other than a nearly pure helium star, it follows that both 'single' and binary stars are pure helium stars approximately of the same mass and luminosity.

8. How 'Rare' are the WR Stars?

The statement has often been made that WR stars are exceedingly rare objects. It is clear that a WR spectrum is probably characteristic of only a part of any given star's evolution. The 'rarity' of WR stars has been taken to imply that only a small fraction of stars have a WR phase. Let us re-examine the evidence.

The Hamburg, Warner and Swasey Catalogue lists 5800 stars of spectral type earlier than B 1 V (see Sim, 1969). Their limits were: $l^{11} = 10°$ to $220°$ and $m_{pg} < 13$ mag. If we guess that this represents completeness to a distance of 3 kpc, we may count the number of WR stars known to this distance; in the same longitude limits, the answer is 15. (This number allows for the change in the distance scale of the WR stars resulting from changes of some of the M_v's adopted in Section 4.) Now a B 1 V star has a mass of about 10 $M_\odot$ and a main sequence lifetime of about 2×10^7 yrs (Iben, 1967). Thus:

$$\left[\frac{\text{The number of WR Stars}}{\text{The number of OB stars with } M > 10 M_\odot} \right] \sim \left[\frac{1}{400} \right].$$

And, if a fraction, 1/W, of all stars with $M > 10 M_\odot$ become WR stars, the lifetime of the WR stars must be of the order of $5 \times W \times 10^4$ yrs. Kippenhahn (1969) [See also Smith, 1968d)] estimates the total lifetime for helium *and* carbon burning of an 8.5 $M_\odot$ nearly-pure-helium star as less than, but of the order of, 10^6 yrs. Thus, we might imagine that $W \approx 20$, and indeed the WR phenomenon would be moderately rare.

However, there is a fallacy in the above logic. To generate an 8 $M_\odot$ pure helium core during its main sequence lifetime, a star must be initially about 25 $M_\odot$ (Kippenhahn, *loc. cit.*) and it is therefore probably only stars of 25 $M_\odot$ or greater that have any possibility to become WR stars. From Sandage's (1957) mass function, only about 1/100 of stars with masses greater than 10 $M_\odot$ are also greater than 25 $M_\odot$; thus the ratio of WR stars to 'progenitor OB stars' is only $\frac{1}{4}$. The lifetime on the main sequence of a 25 $M_\odot$ star is 5.5×10^6 yrs. (Stothers, 1963; Stothers, 1965). Thus, if 1/W of all main sequence stars greater than 25 $M_\odot$ becomes a WR star, the lifetime of the WR phase must be approximately $W \times 10^6$ yrs. In as much as a lifetime greater than about

10^6 yrs appears theoretically impossible for 8-14 $M_\odot$ pure helium stars, it would appear that W must be of order 1, and all stars with initial masses greater than 25 $M_\odot$ must have a WR phase. Thus, far from being rare, the WR phenomenon would appear, among massive stars, to be the rule.

Obviously this calculation can, and should, be checked with the stars in the Magellanic Clouds.

9. Interpretation of the WN Sequence

It may be concluded from the data presented in Sections 2 and 7 that WN stars of all subclasses are pure helium stars with masses between 8 and 14 $M_\odot$ and that the stars in various subclasses differ from one another in their atmospheric composition and, thereby, in their sizes and surface temperatures. To demonstrate more clearly the consequences and uncertainties inherent in this statement, the subclasses are arranged in a two dimensional grid in Table VIII. The mass and bolometric magnitudes, which

TABLE VIII

The WN sequence

$M/M_\odot$	M_{bol}					
17	−9.5		⊤ WN7			⊤ WN5
14	−9.0	⊤ WN8			⊤ WN6	
11.5	−8.5			⊤ WN3	WN4	
8.5	−8.0	⊥	⊥			
7	−7.5			⊥	⊥	
(H/He)		2.3	1.0	0.8:	0.4	0.0
T_{eff} (K)		23000°			(46000° 40000° 29000°	53000°

are assumed to be related as for pure helium, helium-burning models, are displayed vertically. The H/He ratio is displayed horizontally; temperatures, when known, are also indicated. It is implicitly assumed that each subclass may be associated with a comparatively small range of each of these parameters.

Several things are immediately noticeable:

(1) The constancy of M_{bol}, within the uncertainties, is demonstrated. The bars correspond to the observed range of M_v in the LMC or to ±0.4 mag., the rang observed for the WN4 stars; the latter are the most numerous subclass in the LMC, the intrinsic range of their luminosities is better determined than for any of the other subclasses and is probably typical. For the WN6 stars the bars include the uncertainty due to two divergent temperature determinations. The observed range of M_v presumably corresponds to a spread in both M_{bol}-mass and radius-temperature.

(2) The H/He ratio alone is not sufficient to determine the subclass of the star; e.g., WN4 and WN6 stars are found to have comparable H/He ratios. While the H/He ratio may be an important parameter in determining the subclass, a second parameter is obviously necessary. The data give no clear indication what that may be. However, one should recall that the WN6 stars differ from other subclasses of the WN stars in their galactic distribution, being found more often than other types in directions close to the galactic center (Smith, 1968c). Moreover, among the WC stars, there are gross differences from galaxy to galaxy; in the Magellanic Clouds many of the subclasses are completely missing; in M33, the spectra of the two WC stars observed by Lynds (Wray and Corso, 1971) are quite unlike those in either the Galaxy or the Magellanic Clouds. Whatever the parameters are that control the WR subclasses, at least one of them is variable from place to place.

I have previously (Smith, 1968c; Smith, 1968d) suggested that the parameter responsible for the observed distribution of subclasses in the Galaxy and the Magellanic Clouds is initial chemical composition. Data has accumulated supporting the hypothesis that there are indeed variations of the interstellar chemical composition from galaxy to galaxy and from the center to the edge of large galaxies (see Peimbert and Spinrad, 1970). However, no causal relation between initial chemical composition and spectral subclass has yet been established.

(3) The sequence of decreasing H/He ratio is not monotonically related to the excitation subclass. In fact, the subclasses WN3, WN4 and WN5 are inverted. The observed correlation between H/He ratio and temperature strongly implies that the WN3 stars have the lowest surface temperature and the WN5 stars the highest. There is no serious objection to the suggestion, although it is, at first glance, surprising. The excitation temperatures of the atmospheres appear, in general, to be higher than the effective temperatures of the stars. Thus, the atmospheric temperature must be supported by a source other than the radiation field and there is no *a priori* reason why the two temperatures should be monotonically related. It is of interest in this regard to notice that the progression of spectral appearance from WN5 to WN3 consists mainly of progressive reduction of first the N III and then the N IV contributions to the spectrum; N V lines are strong in all of these subclasses. If ionisation decreases outwards (see Kuhi, in this Symposium) the observed spectral sequence could be produced by stars whose atmospheres have basically the same ionization structure, but different density structures, such that densities in the outer layers, where the N III and N IV ions are found are successively reduced as one proceeds from WN5 to WN3.

(4) The relative position of WN7 and WN3 is somewhat arbitrary since the H/He ratios for both are uncertain. However, there is obviously a sudden break in properties of the spectra as we pass from WN3 to WN7 and WN8; the excitation drops radically, He I lines become prominent and violet absorption edges become more frequent (most notably on the Pickering lines). The energy supply to the atmosphere of WN7 and WN8 stars is apparently insufficient to keep the helium doubly ionized. Since helium is presumably the dominant element in the atmosphere, a profound change of atmospheric excitation and structure is not surprising.

10. WC Stars

An attempt to interpret the WC stars in the same framework is possible. Unfortunately, H/He abundances may not be derived with the same facility as for WN stars because of severe blending of carbon lines with the Pickering series. Since the stars are not commonly associated with nebulae, temperatures are not available either. The interpretation depends heavily on the properties of binary systems and is better held until after Kuhi's review of binaries later in this Symposium.

Acknowledgements

I thank James Wray and George Corso for making their data and diagrams available prior to publication. The final manuscript and diagrams were prepared while the author was at the Institut d'Astrophysique, Liège, supported by an ESRO Fellowship; assistance given with typing, diagrams and photography is appreciated.

References

Aller, L. H. and Heap, S. R.: 1971, *Bull. Am. Astron. Soc.*, **3**, 400.
Bashek, B. and Scholz, M.: 1971, *Astron. Astrophys.* **11**, 83; **12**, 322.
Beals, C. S.: 1938, *Trans. IAU* **6**, 248.
Brandt, J. C., Stecher, T. P., Crawford, D. L., and Maran, S. P.: 1971, *Astrophys. J. Letters* **163**, L99.
Castor, J. I. and Van Blerkom, D.: 1970, *Astrophys. J.* **161**, 485.
Conti, P. S. and Smith, L. F.: 1972, *Astrophys. J.*, **172**, 623.
Courtès, G., Georgelin, Y., Monnet, G., and Pourcelot, A.: 1968 in Y. Terzian (ed.), *Interstellar Ionized Hydrogen*, Benjamin, New York, p. 571.
Courtès, G., Georgelin, Y. M., and Monnet, G.: 1969 *Astrophys. J. Letters* **4**, 129.
Cowley, A. P. and Hiltner, W. A.: 1969, *Astron. Astrophys.* **3**, 372.
Crampton, D.: 1971, *Monthly Notices Roy. Astron. Soc.* **153**, 303.
Gebel, W. L.: 1968, *Astrophys. J.* **153**, 743.
Georgelin, Y. P. and Georgelin, Y. M.: 1970, *Astron. Astrophys.* **7**, 133.
Graham, J. A.: 1965, *Observatory* **85**, 196.
Graham, J. A.: 1970, in W. Becker and G. Contopoulos (eds.), 'The Spiral Structure of Our Galaxy', *IAU Symp.* **38**, p. 262.
Hiltner, W. A. and Schild, R. E.: 1966, *Astrophys. J.* **143**, 770.
Iben, I.: 1967, *Ann. Rev. Astron. Astrophys.* **5**, 586.
Johnson, H. M. and Hogg, D. E.: 1965, *Astrophys. J.* **142**, 1033.
Kerr, F. J.: 1970, in W. Becker and G. Contopoulos (eds.), 'The Spiral Structure of Our Galaxy', *IAU Symp.* **38**, p. 95.
Kippenhahn, R.: 1969, *Astron. Astrophys.* **3**, 83.
Kuhi, L. V.: 1966, *Astrophys. J.* **143**, 753.
Kuhi, L. V.: 1968, in K. B. Gebbie and R. N. Thomas (eds.), *Wolf-Rayet Stars*, U.S. Government Printing Office, Washington, D.C., p. 110.
Kuhi, L. V. and Smith, L. F.: 1973, *Atlas of Wolf Rayet Line Profiles*, U.S. Government Printing Office, Washington, D.C.
Mezger, P. G.: 1970, in W. Becker and G. Contopoulos (eds.), 'The Spiral Structure of Our Galaxy', *IAU Symp.* **38**, p. 107.
Morton, D. C.: 1969, *Astrophys. J.* **158**, 629.
Morton, D. C.: 1970, *Astrophys. J.* **160**, 215.
Oke, J. B.: 1954, *Astrophys. J.* **120**, 22.
Peimbert, M. and Spinrad, H.: 1970, *Astron. Astrophys.* **7**, 311.

Popper, D. M., Jørgensen, H. E., Morton, D. C., and Leckrone, D. S.: 1970, *Astrophys. J. Letters* **161**, L. 57.

Sandage, A.: 1957, *Astrophys. J.* **125**, 422.

Sim, M. E.: 1968, *Publ. Roy. Obs. Edinburgh* **6**, No. 5, 123.

Smith, H. J.: 1955, Dissertation, Harvard.

Smith, L. F.: 1968a, *Monthly Notices Roy. Astron. Soc.* **138**, 109.

Smith, L. F.: 1968b, *Monthly Notices Roy. Astron. Soc.* **140**, 409.

Smith, L. F.: 1968c, *Monthly Notices Roy. Astron. Soc.* **141**, 317.

Smith, L. F.: 1968d, in K. B. Gebbie and R. N. Thomas (eds.), *Wolf-Rayet Stars*, U.S. Government Printing Office, Washington, D.C., p. 54.

Smith, L. F.: 1968e, in K. B. Gebbie and R. N. Thomas (eds.), *Wolf-Rayet Stars*, U.S. Government Printing Office, Washington, D.C., p. 72.

Smith, L. F.: 1972, in A. D. Code (ed.), *The Scientific Results from the Orbiting Astronomical Observatory (OAO-2)*, National Aeronautics and Space Administration, Washington, D.C., p. 429.

Smith, L. F. and Bachelor, R.: 1970, *Australian J. Phys.* **23**, 203.

Smith, L. F. and Kuhi, L. V.: 1970, *Astrophys. J.* **162**, 535.

Stothers, R.: 1963, *Astrophys. J.* **138**, 1074.

Stothers, R.: 1965, *Astrophys. J.* **141**, 671.

Stothers, R.: 1966, *Astrophys. J.* **144**, 959.

Van der Borght, R. and Meggitt, S.: 1963, *Australian J. Phys.* **16**, 68.

Wray, J. D. and Corso, G. J.: 1972, *Astrophys. J.*, **172**, 577.

DISCUSSION

Thomas: Regarding the question of the masses, suppose I accept that some central stars of planetary nebulae have a Wolf-Rayet phase, what do you do then? The mass drops to one or two or three solar masses.

Smith: Kuhi has the data relating to masses assembled; the Population I Wolf-Rayets range from 5 to 15 solar masses. So they must evolve from stars which are ten solar masses or greater. The nuclei of planetaries are probably about one solar mass and must be considered separately. Here I am considering only Population I WR stars.

Thomas: But you cannot close your eyes to the other things.

Smith: I am just saying that every star greater than a certain mass or a large fraction of stars greater than some mass go through the Wolf-Rayet phase. The phenomenon is not as rare as we thought. We are not talking about one in a thousand stars becoming a Wolf-Rayet star. We are talking about one in ten.

Thomas: If the central star of a planetary is a Wolf-Rayet phenomenon and if the central star is one or two solar masses, then just taking the kind of numbers you gave (the smaller the mass the less life time of the phase) one would get down to 10^3 yrs or something like this for the life time of a Wolf-Rayet star in a planetary nebula.

Smith: Obviously the planetary nebulae nuclei have a completely different evolutionary situation. To derive their lifetime you would have to know what mass range on the main sequence they evolve from and compare the number of WR nuclei with the number of main sequence stars. It could be done but I have not done it.

Thomas: All I say is that if I have two different kinds of things exhibiting a Wolf-Rayet spectrum then I would just worry if you restricted your attention to one of them as representing all Wolf-Rayet phenomena.

Smith: One should also do the same calculation, comparing numbers of OB and WR stars in the Magellanic Clouds.

Thomas: In regard to the question of the observed visual magnitude and the bolometric corrections, I wish Lindsey Smith to clear one thing up. It seemed to me that what you were going to say was that those stars which have a low visual magnitude, but a high bolometric correction, would do so because of the difference in size. But I expected something different. Here you say that WN5 has a greater extended atmosphere than WN8 and I would expect it the other way because it has a lower visual magnitude.

Smith: What you say is correct. The bolometric magnitude depends on the effective temperature and on the size of the photosphere. It appears to be approximately constant. Thus the radius you would observe in the continuum must decrease dramatically with increasing temperature.

Thomas: So that when you say bigger you imply that the momentum transfer is still larger.

Smith: To explain the properties of the emission spectrum, I suggest that the radius of the line emitting region increases as you go upwards and to the right in the diagram.

Thomas: Now, physically, how does this happen?

Smith: I do not know. Three years ago I would have suggested that pulsational instability increases in that direction. The failure of all efforts to detect light variations due to pulsational instability and the increase in the theoretical mass limits for instability to occur make that suggestion very weak now. So I merely emphasize that the intensities and widths of the lines in the subclasses WN5 and WN6 (both binary and 'single') are much greater than in subclasses WN3 and WN4. It implies that the amount of material in the atmospheres is much greater, i.e., that the atmospheres are probably both bigger and denser. Thus, one requires a greater energy input into the atmospheres of WN5 and WN6 than in WN3 and WN4. In terms of our 1968 discussions, I suggest a greater *mechanical* energy input. The anticorrelation I suggest between T_{eff} and excitation class speaks against radiative control.

Underhill: There are two points which I think it important to clarify concerning your deduction of zero or low hydrogen abundances in WN stars. (1) The unit of flux in your diagram giving line strength vs. upper quantum number of He II is F (5000 Å). The emission line strengths were measured at the wavelengths of the lines. I assume that you used spectral scans such as those of Kuhi to relate the continuous spectrum at the line to that of the unit at 5000 Å; if so, we can infer that the measurements are all on the same scale and any error on this point is proportional to the uncertainty of defining the continuous spectrum; (2) My greatest dismay with your implied knowledge of the H/He ratio is that you have implied that the answer is known to some very serious theoretical problems which are intrinsically very significant for understanding Wolf-Rayet spectra.

Consider a line in a hydrogen like spectrum with upper level n and lower level m. Then the measured flux in arbitrary units is given by

$$F(n \to m) \sim N(m) \, A_{nm} \times hc\lambda^{-1}{}_{nm} \times \text{(transfer effects in line } n \to m).$$

For two lines $n_1 \to m_1$ and $n_2 \to m_2$ measured in the same units,

$$\frac{F_{n_1}}{F_{n_2}} = \frac{N(n_1)}{N(n_2)} \times \frac{A_{n_1}}{A_{n_2}} \times \frac{\lambda_{n_1}}{\lambda_{n_2}} \times \frac{\text{(transfer effects)}_1}{\text{(transfer effects)}_2}$$

Here n_1 represents the upper quantum number of line 1, n_2 represents the upper quantum number of line 2. My criticism lies in that you are assuming that the ratio of the line transfer effects (which is the solution of the equation of transfer) for both lines is unity. This is only in a truly thin nebula. If you take the estimated electron temperature and density in a typical Wolf-Rayet atmosphere of the type WN6 to be 10^5 and 10^{11} respectively, you find that an optical depth 10 is reached in a distance of the order of 1000 cm. Allowing for optical transparency introduced by motion, the effective path length may possibly be extended to 10^8 cm for Hε or 10^9 for the He II line (4–14). The geometric length of a Wolf-Rayet atmosphere is considered to be something like thirty solar radii which means that complete optical thickness will be reached in both lines. If this is so, then it is essential that the transfer effects in the H and He II lines be evaluated before anything can be said about the relative abundance of H and He.

Smith: We measured the equivalent width relative to the continuum in each case. And then we have used Kuhi's published continuum flux determinations corrected for reddening to convert to flux in units of the continuum of the star at 5000 Å.

Observationally the case is based, first of all, on Castor and Van Blerkom's calculations that, for HD 192163, the Pickering lines are thin for $n > 10$. The change in slope of the decrement observed for HD 192163 and HD 50896 dramatically confirms the suggestion that there is a change at about $n = 10$ from thin lines to thick lines. Second, when the Pickering line is blended with N III or He I the optical thickness is not so great that the line strength is not greatly increased. Third, there is marked consistency of the slope of the graph between the high-n sections of HD 192163 and HD 50896 and those other subclasses in which the lines are weaker and a contribution from hydrogen is clearly observed.

Fourth, if you did have an abduance ratio of 10 to 1 between H and He in these stars your H lines might still be thick to high value sof n, but the helium lines certainly would not be. Thus, from observational evidence alone, I think it is clear that the higher Pickering lines are optically thin in al.

the stars observed. I emphasize that we are talking about lines that only exceed the continuum by about ten per cent and their equivalent widths are down to an angstrom by the highest members of the series.

Underhill: How do you account for that formula that I just sketched on the blackboard?

Smith: If you had such high opacity in such weak lines, I wonder what is the opacity in the continuum. Perhaps you had better choose different numbers.

Underhill: I am talking about the opacity just in a line. If I add the continuous absorption in an atmosphere with typical parameters, as deduced by Castor and Van Blerkom, I have a really opaque atmosphere.

Smith: Then you are not talking about the atmosphere, you are talking about the interior of the star.

Conti: May be a few more people would like to contribute to this discussion.

Niemela: In γ_2 Velorum, He I lines and the hydrogen lines display the so-called V/R variations; these variations are not present in the He II and C lines. This suggests that hydrogen may be present.

Bappu: At what phases were the plates taken?

Niemela: At 47 and at 4 days.

Conti: You now say that the V/R variations are purely due to the presence or absence or dominance of H or He. On certain occasions, could it not be due to the fact that the O companion's velocity mutilates the emission. What strikes me is that the effect is not shown by all the He II lines.

Bappu: You find it only in $H\alpha$?

Niemela: I find it in $H\alpha$, $H\beta$, $H\gamma$, etc. and also in He I lines, but not in 5411 Å, 4200 Å and 4570 Å.

Conti: Are there any comments on this particular theme?

Van Blerkom: The optical depth in He II Pickering lines from two WN6 stars were computed by John Castor and myself on the basis of a simple expanding envelope model. In that analysis, the lines became optically thin for transitions out of levels with $n > 9$, which agrees with Lindsey Smith's interpretation.

Underhill: I know your calculations of He II, and I think the result is true. I am not at all certain that you can make the same argument for the $H\varepsilon$ line of hydrogen. You went on to argue that if $H\varepsilon$ was optically thin, your answers would come out as indicated. However, it is an interesting mental experiment to ask yourself what would be the answer with a normal hydrogen abundance and the same N_e and T_e.

Conti: Even despite the fact that you really do not see the hydrogen?

Underhill: Yes, I can see of ways of hiding the hydrogen lines. Possibly you cannot find them because of the free-free emissions.

Morton: Since the expanding atmosphere enhances the optical thinness, can Lindsey Smith tell us whether anyone of the stars you analyzed do show shifted absorption lines in the visual spectra, suggesting the existence of an expanding atmosphere?

Smith: Yes, WN5, WN6, WN8 and WN9 have the strongest violet displaced absorption; however, some violet edges are observed in all spectra, I think. I do not have the velocities now.

Underhill: It depends which lines you look at. Absorption edges are there in all of the Wolf-Rayet stars. You may look at He I, for example; sometimes you can see absorption edges in C III, sometimes in N IV. Sometimes you see strong displaced absorption lines from levels that are not expected to be strongly overpopulated by non-LTE effects. In C IV and N V such lines are extremely strong. That is interesting because you cannot hold the atoms up there unless you get tremendous optical depths.

Conti: Then, I suppose we might say that most WC and WN stars have at least some violet displaced components in some lines, and it is different for different stars.

Smith: And in certain stars you see them in He II; in some stars they are not conspicuous in He II.

Alcaino: Coming back to your absolute luminosity slide Lindsey Smith, you said that most of them are calibrated from the distance moduli of the Magellanic Clouds. Are there any of these stars in globular clusters? Because it could bring up a very important point! If they are globular cluster members they must belong to Population II.

Smith: I know none in globular clusters.

Alcaino: Are there any anomalies in the color excesses? What about the U−B colors?

Smith: Crampton found that the color excesses from the Wolf-Rayet stars are very consistent with the color excesses from the OB stars, so there seems to be no problem with the colors. There are some WR stars in the Galaxy, in the galactic clusters, in Carina and in Sco-Cen OBI; in these cases, his numbers and mine do not differ very greatly. So most of the strong disagreements come from derivations that depend on distances of H II regions derived from one or two O and B stars, and you have all the uncertainty of luminosity assignment for O and B stars as well as assignment of the stars to the same H II regions.

Conti: I have just one question recalling our experience with γ_2 Velorum. You only have four or five stars in the Clouds which have -6.8. Could they not be binaries and could any of the possible companions really dominate the spectrum?

Smith: It is possible, particularly in the more luminous subclasses.

Conti: It really struck me when we made the absolute magnitude of the WC8 component of γ_2 Velorum fainter by one and a half magnitudes.

Underhill: Crampton in sending me the copy of his paper noted that the only stars he had significantly brighter than -5 were the WN7s which came up to -6.5 or so. He notes that these are in exotic places like η Car and Sco OBI. For two of the WN7 stars in Carina, as published in the Annual Reviews, you can see an O type spectrum superposed. Sally Heap told me of an O subdwarf star that has almost an identical looking spectrum. I do not know what is going on with type WN7. Crampton wrote to me and asked whether I thought there are two different kinds of WN7 stars. All I can say is yes, but I do not know the full answer.

Conti: We may end up with the same experience we had in γ_2 Velorum, that the O star, in fact, dominates. Perhaps Lindsey Smith would like to add something.

Smith: I indicated an enormous range for the mass and luminosity of the WN7 stars. This is based on two things. First of all, the observed range in absolute magnitude for the WN7 stars in the Magellanic Clouds is nearly two magnitudes; so, either they do have companions, as has been suggested, or they do have an enormous range in their intrinsic luminosity. Second, they also appear to have an enormous range in their mass. The binary HD 228766 (WN7 + O) has a large mass function and may be edge on; that would put the mass of the WN7 star at about 5 $M_\odot$, one of the lowest mass Wolf-Rayet stars we know. On the other hand, if there really are WN7 stars with absolute magnitude as high as -7.5, their masses must be about 15 $M_\odot$.

Bappu: I was just wondering whether the calibrations of the WN7 stars you have picked up from the Magellanic Clouds, are affected by inadequate sky background corrections or field stars below the limit of visibility. Do the WN stars have any preference for the bar of the Clouds?

Smith: The WN7 stars are in 30 Doradus, η Carinae, etc. The problem is not so much about faint stars, but the nebulosity. I did the obvious things like taking the nebulosity on two sides and so on. I do not think it is a serious problem.

Bappu: Regarding HD 151932, if you look at the spectrum of this star in the near infrared, there seems to be positively something odd in the energy distribution. The energy increases towards longer wavelengths. I wonder if Kuhi has any scans of this star, for it certainly is an object which one should look with a scanner as early possible. It is in NGC 6231. I shall show later today a slide which will give a rough idea of the energy distribution.

Kuhi: I do not have any scans of the star but I hope that I shall soon come to Chile when the scanner at Tololo gets into operation.

Bappu: I am very happy to see that all WC stars have come down within the same domain of absolute magnitudes. However, I still find WC8 brighter by about 0.4 of a magnitude. Rajamohan at Kodaikanal has looked into this problem by using Mount Stromlo spectra of γ_2 Velorum at 6A mm^{-1} dispersion. He has utilized the hydrogen absorption lines, with the old-fashioned way of making corrections. He has also determined the distance of γ_1 Velorum, using Petrie's technique, but with the improved calibration, and the absolute magnitude of γ_2 comes out to -3.1. With a difference of 2.4 mag., the entire γ_2 system becomes -5.5. The absorption line intensities give you the fact that the O star is brighter than the Wolf-Rayet star by 0.6 magnitudes while the emission line intensities in infrared give a difference of 1.4 mag. Assuming then that the O star is brighter by 1.0 mag., the final values of the absolute magnitudes are -4.2 for the WC8 star and -5.2 for the O star.

Smith: That is awfully faint for stars that are 15 and 60 solar masses. Conti and I have summarized the argument in our current paper which Kuhi will review later. My inclination is to stick to Graham's distance of 460 pc which has been recently confirmed by Brandt *et al.*

Conti: Lindsey Smith and I have had a re-analysis of γ Velorum and have come out with somewhat different numbers all around. We derived an absolute magnitude of -4.6; we have the O star brighter by 1.4 mag., and the distance is taken from Hβ photometry.

Kuhi: I would like to remark that Lindsey Smith's comments as far as the H/He ratio, apply only to the WN stars. It is almost impossible to do the same sort of analysis for the WC stars because their spectra are so badly blended that it is almost completely hopeless to trace out the Pickering series or any series throughout the normal photographic spectral region. I have also obtained scans out to 1μ or so, of a number of the brighter WC stars in the hope of finding relatively free spectral regions and have not succeeded. The WC spectrum is full of emission lines, all the way out to about 1.2μ; the WN's, on the other hand, out to about 1.3μ, are dominated by helium lines and virtually nothing else.

Sahade: I would just like to make two comments. In regard to γ Velorum; I wish to remark that γ_1 Velorum is a spectroscopic binary and that its velocity curve is somewhat peculiar, atleast we have not been able to settle on a definite velocity curve so far. The second comment is that the so-called V/R variations of the intensities of the emission lines which Virpi Niemela showed this morning to be phase dependent in stars like γ_2 Velorum, are also present in WN objects. They are seen very nicely on the illustrations of Giudo Munch's paper on V444 Cygni, and they also appear to be phase dependent.

Kuhi: I think that those changes in V444 Cygni or any of the other stars that are binaries really just confuse the problem of determination of the H/He ratio and of everything else. I am sure that they arise from material between the two stars. In fact, changes occur in geometrical properties as one looks at the system from different aspects, as well as changes in the physical properties of the material associated with the stars, especially the material located in the regions between them. So, I think one should stick to the single stars, in all cases, if one is going to determine anything about hydrogen and helium.

Sahade: I agree with Kuhi that these effects are connected with the matter which is between the two stars. But I think we should make a point in trying to see whether there are any changes in the spectra of those stars which we think are single.

Underhill: There is a point that Kuhi covered rather quickly. If it is a binary you can see the material at certain phases and you can see changes. Now, you should ask yourself the question why do these changes happen? Is it something originating in the Wolf-Rayet star? If so, then, the same thing may originate in a single Wolf-Rayet star and you should ask yourself, how could I observe it? It may be that a particular geometric arrangement in specially lined up binaries enables us to see it, but if these changes are going on anyway, it may have a lot to do with the ratio of transfer effects 1 to 2. Just glibly putting that ratio equal to 1, still remains a problem to me.

Kuhi: It is very likely in the binaries that we do see and detect this kind of effect if the material that we are looking at has large motion in the line of sight, which obviously, can happen very easily with systems seen edge on. However, if you are looking at a system face on, you are likely to miss everything.

Thomas: I cannot help sitting back and trying to just see overall what Lindsey Smith has done. And I must say, you make an awfully convincing case for the self-consistency of your argument, which seems to be based on three separate things; and you will forgive me if I do not worry about details. (A) you worry about your line ratios; (B) you worry about some absolute magnitudes (and that involves an observed visual magnitude and a computed bolometric correction), and (C) you come up with some stellar interior calcualtions resting on mass. In (A), I can have all kinds of uncertainty in the direction that Anne Underhill is worried about, but for the moment let us forget about it since it is simply an observational question. In (B) and (C) it is a question of the model, with all kinds of uncertainties. What you come up with is self-consistent for a mass in the range 10 to 15 solar masses. Now, if I go back and ask the same kind of question I asked before, namely, if I look at the Wolf-Rayet spectrum and ask, as at the last Symposium, is it a phenomenon or is it a kind of object, in my answer, I also observe that central stars of planetary nebulae are objects that show a Wolf-Rayet spectrum and have masses like one solar mass. How, then, can I reconcile this with the conclusion from (A) to (C)? Lindsey Smith's argument is beautifully self-consistent until I introduce something with a mass like 1, unless we say there exists several kinds of objects capable, for different reasons, of producing a Wolf-Rayet spectrum.

Smith: My intuitive answer to that would be, at the moment, give me a pure helium star and I will give you a Wolf-Rayet star.

Thomas: But I may also get a Wolf-Rayet star from something which is not a pure helium star. That is the basic thing.

Smith: No, I would suspect that the situation regarding the planetary nuclei is probably also one where we have a pure helium star of mass 1. I do not see why there should be any trouble with the same sort of arguments for this case. (I use the term 'helium star' rather loosely. Lack of hydrogen is required. Carbon cores or other more complicated models are not excluded).

Thomas: Yes, but you put a lot of emphasis on this being in the 10 to 15 mass range.

Smith: The observations suggest that the stars I have discussed are 10 to 15 solar masses, they do have the bolometric magnitudes and hydrogen to helium content I have shown.

Thomas: I thought your argument was that the interior models for a helium star, with mass 10 to 15 solar masses, would give me the kind of absolute magnitude to which all the WN sequence reduce.

Smith: I am saying that the observed stars must be pure helium stars because they are observed to have a certain bolometric magnitude and a certain mass, which eliminates them from being pure hydrogen stars. Population I stars are observed to be 5 to 15 solar masses.

Thomas: Can you get the same consistency with one solar mass?

Conti: I do not think we know the masses of the nuclei of planetary nebulae.

Thomas: Would you accept that they have a mass not much greater than one?

Conti: No, but your point is whether we can get Wolf-Rayet stars from two different kinds of masses, and I think both Lindsey Smith and I would agree with it.

Thomas: No, all I am saying is that you have a beautifully self-consistent argument, each link of which has a number of weak points. Either it is unique to mass 1 or 15 or it is not. If it is not, you had better examine each weak point.

Conti: I think the bolometric correction is a little shaky.

Thomas: I am not disputing a single thing. All I am saying is that there is a lot of uncertainty and the coherence of this very nice argument depends on the absolute magnitude being of the value (forget about fluctuations here) that a 10 to 14 solar mass pure helium star would produce. So, I say now, go to whatever the central star of the planetary nebuale are (helium, hydrogen), whatever their mass is; can I, then, make the same kind of a nice, self-consistent argument from them, and come up with some kind of a conclusion?

Bappu: I would like to ask Lindsey Smith how she explains the absence of λ 5696 in some of the WN stars, particularly, when you have an abundant supply of C IV ions. Selective excitation I could understand normally, but, how do you have a sort of selective supression?

Smith: We see C IV in the λ 5810 line in nearly all of the WN stars. You see the whole of the C IV spectrum, as I showed you in HD 50896, and we do not know if this situation applies also to the others. The only place I have seen C III, is in the WN8's, where it appears quite well developed. I have never seen λ 5696 in a WN spectrum. Henry Smith does claim to see it in HD 50896 but it certainly was not on our spectrogram. I guess the implication is that most of N stars have C IV and it is only when you get down to the WN8s that the excitation is low enough that you see C III. After all, the ionization is comparable. C IV is the ion you expect to be the strongest in HD 50896.

Bappu: Yes, but C III could not be that weak.

Smith: I agree. By strict analogy between the C and N spectra we should have been able to see the stronger C III lines.

Underhill: That is still in question. You have to work it out. I can show some graphs, giving some rough calculations that are not applicable to Wolf-Rayet stars, but if you are brave you can interpolate and get an idea of the ionization fractions in non-LTE situations.

UV AND RADIOFREQUENCY OBSERVATIONS
OF WOLF-RAYET STARS

HUGH M. JOHNSON

Lockheed Missiles and Space Company, Palo Alto, Calif., U.S.A.

1. UV Observations with OAO-2

Very few observations are available. Of early rocket work on γ Vel (Stecher and Milligan, 1962; Smith, 1967) the most detailed spectra are by Stecher (1968) in the range λ 1200–3100 Å and by Carruthers (1968) in the range λ 1050–1216 Å. West's (1971) spectrum of γ Vel with the OAO-2 spectrometer (Code *et al.*, 1970) in the range λ 1050–2000 Å appears to be somewhat different from Stecher's. Comparison must await full details.

The OAO-2 spectrometer has scanned apparently only one other W star, HD 50896. It is more informative than γ Vel because it is possibly a single star without contamination by a dominant O-star spectrum. It is also one of the WN5 stars in a symmetric nebula (cf. Section 2). Smith (1971) has reported the OAO-2 spectrum of HD 50896. Let us make a few more remarks about the same scans. First, the measured FWHM of the strong He II 1640 emission is $\Delta\lambda = 20$ Å, or about 17 Å after correction for instrumental broadening, in agreement with the ratio $\Delta\lambda/\lambda$ of fairly unblended lines at wavelengths several times larger. Thus Doppler broadening is confirmed. Second, interstellar hydrogen absorption at λ 1215 Å may be estimated from the 21-cm survey of McGee and Murray (1961): $N_H = 2.0 \times 10^{21}$ H atoms cm^{-2} toward HD 50896. Since the star is at $l = 234°8$, $b = -10°1$, and at the distance 1.59 kpc (Smith, 1968b),

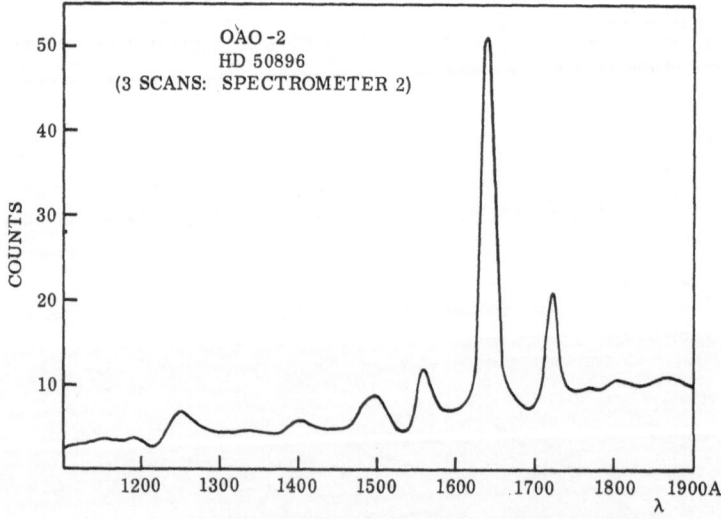

Fig. 1. Scanned spectrum of the WN5 star HD 50896.

M. K. V. Bappu and J. Sahade (eds.), Wolf-Rayet and High-Temperature Stars, 42–53. All Rights Reserved.
Copyright © 1973 by the IAU.

most of the hydrogen should be in front of it. Morton's (1967) formula for the equivalent width $EW(L\alpha) = 7.31 \times 10^{-10} N_H^{1/2}(A)$ due to radiation damping of the interstellar line, then gives $EW(L\alpha) \leqslant 33$ Å in absorption. As estimated from Figure 1 the net equivalent width of the features around $L\alpha$ is $\leqslant 30$ Å in emission, so the net equivalent width of the features before interstellar absorption should be < 60 Å in emission. Presumably the dominant emission near λ 1215 Å would be He II 1215, the second line in the series which He II 1640 heads. We estimate $EW(\lambda 1640) = 140$ Å, a strength which appears to confirm the absence of continuum from any hot companion star.

The OAO-2 photometers (Code *et al.*, 1970) have been used in the OAO-2 guest-investigator program to observe HD 192163. This WN 6 star was selected because it is apparently a single star and it is in a symmetric nebula, NGC 6888 (cf. Section 2). In respect to the latter it may not be normal. The four photometers are each equipped with three filters. The range of effective wavelengths $\lambda\lambda 1330$–3320 Å is covered with passbands of FWHM = 240–860 Å such that practically no gaps are left in the observed continuum. However, the data will tell little about a line spectrum. Figure 2 shows the observations. Figure 3 corrects them for interstellar extinction by means of the color excess $E_{b-c} = 0.33$ mag. for HD 192163, and the ratio $E(B-V) = 1.6 E_{b-c}$ (Smith and Kuhi, 1970), and the average ultraviolet extinction $E(\lambda - V)/E(B - V)$ given by Bless and Savage (1971). There is still a question about the calibration of some of the photometer-filter responses (Bless, 1971), but it is apparent that the run of the data corrected according to this differential extinction agrees rather well with an interpolated model atmosphere of $T_{eff} = 30700$ K. This is the effective temperature for HD 192163 which Morton (1970) derived in a completely different way.

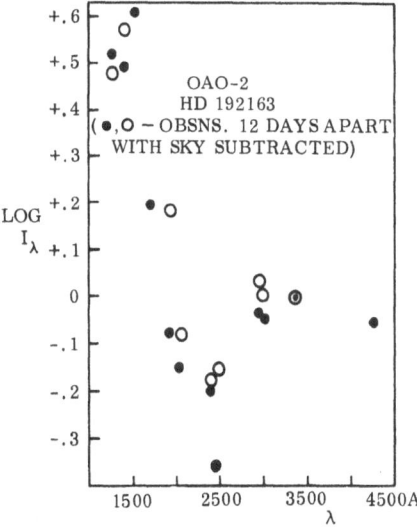

Fig. 2. Photometered spectrum of the WN6 star HD 192163, with intensity per unit wavelength I_λ normalized to an arbitrary zero-point at λ 3320 Å. The first set of data has 12 solid points; the second has 9 open circles.

Ultraviolet spectral lines might have some effect on Figures 2 and 3, for example the strong line of EW(λ 1640) = 140 Å such as HD 50896 exhibits. It would fall 40 Å off center in the passband with $1/\lambda_{eff} = 5.95\mu^{-1}$ and FWHM = 270 Å. It should therefore add close to $140/270 \times 100\% = 52\%$ to the light of the continuum, but in fact this point falls low in the plots. We must conclude that He II 1640 is not prominent in HD 192163. C IV 1548–50 and N V 1238–42 fall in passbands with relatively high points in Figure 3, but we cannot claim the presence of these lines from the data.

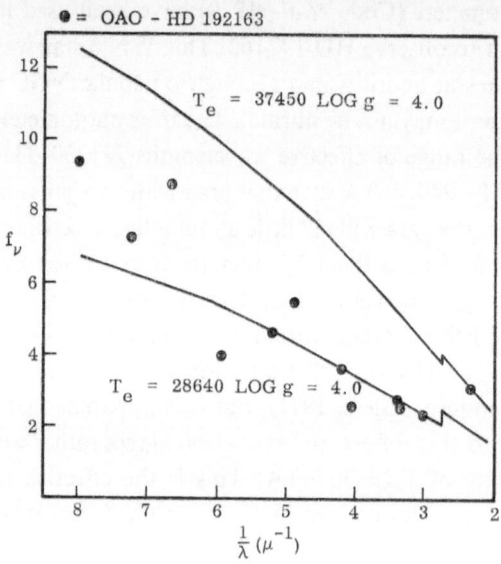

Fig. 3.   Mean data of HD 192163 (points) in Figure 2 reduced to intensity per unit frequency, corrected for interstellar extinction, and compared with two model O-type atmospheres (labeled curves) by Bradley and Morton (1969). Dr Morton suggested the comparison with these models. The zero-point of observed intensity is fitted arbitrarily to the scale f_ν of the curves.

Houck (1971) has made 200 OAO-2 spectrometer scans of β Lyr at 10 Å resolution in the range $\lambda\lambda$ 1100–1800 Å. He has found emissions of C IV, Si IV, and 'Lα' (the latter periodically shifted from interstellar absorption by orbital motion) in a spectrum which compares with a star such as γ Vel.

Davis (1971) has presented preliminary photometry of about 500 stars observed with the Celescope on OAO-2. Two of them are W stars, CD - 45°4482 and HD 76536. The new data are magnitudes called U1 and U2, respectively taken with passbands of 2100–3200 Å and 1550–3200 Å. Despite the gross passbands, additional information on effective temperatures of a few W stars should come out of the final catalog.

2. Observations of W Stars With Symmetrical Nebulae Around Them

Johnson and Hogg (1965) used the Green Bank telescopes to detect NGC 6888 around HD 192163 at 750 MHz and 1400 MHz; also NGC 2359 around HD 56925 at 750,

1400, and 3000 MHz. The W stars are included with the nebulae in the available tele-scope beams but it has been assumed that the nebulae account for the observed flux densities. Johnson and Hogg (1965) reported the privately communicated independent discoveries by Herbig and by Minkowski of a third nebula 35′ in diameter around HD 50896, which may be called S 308, but it was not detected in radiofrequency until later (Johnson, 1971). Smith (1967; 1968a) searched Palomar Sky Survey charts and other material, discovered four more symmetric nebulae around W stars, and noted that all seven stars were WN 5, 6, or 8 types. Smith and Batchelor (1970) pro-ceeded to observe three of them, NGC 3199, RCW 104, and RCW 58 at 11 cm. Lozinskaya (1970), Terzian (1970), and Johnson (1971) observed NGC 6888 again at 8500, 318, and 7795 MHz, respectively. The radiofrequency spectrum of NGC 6888 is

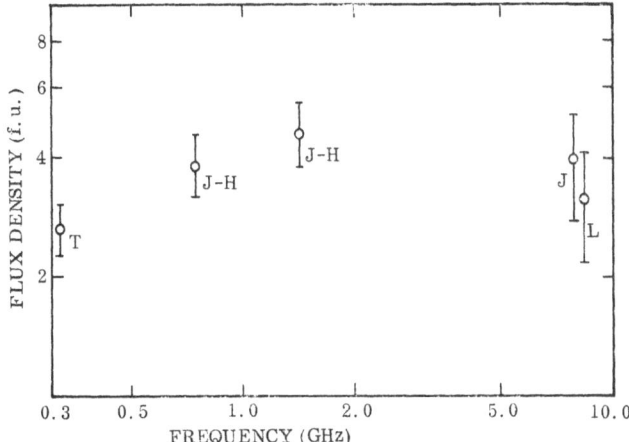

Fig. 4.   Radiofrequency spectrum of NGC 6888. Unit flux density (f.u.) = 10^{-26} Wm^{-2} Hz^{-1}. The observed data are by Johnson and Hogg (1965) J-H, Lozinskay (1970) L, Terzian (1970) T, and Johnson (1971) J.

shown in Figure 4. It is apparently thermal Bremsstrahlung. These radiofrequency data, or alternatively the integrated flux of the nebula in a Balmer line, corrected for extinction, such as Parker (1963) has estimated only for NGC 6888, with a distance estimate of the W star, lead to the mass of the nebula and to the 'excitation parameter' U of the star, or to the equivalent spectral type and effective temperature. The 'Zanstra temperature' of W stars is also obtained if the stellar magnitude, rather than the dis-tance, is employed with the nebular flux density. Morton (1970) refined the Zanstra method by calculating model atmospheres in place of blackbodies, and he found effective temperatures as a function of W-star types. By including the observed angular size we also get the linear dimensions and a confirmation of the mass distribution (thin-wall shell appearance on photographs). Finally, observations of the internal motions of the nebula may be combined with observations of the rate of mass loss and ejection velocity from the W star, and with an estimate of ambient interstellar density, in order to obtain the age of the system.

Conservation of momentum and continuous ejection were assumed in the simple theory (Johnson and Hogg, 1965). Spherical symmetry was also assumed, tacitly contrary to the observed ellipticity of the nebulae, eccentricity of stellar site, and correlated azimuthal asymmetry of nebular perimeter (brightest nearest the star). In apology one can only say that these nebulae are quite symmetrical in comparison with ordinary diffuse nebulae, and perturbations that are attributable to a stellar or interstellar magnetic field, or to irregularity of ambient interstellar density, appear to be negligible in the first objectives of the theory. The higher degree of central-star concentricity which prevails in planetary nebulae may be explained by their smaller radii, statistically higher z-distances, and consequent lack of interaction with the ambient medium. However, the non-circular projection of many planetary nebulae shows that central stars do not eject mass isotropically.

The full application of these ideas has been made only to NGC 6888, the W-star nebula in which large nebular velocities are observed (Courtès, 1960); Lozinskaya and Esipov, 1968; Lozinskaya, 1970; Georgelin and Monnet, 1970). The result is a self-consistent picture of a single WN 6 star with the effective temperature of about 31 000 K, which for unknown reasons ejects mass at a velocity of 1400 km s^{-1} and rate of 10^{-5}–10^{-6} M$_\odot$/yr into collision with an ellipsoidal nebular shell of projected dimensions $12' \times 18'$. If the distance is 1.2 kpc the mean radius of the shell is 2.6 pc, which is considerably larger than the shells of planetary nebulae. Smith (1968b) finds the distance to be 2.29 kpc, but this summary is based on the smaller distance. The mean linear thickness of the nebula is only 10^{-2} pc (shell walls); its electron density is 400 cm^{-3} and electron temperature is 15–19×10^3 K. The shell is expanding 50–80 km s^{-1} and sweeping up interstellar matter with a density of about 1–2 cm^{-3}. The age is about 2×10^4 yrs if the ejection and transfer of momentum has been steady. The shell mass is about 4 M$_\odot$ of which only 3% has been contributed by the ejecta of the W star. Thus the element abundances inferred from the stellar spectrum of HD 192 163 need not agree with those inferred from the nebular spectrum (normal). The excitation of the shell appears to be radiative, not collisional.

Before these investigations NGC 6888 was classified as a supernova remnant (e.g. Pikelner, 1959; Lozinskaya and Esipov, 1968) or as a 'giant planetary nebula' (e.g. Parker, 1964). Minkowski (1965) has suggested also that S 308, the shell around HD 50 896, is possibly a planetary nebula. Lozinskaya and Esipov (1968) and Georgelin and Monnet (1970) agree that the mean radial velocity of NGC 6888 is -50 km s^{-1}; no rms error is given. (Lozinskaya and Esipov also estimated the radial velocity of HD 192 163 equal to -120 ± 20 km s^{-1}, despite the difficulty of the broad spectral lines.) The galactic coordinates are $l = 75°.5$, b $= +2°.4$, the component of differential galactic rotation is $+7$ km s^{-1}kpc^{-1}, and the component of solar motion with respect to the local standard of rest is 18 km s^{-1} toward HD 192 163. Hence one component of the velocity of the nebula with respect to its local standard of rest is probably -41 to -49 km s^{-1} at distances of 1.2–2.3 kpc. Of course, the near side of the nebula may contain most of the observed filaments. If not so, the mean velocity is appropriate to a 'runaway' O star or a planetary nebula.

In view of the high dispersion of velocities in NGC 6888, it is interesting that three other examples do not show it. They are given in Table I and their velocity dispersions do not differ significantly from those in ordinary nebulae. Successive columns give the nebula, its radius in arc minutes and in pc according to the spectroscopic distance, its mass (Smith and Batchelor, 1970), the rms σ of n velocities in the nebula (Georgelin and Georgelin, 1970), the included star, its spectral type, its spectroscopic distance R(sp) (Smith, 1968b), and the kinematical distance of the nebula R(kin) (Georgelin, 1970). If the theory which has been applied usefully to NGC 6888 is applied to the objects of Table I, one must conclude that these nebulae have become massive and the velocities of the shells have slowed down to ordinary internal motions in nebulae,

TABLE I

Symmetric nebulae with small dispersions of velocity

Nebula	Radius		M	σ	n	Star	Sp.	R(sp)	R(kin)
	('arc)	(pc)	($\odot$)	(km s^{-1})				(kpc)	(kpc)
NGC 2359	2[a]	4[a]	330[a]	8.0	29	HD 56925	WN5	6.92	4.57
NGC 3199	8	8	400	6.5	33	HD 89358	WN5	3.63	[b]
RCW 104 + 106	10	18	650	8.0[c]	60[c]	HD 147419	WN6	6.31	3.75

[a] Pertains to the sharp inner ring of the nebula.
[b] Differential galactic rotation at $1 = 283°5$ is too small for a significant kinematical solution.
[c] Data for both nebulae.

because they are older than NGC 6888. For conventional values of the parameters, e.g. rate of mass loss $= 10^{-5}$ M$_\odot$/yr, velocity of ejection $= 10^3$ km s^{-1}, and interstellar density $= 1$ hydrogen atom cm^{-3}, the derived ages are $2-20 \times 10^5$ yrs, or $10-100 \times$ the age of NGC 6888. The difficulty is that WN 6 stars apparently survive 2×10^6 yrs if HD 192163 and HD 147419 are equal members of the class WN 6. But the total mass loss is 20 M$_\odot$ at this age, and the mass of WN 5-6 stars in binary systems is about 11 M$_\odot$ (Smith, 1968a). A reduction in rate of mass loss by a factor of 10 increases age by about $\sqrt{10}$ so that the total mass loss is then 6 M$_\odot$ rather than 20 M$_\odot$. An increase of ambient density outside the shell by a factor of 10 also increases age by about $\sqrt{10}$.

The mass of RCW 58 is estimated to be small, 5.7 M$_\odot$ (Smith and Batchelor, 1970), and likewise the mass of S 308 to be only 3 M$_\odot$ (Johnson, 1971). They are candidates for large expansion velocities but they have not been observed with the interferometer. Neither has any symmetric nebula around a W star been reported in the radiofrequency H recombination lines. It is reasonable to consider measuring proper-motion expansions in NGC 6888 and S 308, but RCW 58 may be too small for the scale of available plates. At present we tentatively conclude that the less massive nebulae of the class we have discussed are the younger, and that the best-studied member of the class, NGC 6888, may be peculiar as well as young.

If we accept the theory of ejected stellar mass sweeping out a volume of ambient interstellar gas around some W stars, and note the result that the star contributes only a few per cent of the total mass in each case, we may derive the ambient interstellar density N_H of the swept volume. This is, first, a check on self-consistency of the theory for which Johnson and Hogg (1965) originally had to estimate N_H independently in order to establish the theory; and, second, it makes possible some comparisons with the rival suggestion that the gas in a symmetric nebula has been ejected from the star at some earlier stage of evolution (e.g. red giant) and said gas might have no dependence on current mass-loss in the W star. The four nebulae of Table II are the only ones for which the symmetric-nebular mass M has been derived from the spectroscopic distance R, the observed flux density, and the observed nebular gas density N_e via the method of the [O II] 3726–29 doublet-ratio. If the gas is non-uniform, N_e is overestimated and M and N_H are underestimated. We draw the following conclusions from the table: first, there is an apparent dependence of N_H on the z-distance from the galactic plane,

TABLE II

Interstellar densities near W stars

Nebula	Star	Sp.	R(sp) (kpc)	z (pc)	r (pc)	N_e(sp) (cm^{-3})	M ($\odot$)	N_H (cm^{-3})
NGC 2359	HD 56925	WN5	6.92	12	4[a]	100[a]	330[a]	50
NGC 3199	HD 89358	WN5	3.63	63	8	240	400	5.7
NGC 6888	HD 192163	WN6	2.29	96	5	400	21	1.2
RCW 104	HD 147419	WN6	6.31	165	18	190	650	0.8

[a] Pertains to the sharp inner ring of the nebula.

in the sense expected of swept interstellar gas rather than of a mass entirely derived from the star during stellar evolution. Second, N_H is a factor of 10 greater than mean interstellar densities, so that an association of W stars with denser clouds is implied. Alternatively, we could say that N_H decreases from type WN 5 to type WN 6, and the absolute density is governed by factors other than ambient interstellar mean density to make it greater than the mean. At present the interstellar density near specific W stars is not well enough known independently to decide.

It is interesting to ask about the visibility of the ejecta of W stars before any effects of collision with ambient interstellar matter. If the velocity of the mass lost at radii $r \geqslant 30\,R_\odot = 7 \times 10^{-7}$ pc is constant, then electron density $N_e(r) \propto r^{-2}$. If $N_e = 10^{12}$ cm^{-3} at $r = 30\,R_\odot$ the hydrogen emission measure EM $= \int N_e(s)\,N_H(s)\,ds$ may be computed along lines of sight which pass ϱ pc from W stars. In the approximation that ϱ is small compared with the distance of the star, EM $= 0.12\pi\varrho^{-3}$ pc cm^{-6}. For example, at $R = 1200$ pc, EM $= 1.9 \times 10^6$ pc cm^{-6} at 1″ arc from the star, or 1.9×10^3 pc cm^{-6} at 10″ arc. Nebulae for which EM $\geqslant 400$ pc cm^{-6} are commonly visible on Hα photographs. However, the star image would be competitive on ordinary photographs, and the hydrogen emission measure would not be appropriate for hydrogen-deficient W stars. According to Pengelly (1964) the intensity of recombination He II 6560 is 0.6 × the intensity of Hα in the limit of a low electron density and at $T_e = 10^4$ K. Likewise the intensity of He II 4686 is 4.4× the intensity of Hα. We should probably find $N_e = 2N_{He}{}^{2+}$ rather than $N_e = N_H{}^+$ in the envelopes of some W stars. Note also that, according to the adopted density law, $N_e \leqslant 10^{10}$ cm^{-3} at $r \geqslant 300$ $R_\odot$, and forbidden lines can be emitted. They may compete with He II lines in intensity as in planetary nebulae.

But one more question remains about the expanding envelope of a W star regarded as a special H II region. Is the Strömgren radius r_s significantly larger than the star? Can the continuum photons of the core below λ 912 Å escape to photoionize the symmetric nebulae? If the envelope density $N_e = 10^{12}$ cm^{-3} is constant inward from $r = 30$ $R_\odot$ to the photosphere of a star of radius $r_* = 6$ $R_\odot$ and effective temperature of 30 000 K, $r_s = 29$ $N_e^{-2/3}$ pc $= 2.9 \times 10^{-7}$ pc $= 13$ $R_\odot$, as given by Spitzer (1968) for a standard O8 star. The actual radius r_s will be larger if the W-star core is larger or hotter or if N_e is less. However, N_e may be greater since the possible range is 10^{10}–10^{14} cm^{-3} (Underhill, 1968), and the density may follow the r^{-2} law for constant-velocity ejection down to the photosphere. Doubly-ionized He regions are smaller than ionized H regions. Detailed models are in order, but an immediate conclusion is that some W envelopes may smother the stellar uv radiation, unless the envelopes are confined to the equatorial plane or some other non-isotropic configuration. We should be tempted to explain in this way the absence of symmetric nebulae or other nebulae around some W stars except for the complication that they may be doubles with companions that are independently able to photoionize surrounding nebulae.

Are W stars or their expanding envelopes detectable as radio sources? Davies *et al.* (1967) reported HD 16523 and HD 193793 at 2695 MHz in a beam of 15′ HPBW. However, the former source is probably identifiable with 4C56.04 rather than the W star (Johnson, 1971), and no W star has been detected in a search with the Green Bank interferometer (Hjellming, 1971). Johnson (1971) looked at HD 9974, HD 168206, HD 177230, HD 187282, HD 190918, HD 191765, HD 193793, HD 211853, and HD 214419 at 7795 HMz in a beam of 4′.4 HPBW. One source was found in the beam at HD 211853, but this is probably part of a (non-symmetric) nebula in the area. The answer to the question appears to be no, at present.

Acknowledgements

This report has been supported partially by the Lockheed Independent Research Program and partially by the NASA under contract NASW-1977. The OAO-2/WEP investigators have provided primary data under the guest-observer program. The Argentine Consejo Nacional de Investigaciones Científicas y Técnicas defrayed much of the cost of travel to the symposium. The hospitality of Dr J. Sahade and his colleagues in the Instituto de Astronomía y Física del Espacio is very much appreciated.

References

Bless, R. C.: 1971, private communication.
Bless, R. C. and Savage, B. D.: 1971, in A. D. Code (ed.), *Symposium on Scientific Results from OAO-2*, Amherst, p. 175.
Bradley, P. T. and Morton, D. C.: 1969, *Astrophys. J.* **156**, 687.

Carruthers, G. R.: 1968, *Astrophys. J.* **151**, 269.

Code, A. D., Houck, T. E., McNall, J. F., Bless, R. C., and Lillie, C. F.: 1970, *Astrophys. J.* **161**, 377.

Courtès, G.: 1960, *Ann. Astrophys.* **23**, 115 (Table IX).

Davies, J. G., Ferriday, R. J., Haslam, C. G. T., Moran, M., and Thomasson, P.: 1967, *Monthly Notices Roy. Astron. Soc.* **135**, 139.

Davis, R. J.: 1971, *Preliminary Catalog of Celescope Ultraviolet Observations*, Smithsonian Institution, Ap. Obs., Cambridge.

Georgelin, Y. P. and Georgelin, Y. M.: 1970, *Astron. Astrophys.* **6**, 349.

Georgelin, Y. P. and Monnet, G.: 1970, *Astrophys. Letters* **5**, 239.

Hjellming, R. M.: 1971, private communication.

Houck, T. E.: 1971, in A. D. Code (ed.), *Symposium on Scientific Results from OAO-2*, Amherst, p. 479.

Johnson, H. M.: 1971, *Astrophys. J.* **167**, 491.

Johnson, H. M. and Hogg, D. E.: 1965, *Astrophys. J.* **142**, 1033.

Lozinskaya, T. A.: 1970, *Astron. Zh.* **47**, 122.

Lozinskaya, T. A. and Esipov, V. F.: 1968, *Astron. Zh.* **45**, 1153.

McGree, R. X. and Murray, J. D.: 1961, *Australian J. Phys.* **14**, 260.

Minkowski, R.: 1965, in A. Blaauw and M. Schmidt (eds.), *Galactic Structure*, Univ. of Chicago Press, Chicago and London, p. 337.

Morton, D. C.: 1967, *Astrophys. J.* **147**, 1017.

Morton, D. C.: 1970, *Astrophys. J.* **160**, 215.

Parker, R. A. R.: 1963, Thesis, California Inst. of Technology.

Parker, R. A. R.: 1964, *Astrophys. J.* **139**, 493.

Pengelly, R. M.: 1964, *Monthly Notices Roy. Astron. Soc.* **127**, 145.

Pikelner, S. V.: 1959, *Physics of the Interstellar Medium*, Acad. Sci. U.S.S.R., Moscow, Figure 30.

Spitzer, Jr., L.: 1968, *Diffuse Matter in Space*, Interscience Pub., New York, Table 4.5.

Smith, A. M.: 1967, *Astrophys, J.* **147**, 158.

Smith, L. F.: 1967, *Astron. J.* **72**, 829.

Smith, L. F.: 1968a, in K. B. Gebbie and R. N. Thomas (eds.), *Wolf-Rayet Stars*, U.S. Government Printing Office, Washington, D.C., p. 21.

Smith, L. F.: 1968b, *Monthly Notices Roy. Astron. Soc.* **141**, 317.

Smith, L. F.: 1971, in this Symposium; and in A. D. Code (ed.), *Symposium on Scientific Results from OAO-2*, Amherst, p. 429.

Smith, L. F. and Kuhi, L. V.: 1970, *Astrophys. J.* **162**, 535.

Smith, L. F. and Batchelor, R. A.: 1970, *Australian J. Phys.* **23**, 203.

Stecher, T. P.: 1968, in K. B. Gebbie and R. N. Thomas (eds.), *Wolf-Rayet Stars*, U.S. Government Printing Office, Washington, D.C., p. 65.

Stecher, T. P. and Milligan, J. E.: 1962, *Astrophys. J.* **136**. 1.

Terzian, Y.: 1970, *Astron. J.* **75**, 1155.

Underhill, A. B.: 1968, in M. Hack (ed.), *Mass Loss from Stars*, D. Reidel Pub. Co., Dordrecht-Holland, p. 17.

West, D. K.: 1971, in A. D. Code (ed.), *Symposium on Scientific Results from OAO-2*, Amherst, p. 441.

DISCUSSION

Underhill: Regarding the *OAO* scans, what is the total filter bandpass at half intensity?

Johnson: Well, they are different from band to band, but they are roughly 200–400 Å. The passbands are such that they practically overlap at all places along the spectra. In other words, if there is any monochromatic line in the spectrum, it will be captured by one or more bands. So that it is true that the lines do affect the spectrum, but I do not think that we can say how from these data. In the case of HD 50 896, I found no suggestion of the presence of any very strong lines in the ultraviolet. But then may be I am wrong. I am really giving this report to show you the lack of information rather than give you much more information.

Morton: One problem in trying to estimate the interstellar contribution is that the interstellar Lα line, in many hot stars, is much weaker than one would expect from the 21 cm data. HD 50 896 is close to Orion where we found at least a factor 5 less neutral hydrogen than we might expect from the

21 cm data. If that is the case, we would conclude that the contribution from the emission lines would be less.

Thomas: In the case of NGC 2359, how can one say that one has a nice spherical nebula when one sees a fountain-like ejection coming out from the star. What is your picture?

Johnson: In all of these nebular problems one is never able to get the perfect symmetry that one might have in a stellar problem. Consequently, when I say this is a nice shell around the star I am taking a little artistic license here and of course we have no direct evidence that this star actually is sitting inside a loop or a shell of nebula. It could be a foreground or background object but the coincidence looks a little too inviting to pass it by. Now it could be that this is not a shell but rather a spiral, but I call it a shell for convenience.

Thomas: I ask, whether it is conceivable that we do not have spherical symmetry, but may be something localized over certain points in the ejection.

Johnson: As a nebular astronomer I would like to call NGC 6888 'a neat shell' because in comparison to most nebulae, it is a neat shell. It is an elliptical feature perhaps lacking part of its perimeter, but nonetheless almost complete, and in stronger exposures, it is complete though weak on the one side. I draw your attention to the possible appearance of a general darkening of the inside of the shell, but outside we do not see that nebulosity. Now, as to the mechanism of ejection, of course we assume spherical symmetry. Certainly, stellar astronomers cannot blame us too much for that; it is the natural thing to do and we did believe that the star in the center does eject matter equally in all directions. We have not tried to explain the ellipticity or the finer points of departure from a spherical shell. For example, one might believe that a magnetic field did not permit the stuff to expand quite equally in all directions at the same speed. However, that is a relatively secondary feature of the argument, I think, at the present time.

Thomas: Could you not if you would want to be speculative and accept your model? In essence it says that mass goes out rapidly until it has swept up an interstellar medium mass equal to the ejected mass and then the velocity drops rapidly and reaches a small, asymptotic value where $4\pi R^2_* m_* v_{\text{eject}} = 4\pi R_{\text{neb}}^2 m_{\text{ISM}} v_{\text{neb}}$. This gives a spherical nebula, for uniform interstellar medium. But suppose you do not require such uniformity, could you not use the observed ellipticity of the nebula to infer a non-uniform distribution in the interstellar medium around the star, or is that too speculative? It can be made quantitative by exactly your approach.

Johnson: I think that is a pretty good speculation, because if the density of the ambient medium agrees with this correction you can explain both why the star is closest to one edge and why the stuff gets brighter where it does because the distance of travel is smaller and it piles up faster. So I do not object to that kind of an interpretation at all. I suppose in fact that the ambient medium outside the star may be able to control the shape of the shell whether it be a magnetic field or inhomogeneities of the ambient medium or whatever. The star can still be expelling the matter isotropically.

Smith: Just in support of what Johnson has just said, let me refer to NGC 3199 (see Proceedings 1968 WR Symposium, p. 41). It is a very nice example of the situation where the brightest parts of the nebula are closest to the star and faint extensions on the opposite side are at much greater distance. So I would support Johnson in the interpretation that the shape depends on the distribution of the interstellar medium. Let me summarise the general situation of rings. In 1968 there were 6 known, all associated with WN5, WN6 or WN8 stars, plus NGC 7635 with its Of star.

There is one recent development. Crampton has suggested the addition of two objects to the list of ring nebulae. One is the WN6 star HD 191765 which falls nicely into the previous generalization about the nebulae. However, his second one is a WN7 star which breaks the pattern. Neither are as well defined rings as the previous six and their reality needs confirmation. For the moment it looks like the association of ring nebulae with single WN5, WN6 and WN8 stars stands.

Paczyński: It is really very nice to know that ring nebulae are observed around single stars only. I am surprised, though, that all these are WN stars, and that there is no WC star within ring nebulae, while there are single WC stars, I believe. As far as I know there is no systematic difference in the presently observed mass loss from single and binary WR stars. The outflow velocity is typically about 1000 km per second, considerably more then the velocity of orbital motion. Therefore, the mass outflow that is observed now is not seriously affected by the duplicity of some WR stars. It looks like ring nebulae were not created by the observed mode of mass loss, as in such case we should expect to see those nebulae around binary WR stars too. It seems that we should consider the possibility of a large mass loss that occurred in the past in single WR stars only. We know that there was a large mass loss in the past in the binaries: almost entire hydrogen-rich envelopes were transferred from the progenitors of present day WR components to their companions. It is possible that the progenitors of single WR stars have lost their massive hydrogen-rich envelopes in the past, and those envelopes are seen now as ring nebulae.

Thomas: Having been supported quickly by both Paczynski and Lindsey Smith on this idea of differential distribution of the interstellar medium, I hate to start backing down from it. But suppose you accept that every time I have a ring like this, it simply reflects the interstellar medium density distribution. Then, I have, it seems to me, an enormous variation in the interstellar medium around individual stars. Is that something that one can accept happily? But suppose, just for fun, you accept it, then what?

Paczyński: But all those stars are believed to be single? I think there should be no relation between the distribution of interstellar matter and the duplicity of WR stars.

Thomas: Are you sure? If you think back on the question of how the stars evolved, of how they came together, what determines the distribution of the material in the interstellar medium around the star? You see, you get into all sorts of really speculation here, but if you want to start some place, you might as well be wild.

Johnson: I take it from Paczynski's remarks that he has given up the idea that the secondary star influences in an important way the loss of mass from the Wolf-Rayet star.

Paczyński: No.

Johnson: You think the secondary star can capture a good part of the mass?

Paczyński: There are two different things. One is the mass exchange in a close binary which is due to the fact that one of the components is so large that mass can freely flow to the companion. This does not require any ejection velocities, but if the ejection velocities are considerably larger than the velocity of the orbital motion, then the presence of the companion will produce a perturbation.

Johnson: A small perturbation?

Paczyński: It depends upon the ratio of the expansion velocity to the velocity of the orbital motion.

Conti: Is there any expansion velocity that is in fact less than the orbital motion? For example, the orbital motion in V444 Cyg is small as the period is about 4 days. That is a lower limit to the expansion velocity, so it seems that the expansion velocity is always larger than the orbital velocity.

Paczyński: In that case the mass loss from the Wolf-Rayet stars, which is observed by means of violet displaced absorption lines, should hardly be affected by the presence of the companion. Of course it does not mean that the star which is observed to have a Wolf-Rayet spectrum has not lost a large amount of mass to the companion some time in the past. Without that mass exchange in the past one cannot explain the observed mass ratios and the observed over-luminosities. We have here two different kinds of mass loss processes.

Johnson: While we were walking to lunch, Paczynski very kindly explained to me how he thought planetary nebulae nuclei, some of which have WR-type spectra, came from Mira stars, and I wonder if he has any analogous process involving something other than a Mira star to explain the origin of population I WR stars? In other words, do you have any idea of what kind of star goes into a single Wolf-Rayet star?

Paczyński: Well, there is a very recent suggestion due to Bisnovaty, Kogan and Nadyozhin, that a massive star cannot become a red supergiant because of mass outflow during the preceding evolutionary phases. Unfortunately, their calculations are not very precise. Still it is possible that the observed deficiency of very luminous red supergiants may be due to a severe mass loss from these massive stars. In fact the mass loss from supergiants is observed, and the rate is the highest for the most luminous objects. Only a single star may develop an extended envelope as it has no companion that could capture the envelope too soon. Perhaps such an extended envelope is lost to the interstellar space and can be observed as a ring nebula. This is just a speculation.

Johnson: If the nebular shells, such as Hogg and I observed around some WR stars, are left over from slow mass loss during an earlier stage of the evolution of a single supergiant, Hogg and I are wrong to use the current 1000 km s^{-1} mass loss of WR stars to explain the nebular shells.

Underhill: That is right. It also ruins the age estimates of the star.

Johnson and Underhill: It takes away quite a few things from previous arguments.

Paczyński: I believe the assumption of large previous mass loss is the easiest explanation for the fact that all the ring nebulae are observed around single stars only.

Johnson: I agree with you.

Thomas: It depends upon the phase of evolution. Suppose you argue that Wolf-Rayet stars occur early in the phase of evolution, then it would be in terms of the contraction hypothesis that we are fixing the surrounding medium. If it were late, which I gather that most of you believe, then maybe what most of you say is fine. But then, you talk about the red giants and there the velocities are low, so low that you cannot assume that the matter has really escaped unless you have quasi-slow previous ejections. Then the bigger ejection later catches up with it.

Paczyński: Yes.

Conti: But may I remind us all that we have Of stars and very luminous OB supergiants in which the rocket UV observations themselves indicate very rapid mass loss. One might think that before a star is a red supergiant it is a blue supergiant and while it is a blue supergiant it is losing mass, lots of it, at a very rapid rate. The star is first a blue supergiant, then a red supergiant and when it loses all its envelope, you are down to the helium core and you have a Wolf-Rayet star.

Underhill: I would like to know what is the linear size of one of those rings, relative to Strömgren spheres. Now, if you have an O5 star with an effective temperature of 40 to 50 thousand degrees, the simple theory gives, if I recall correctly, a Strömgren sphere of 20 parsecs. Everything within 20 parsecs of the star will be completely ionized so you will not see a hydrogen recombination spectrum which I presume is making your nebula visible. Therefore, we must assume that these nebulae are recombining and giving you Hα and that is the end of their Strömgren spheres.

Johnson: Around NGC 6888, if the distance is 1200 parsecs, it is about 2.6 parsecs in radius. It is quite a small H II region but probably the gas in the shell is somewhat denser than the gas in a typical H II region of ambient interstellar matter, so it is able to absorb the photons a little better.

Underhill: So, the shells are smaller and to make them visible you have to deduce a bigger density.

Smith: The densities, in several cases, were determined from the intensity ratio of O II at 3727 Å.

Thomas: Do I understand that there are no WC stars that illuminate shells?

Smith: That is right, no population I WC stars have ring nebulae.

Paczyński: With the possible exception of a very controversial object, BD + 30° 3639, which is sometimes classified as a planetary nebula and sometimes as a Wolf-Rayet star.

Thomas: If I really accept your analogy and exclude the case you have just mentioned, then the evolutionary life of a WR star can be described as follows: first, it is a WC star; then, it becomes a red giant and loses some mass; and then, it becomes a WN star; furthermore, I see the mass loss because of the mass that came out of the giant phase. I am just reasoning phenomenologically following your logic; I do not see shells because I do not have any prior mass loss which gets swept up.

EFFECTIVE TEMPERATURES OF WOLF-RAYET STARS

DONALD C. MORTON

Princeton University Observatory, Princeton, N.Y., U.S.A.

We have learned from Lindsey Smith's talk how useful it would be for our understanding of Wolf-Rayet stars to have reliable effective temperatures and bolometric corrections. The ring nebulae surrounding a few WN stars provide one method for estimating the stellar temperatures. I have assumed each nebula is a normal H II region excited primarily by the Lyman continuum radiation of the central star. Then, if the nebula is optically thick shortward of 912 Å, each stellar photon with more energy than 13.6 eV is absorbed by the nebula and re-radiated in a predictable way as hydrogen lines and free-free radio emission. Consequently, nebular radio measures such as those recently obtained by Hugh Johnson (1971) provide a direct indication of the number of Lyman-continuum photons emitted by the central star. The exact equations have been described in the *Astrophysical Journal* (Morton, 1969; Morton, 1970).

For the exciting stars a measure of the visual flux in the continuum is available from Lindsey Smith's UBV photometry of narrow bands between the emission lines. I have used the relation

$$V^* = v - 0.02 - 0.36 \, (b - v)$$

to convert each v magnitude to an equivalent one on the Johnson UBV system such that V^* represents the visual magnitude the WR star would have if there were no emission lines. This transformation is necessary to convert the observed magnitudes to absolute visual fluxes incident at the Earth, since the calibration factor is not known for v. I adopted a flux of 3.8×10^{-9} erg s^{-1} cm^{-2} Å^{-1} at 5460 Å for a star with $V = 0.0$. A recent measurement by Oke and Schild (1970) was 3% lower. The total visual extinction was derived from the observed colour excess by the formula

$$A_V = 3.0 \, \mathrm{E}(B - V) = 3.6 \, \mathrm{E}(b - v).$$

Fortunately there is no interstellar absorption of the radio emission.

Table I lists the photometric data on the central stars and Table II gives the nebular radio fluxes, the derived effective temperatures, and the bolometric corrections. The first six entries of Table II are reproduced from Morton (1970) and depend on radio measurements by Johnson and Hogg (1965), Gebel (1968), and Smith and Batchelor (1970) while the last three rows are based on the new data by Johnson (1971). The Gaunt factor is accounted for through the term

$$g(T, v) = 1 + 0.13 \log \, (T^{3/2}/v),$$

where v is the frequency in Hertz and $T = 10^4$ K.

The observational data provided $N_L/\pi F_V$, the ratio of the number of Lyman continuum photons to the energy flux in the V bandpass. It was assumed that the relation

M. K. V. Bappu and J. Sahade (eds.), Wolf-Rayet and High-Temperature Stars, 54–56. All Rights Reserved.
Copyright © 1973 by the IAU.

TABLE I

Photometry of Wolf-Rayet stars in nebulae

Nebula	Star	Spectrum	v	$b-v$	$E(b-v)$	$V*$	A_V
NGC 2359	HD 56925	WN 5	11.74	$+0.33$	0.47	11.60	1.69
NGC 3199	HD 89358	WN 5	11.20	$+0.54$	0.68	10.99	2.45
RCW 58	HD 96548	WN 8	7.85	$+0.11$	0.26	7.79	0.94
RCW 104	HD 147419	WN 6	11.42	$+0.63$	0.80	11.17	2.88
NGC 6888	HD 192163	WN 6	7.73	$+0.25$	0.42	7.62	1.51
S157	HD 219460	WN 4.5 + B0	10.03	$+0.52$	0.71	9.82	2.56
S308	HD 50896	WN 5	6.94	-0.07	0.07	6.95	0.25
	HD 211853	WN 6 + B0:I:	9.20	$+0.32$	0.49	9.06	1.76

TABLE II

Effective temperatures of Wolf-Rayet stars in nebulae

Nebula	Spectrum	f_v (flux units)	v (Hz)	log $f_v/g(T, v)$	log $N_L/\pi F_V$	θ_e	T_e(K)	B.C.
NGC 2359	WN 5	5.9	1400	$+1.00$	12.50	0.094	53 600	-4.7
NGC 3199	WN 5	20	2650	$+1.56$	12.52	0.093	54 200	-4.7
RCW 58	WN 8	0.2	2650	-0.44	9.84	0.20	25 000	-2.5
RCW 104	WN 6	8.6	2650	$+1.19$	12.04	0.120	42 000	-3.9
NGC 6888	WN 6	4.7	1400	$+0.90$	10.88	0.164	30 700	-3.0
S157	WN 4.5 + B0	40	2650	$+1.86$	12.30	0.105	48 000	-4.3
S308	WN 5	1.3	5010	$+0.40$	10.62	0.171	29 500	-2.9
NGC 6888	WN 6	4.0	7795	$+0.91$	10.89	0.164	30 800	-3.0
HD 211853	WN 6 + B0:I:	0.46	7795	-0.03	10.43	0.176	28 600	-2.8

between this ratio and the effective temperatures of WN stars is given by the theoretical relation derived from a series of model atmospheres with gravity $g = 10^4$ cm s^{-2} representing hot main-sequence stars. The models were derived under the usual simplyfing conditions of hydrostatic equilibrium, radiative equilibrium, and local thermodynamic equilibrium (LTE). Auer and Mihalas (1972) have shown that the continuum energy distribution for O type models is little changed by consideration of the effects of non-LTE and a similar situation may hold for the WN stars. I am less confident about the conditions of hydrostatic and thermal equilibrium which are likely to fail in the region of the emission lines. However the models may not be so bad for relating the effective temperature to the energy distribution in the continuum. Until we have self-consistent models for WN atmospheres there is nothing better to do. The last column in Table II gives the bolometric correction based on the T_e in the previous column and the B.C.–T_e relation derived from the model atmospheres of O and B stars by Bradley and Morton (1969) and Van Citters and Morton (1970).

For the first six stars in Table II, there is the expected trend of decreasing temperature from WN 4.5 to WN 8, though the two WN 6 stars give considerably different values. The new measurement on NGC 6888 nicely confirms the earlier result for HD 192163. An effective temperature of 30 800 K is reasonably consistent with the far-ultraviolet energy distribution obtained from OAO-2 just described by Hugh

Johnson. However, the two new values for HD 50896 (WN 5) and HD 211853 (WN 6) seem a little cooler compared with the earlier estimates, especially the WN 5 star. In that case I wonder if the nebula is optically thin or if the radio telescope missed some of the nebular emission. If any of the photons escape either the nebula or the telescope, the effective temperature will be underestimated.

References

Auer, L. and Mihalas, D.: 1972, *Astrophys. J. Suppl.* **24**, 193.
Bradley, P. T. and Morton, D. C.: 1969, *Astrophys. J.* **156**, 687.
Gebel, W. L.: 1968, *Astrophys. J.* **153**, 743.
Johnson, H. M. and Hogg, D. E.: 1965, *Astrophys. J.* **142**, 1033.
Johnson, H. M.: 1971, *Astrophys. J.* **167**, 491.
Morton, D. C.: 1969, *Astrophys. J.* **158**, 629.
Morton, D. C.: 1970, *Astrophys. J.* **160**, 215.
Oke, J. B. and Schild, R. E.: 1970, *Astrophys. J.* **161**, 1015.
Smith, L. F. and Batchelor, R. A.: 1970, *Australian J. Phys.* **23**, 203.
Smith, L. F.: 1968a, *Monthly Notices Roy. Astron. Soc.* **138**, 109.
Smith, L. F.: 1968b, *Monthly Notices Roy. Astron. Soc.* **140**, 409.
Smith, L. F.: 1968c, *Monthly Notices Roy. Astron. Soc.* **141**, 317.
Van Citters, G. W. and Morton, D. C.: 1970, *Astrophys. J.* **161**, 695.

DISCUSSION

Smith: HD 211853 is a binary. And the ring around HD 50896 is very faint and incomplete.

Thomas: Do you really mean it when you say that there are no non-LTE effects in your calculations?

Morton: Of course there are always non-LTE effects, but in these models they are not serious as they might be with a middle B star.

Thomas: Auer and Mihalas calculations are all for radiative equilibrium; if there is any kind of mechanical energy transport, their calculations go out the window. That is a major objection. My second is what does one get for the ratio of the Lyman continuum to the visible if one uses their hottest model? What is the difference in the ratio of the Lyman continuum to the visible when one considers LTE and non-LTE effects?

Morton: It is not very big, at most a factor two at 25000° and less at higher temperatures.

Thomas: What does it do to the temperature distribution in the model?

Underhill: It does affect the line strengths quite seriously. However, we try to identify the models via the continuum and you identify the model with a star using parts of the Paschen continuum which is almost insensitive to anything.

Thomas: I just do not see how you can get away with it unless you have mechanical heating.

Morton: Since we have only models in radiative equilibrium, it is the best we can do at present.

Thomas: The best you can do in a sophisticated way, but I am not sure that in a fairly rough way you cannot do better.

Conti: If a Wolf-Rayet star is a star that has lost its hydrogen into the interstellar medium by one way or another, the density around the star may be very large and that may be one reason why the Strömgren sphere is not very large, because the density of this sphere is much larger than normal.

Thomas: I personally like better the response having to do with the lack of nebular shells. I am getting more intrigued with the idea of the environment being fixed by previous stages.

Smith: There may be a systematic error in Morton's temperatures because of his assumption that the stars radiate like models; however, the difference between his temperatures for different subclasses may have some validity. It certainly provides a beautiful explanation for the surprising anticorrelation between excitation class and visual magnitude.

Thomas: I just do not believe it, not only out of prejudice, but also from the standpoint of your own arguments this morning when you were asking for much differential mechanical heating between WN6 and WN3.

SECTION III

CHAIRMAN: B. WESTERLUND

THE SPECTRA OF WOLF-RAYET STARS
AT HIGH DISPERSION

M. K. V. BAPPU

Indian Institute of Astrophysics, Kodaikanal, India

1. Introduction

The Wolf-Rayet stars have perhaps the most spectacular spectra among the various celestial species that have been examined with the aid of the spectrograph. The wide emission lines that provide a striking display continue to be the enigma they have been for decades. Progress, however, in evaluating the contributions to the spectrum by different ions has been fairly complete, thanks to the splendid efforts in the laboratory by Edlen and his collaborators. Many lists of wavelengths of individual features exist from the studies of Plaskett, Beals, Cecilia Payne and Swings. Efforts at identification have been such as to provide a list of likely contributors, that by wavelength position and plausible intensity could be present in the emission band at a specified wavelength. The large width of the lines form the principal limitation. For, one can have a wide limit of coincidence in wavelength with resultant emission features that are complexes covering well over a hundred ångströms. The wide nature of these complexes have perhaps encouraged in the past the use of low spectral resolution only, for the many studies that have been carried out. Seldom has one used the resolutions and dispersions that have been usefully employed for the study of the more common relatively narrow absorption lined objects. This situation pertaining to the Wolf-Rayet stars is happily undergoing a change in recent years when coudé spectra with the larger telescopes are becoming increasingly available. The southern hemisphere has been particularly rich in these objects both in variety of behaviour and in having many bright ones, and one can, therefore, justifiably hope that our information on these objects will progress henceforth with remarkable rapidity.

One might well ask the question whether such objects that are characterized by enormously wide emission features could really reveal any additional information when subjected to scrutiny with high dispersion. My answer to this is in the affirmative. The study of microphotometer tracings of high to moderate dispersion spectra does indeed show up details that are often lost in efforts with lower dispersions. I hope, in what follows, to be able to convince you of some of the advantages of higher dispersion spectroscopy of Wolf-Rayet stars.

The first efforts at comprehensive identifications have been of Payne (1933) and Edlen (1933). Later in 1956, Edlen (1956) was able, using new laboratory data for C II, C III, and predicted transitions between high quantum states in C IV, to reach at an almost complete identification of the emission features in the stars of the carbon sequence. The observations that were used for this purpose were principally those

M. K. V. Bappu and J. Sahade (eds.), Wolf-Rayet and High-Temperature Stars, 59–92. All Rights Reserved.
Copyright © 1973 by the IAU.

made by Swings (1942) and in the near infrared region by Swings and Jose (1950). Later efforts by Bappu (1957), Bappu and Ganesh (1968), Underhill (1959, 1962, 1967), Smith and Kuhi (1970) and Smith and Aller (1971) have only added details of contributing wavelengths, profiles and line intensities to the Edlen study.

We have heard this morning a detailed discourse on the dichotomy of spectral display and the basis of the categorization into the WC and WN classes. We are certain today that HeI, HeII, CII, CIII, CIV, OIII, OIV, are present in a typical WC8 (Lindsey Smith's classification) star like HD 192103. Probably present in this star are lines of H, NIII, NIV, NV, OV, OVI and SiIV. In the case of a WN6 star, HD 192163, we are definite of the presence of HeI, HeII, NIII, NIV, NV and Si IV. Many including myself would add to this list CIV, first identified independently in the WN sequence by Swings (1942) and Aller (1943). Probably present are H and O V. The dichotomy of spectral behaviour is an accepted characteristic that needs quantitative theoretical interpretation. The dominant role of nitrogen ions and carbon ions in their respective classes is seen so obviously. The question of interest has been whether a pure carbon sequence free of nitrogen exists. I believe that the moderate to high dispersion spectra obtained give evidence that nitrogen exists in the carbon sequence. Independent support for this statement comes also from the rocket ultraviolet data of Gamma Velorum wherein NV absorption lines and possibly NIV emission lines can be seen. The observed effects are nevertheless marginal and might find interpretation only from the standpoint of stellar evolution.

Our ultimate aim of observation of the spectra of these objects is to build up a physical picture from a theoretical analysis of intensities, profiles, etc. It may not be long before we can compute synthetic spectra that will match in every detail the fluctuations in a line contour caused by the kinematics of the situation or by a contributing blend. To do so, the base necessarily has to be the availability of a minimum set of data of good quality and completeness of information. It is possible to achieve this only by high dispersion studies of selected objects that are amenable to such analyses. My aim in recounting the current status of such information is to give you a picture of the problems we encounter by virtue of the heavy blending of wide lines to form emission complexes. We would have indeed, preferred to have instead of these complexes, single emission features that are likely to be more easily amenable to theoretical analysis!

2. Line Identifications in WC and WN Stars

Let us examine first of all the general problem of identification as applied to a typical star of the carbon sequence and another of the nitrogen sequence. I have chosen for this purpose the direct intensity tracings of HD 192103 and HD 192163 from amongst a collection of such tracings that Mr. Scaria and I have in print in the Kodaikanal Observatory Bulletins. We have used for this purpose coudé spectra that I had obtained some years ago on the Mount Wilson 100-inch at 10 Å/mm in the blue and 20 Å/mm in the red. We have chosen these two objects essentially because they have the narrowest lines amongst the spectra that we have of both sequences. The illustrations of HD 192103 also contain in the lower half the equivalent region in HD 184738, or Campbell's hydrogen envelope star. This tracing is from a blue plate of 10 Å/mm and of 8 hr exposure, obtained by Olin Wilson. The red plate was also obtained by him at 20 Å/mm in order to provide completeness to the study.

The identifications in WC spectra are marked in Figures 1 to 9. It is possible to make an almost complete identification for Campbell's star because (a) the lines are narrow and provide little cause for ambiguity, (b) laboratory wavelength data are

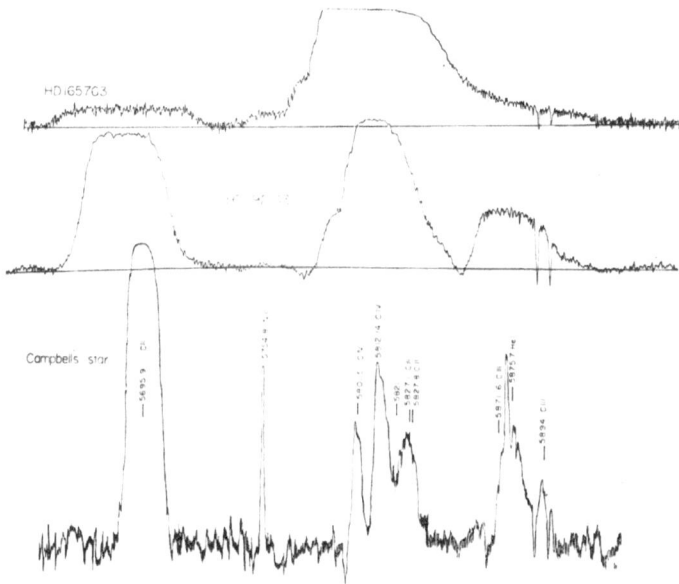

Fig. 1.   The 5800 Å region in WC stars. Note the violet absorption edges of C IV 5801, 5812, the flat-topped nature of the C III 5696 profile, and the blends at 5876 Å.

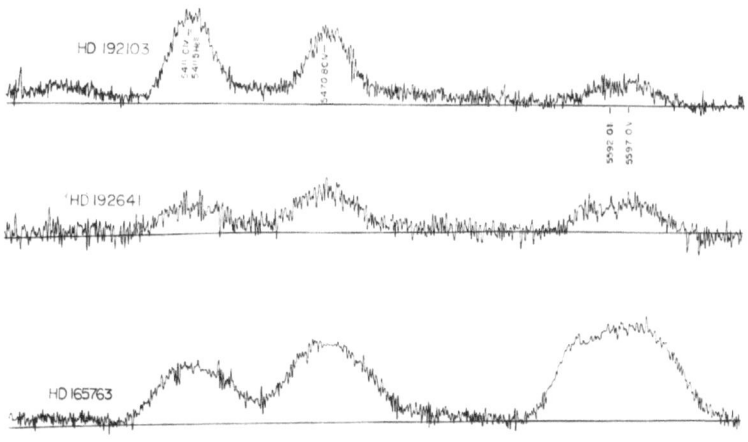

Fig. 2.   The 5500 Å region in WC stars.

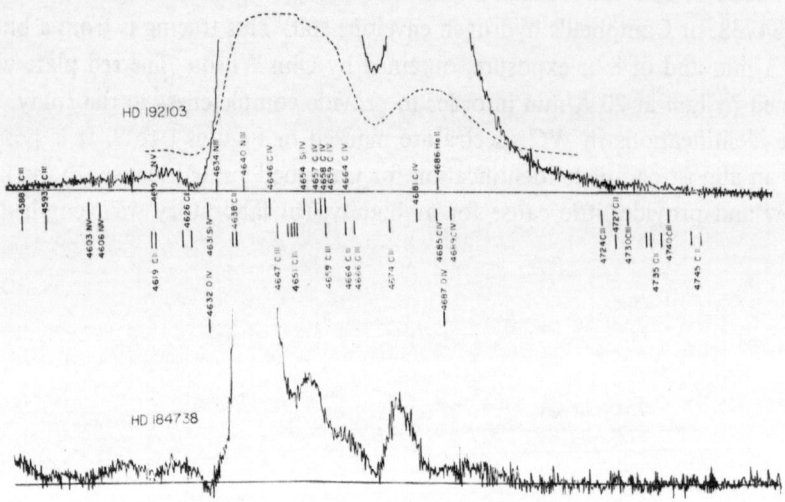

Fig. 3.   The 4600 Å region in HD 192103 and HD 184738.

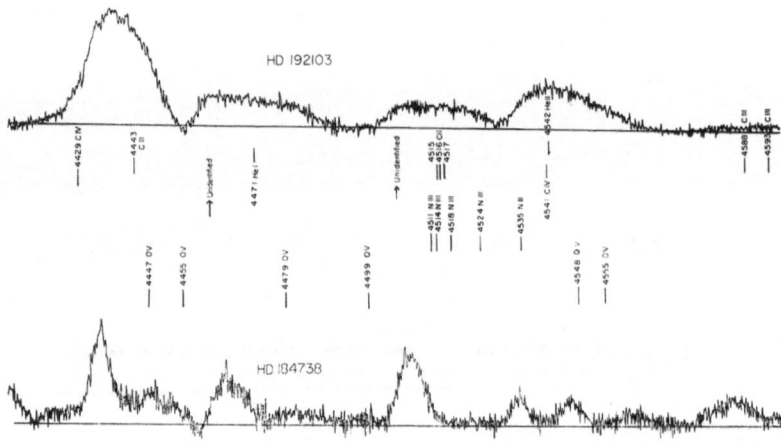

Fig. 4.   Region 4429–4590 Å in HD 192103 and HD 184738.

comprehensive. Allowing for excitation differences between the WC8 and the equivalent of a WC9 star, the identifications in Campbell's star aid in our effort to determine the blending in HD 192103. The details are as follows:

He II *5-n series*: One can see with certainty on the original tracings not reproduced here, the He II lines at 6234 Å (5–17), 6171 Å (5–18) and 6118 Å (5–19). One can follow this series upto the transition (5–21).

5876 Å: He I is the principal contributor. C III exists at 5872 Å, 5894 Å while C II is at 5889 and 5892 Å. C II contribution is possible at 5907 Å and 5915 Å also, while C IV 5866 Å (8–13) can be of appreciable intensity.

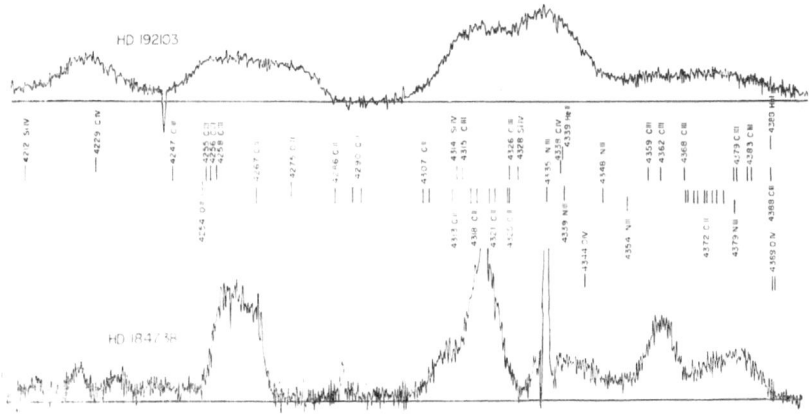

Fig. 5. The region 4210–4390 Å in WC stars.

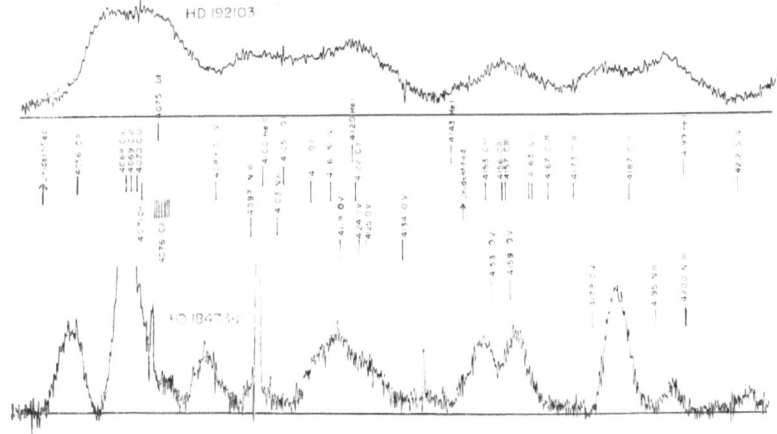

Fig. 6. The 4100 Å region in WC stars.

5816 Å : C IV 5801.3 Å, 5812.0 Å and C III 5826 are certain. There are also the two violet edges corresponding to 5801 Å and 5812 Å. Also, O V 5836 Å seems to be a contributor.

Weak emission between 5696 Å and 5806 Å : From wavelength coincidences alone the likely contributors are N II 5747 Å, 5767 Å and weak C III at 5772 Å. The emission is a definite feature; its identification is not satisfactory.

5696 Å : This is principally due to C III 5696 Å. Possible N II contamination could exist from the strong transitions at 5667 Å, 5676 Å, 5680 Å, 5686 Å and 5711 Å. The identification of N II is very uncertain because the infrared transitions of N II do not appear in the star spectrum.

5592 Å : This is a double humped feature. The humps consist of O V 5572 Å, 5580 Å and 5583 Å as well as 5598 Å. O III 5592 Å also contributes. A possible contributor is 5602 Å of O VI.

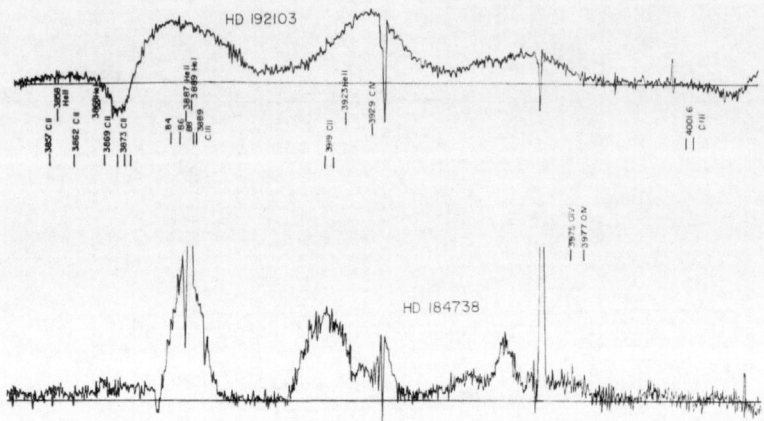

Fig. 7.   Region 3860–4000 Å in HD 192103 and HD 184738.

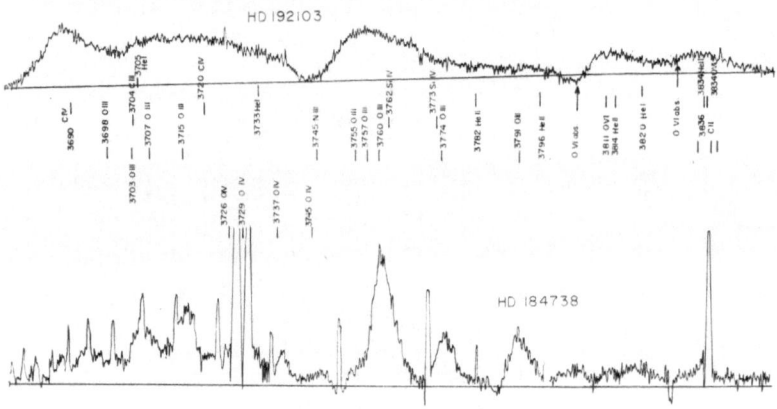

Fig. 8.   The spectral region 3690 Å–3840 Å in WC stars.

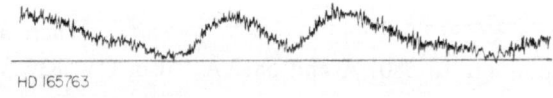

HD 165763

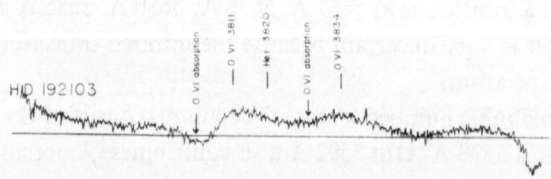

Fig. 9.   O VI absorption in WC spectra.

5469 Å : The most dominant ion is C IV 5070.8 Å (7–10). On the longward side some weak blending which is noticeable on the profile can be ascribed to 5479 Å and 5486 Å of C II. Farther away in the red wing, a slight hump above the continuum can be ascribed to 5508 Å O III. Weak emission caused by C II 5535 Å and 5538 Å can be seen at 5536 Å. This line is another of those rare unblended lines that could be used for derivation of reliable physical parameters of the star.

5411 Å : He II is the principal contributor. C IV 5411 (8–14) has an appreciable share of the intensity considering that C IV 5093 Å (8–15) has about 8% intensity of 5411 Å. Figure 2 illustrates tracings of this region.

Red wing of 4686 Å : C II lines fall at 4745 Å, 4735 Å and 4727 Å. The strongest among these is 4745 Å. C II is unlikely to be dominant. The C III lines at 4739 Å, 4730 Å and 4724 Å probably contribute more than C II even though they are weak.

4686 Å : The principal contributor is He II. The (7–11) transition of C IV at 4689 must be a contributor. The intensity of C IV will be substantial because the next line in the (7-n) series earlier to it is 5470 Å which is quite intense. Other contributors of C IV are 4685 Å (6–8) and 4681 Å (8–17). The transition (6–8) will also be substantial since (6–9) at 3689 Å is quite intense and (6–7) 7726 Å is very noticeable despite its location in the vicinity of the oxygen 'A' band.

4652 Å : Principal contributors range from C III 4674 Å to 4647 Å, with the most intense ones being 4647 Å, 4650 Å and 4651 Å. The profile of the line indicates that its maximum intensity is between 4647 Å and 4653 Å in accordance with the C III dominance. C IV also falls in this region with appreciable intensity at 4646 Å (5d–6f), 4658 Å (5gf–6hg) and 4664 Å (5f–6d). C II has lines in this range from 4638 Å to 4618 Å with the most intense ones at 4618 Å. One has to explain the contribution of intensity over a range 4630–4647 Å before the violet edge comes in at 4627. It is here that one must depend entirely on N III at 4634 Å and 4640 Å to play a role, in the absence of any other contributor. Si IV 4631 Å and 4654 Å also exist and have a share in the intensity profile.

The strongest lines of N III, other than 4634 Å and 4641 Å, are at 4097 Å and 4103 Å. When we look at the spectrum of HD 192 163 and compare 4200 Å with 4100 Å, it is obvious that the complex at 4100 Å is many times more intense than what is expected from He II alone. This is caused by a principal additive contributor to the blend by N III and in a minor way by Si IV. In the case of HD 192 103, the complex at 4100 Å on the long wavelength side has much added to it by C III 4122 Å and the O V lines. The peak intensity of 4100 Å is higher than in the case of 4200 Å. Hence we conclude that N III 4097 Å and 4103 Å are present and definitely distort the He II intensity and profile. But we do not notice such contribution to the extent that we observe at this wavelength in a WN 6 star.

Another line of evidence of the presence of N III in the spectrum of HD 192 103 is the following. The large complex from 3748 Å to 3792 Å contains many contributors, amongst which O III is a good contributor. Shortward of 3748 Å and until 3725 Å, the primary effects are of O IV. Amongst the lines of O III, 3774 Å and 3791 Å are usually of equal intensity. The profile near 3774 Å picks up in intensity because of N III 3771 Å.

Also while the maximum intensity at 3757 Å in the complex can be due to O III 3755 Å, 3757 Å and 3760 Å, its overall increase above 3774 Å and 3791 Å calls for an additional reinforcement from N III 3754 Å. Also N III 3745 Å falls on O IV 3745 Å which is of sufficient intensity to record its existence. The fact that instead of an increase in this position, the contour sweeps down to the continuum level can be explained only by virtue of a violet edge to the O III complex 3755 Å, 3757 Å, 3761 Å. A mild depression in the contour at 3768 Å may represent the violet edge of O III 3774 Å. Both these violet edges are seen in HD 184738. Hence we need to recognize the contribution of N III to this complex also, thus reinforcing our conjecture that the N III ion can be seen in the spectrum of HD 192103.

Complex shortward of the C III *violet edge at 4627 Å* : The possibility exists that this could be the extended wing of the principal line near 4634 Å, and that what seems to be a complex of lines is actually the violet wing of 4634 Å into which the violet edge has made a deep cut. One can consider in addition that N V contribution at 4603 Å, 4619 Å is present. But 4603 Å by itself may not be very bright and 4619 Å will contribute less. Another possibility is N IV 4606 Å. But judging from possible N IV 4058 Å, the contribution of which, if present, is very small, the role of 4606 Å towards the blend is unlikely. Weak contributors may be C III 4593 Å and 4588 Å.

4542 Å : The intensity of this band is shared principally by He II followed by C IV 4541.3 Å (8–18) and two O V lines at 4548 Å and 4553 Å. C IV 4541 Å must necessarily be fainter than 5093 Å (8–15). Hence C IV 4541 Å, while detectable, must be having a small contribution. Two humps on the longward side of the profile portray well the contribution of O V and make the identification beyond doubt. The gap between 4542 Å and the emission band at 4516 Å can be identified as due to a violet edge of 4542 Å of the Pickering series. A similar violet edge is seen in HD 192163 as well as HD 184738.

4516 Å : The appearance of the band indicates its composite nature. C III 4515 Å, 4516 Å and 4517 Å are present. By itself the shortward extension seen can be formed only with the aid of another contributor. Two N III lines at 4514 Å and 4511 Å form this possibility. O V 4498 Å can be seen as a weak emission on the shortward side, even though it seems much weaker than its counterpart 4479 Å, which seems to dominate in the emission band about 4471 Å.

Emission at 4471 Å : He I 4471 Å occurs at the centre of this band with a steep violet absorption edge. O V 4479 Å contributes on the longward side. An unidentified feature at 4465 Å makes a significant contribution.

4441 Å : The principal constituents are C IV 4441 Å and 4429 Å with the former as the dominant contributor. A hump on the longward side agrees with O V 4447 Å. Another O V line at 4455 Å of intensity comparable to 4447 Å falls in the midst of the He I violet absorption edge, and hence is invisible.

4355 Å–4400 Å complex : C III contributors are at 4359 Å, 4362 Å, 4368 Å, 4380 Å, 4383 Å and 4388 Å. C II exists at 4372 Å, 4374 Å, 4375 Å and 4376 Å with fair intensity. O IV 4389 is a likely contributor, but is weak. There needs to be a feature at 4395 Å to give rise to the weak hump in the profile, and this could be O II 4396 Å. He I 4388 Å and N III 4354 contribute weakly.

4339 Å : The principal contribution is of He II. A slight hump on the longward side is caused by O IV 4344 Å. C IV 4338 Å (8–20) must be present, but cannot be separated from He II. On the shortward side the line blends with C III 4326 Å and 4317 Å. In particular, C III 4313 Å, 4317 and 4318 Å merge to form a conspicuous feature at 4316 Å. The central hump can be ascribed to 4326 Å. At this position C II 4325 is also present. Appreciable contribution to the overall band also comes from Si IV 4314 Å and 4328 Å.

4267 Å : C II at 4267 is undoubtedly the principal contributor. Weak contribution at 4247 Å, 4255 Å, 4256 Å and 4258 Å comes from C III. But these do not by themselves explain the peculiar profile. O II at 4254 Å and 4275 Å contributes.

4229 Å : The entire profile is mostly due to C IV 4229 Å (7–12) and is one of the few unblended lines in the entire spectrum. Its longward wing at the very extreme merges with C III 4247 Å. On the shortward side the wing blends with Si IV 4212 Å. The hump is so clear cut that it definitely establishes the presence of Si IV. N III 4216 Å may contribute weakly.

4200 Å : This band is mostly due to He II. N III 4196 Å and 4200 Å may contribute weakly. On the shortward side C III 4186 Å is most definite and this along with O V 4178 form the principal hump on the shortward side.

4157 Å : The emission band consists principally of C III 4153 Å, 4157 Å and 4163 Å. In addition there must be O V 4159 Å and O II 4153 Å. An unidentified contributor at 4149 Å has an appreciable effect.

The 4100 Å–4122 Å complex : Contributors are O V 4134 Å, 4124 Å, O II 4133 Å, 4120 Å, C III 4122 Å, Si IV 4116 Å and O II 4111 Å. He II 4100 Å is the principal contributor to the complex. This is flanked by N III 4097 Å and 4103 Å appreciably. While the net intensity indicates the reality of the N III overlap, its contribution is not consistent with the intensity it should have relative to 4634 Å and 4640 Å. Si IV 4089 Å is also present. O II 4092 Å, 4097 Å and 4105 Å have detectable effects.

4070 Å : The centre of the band is made up of C III 4070 Å, 4069 Å and 4068 Å with sizeable contributions from C III 4056 Å on the shortward wing and from C II 4076 Å on the longward side. O II 4071 Å and 4076 Å are the most prominent of the O II lines whose presence is denoted by distortions at these wavelengths. N IV 4058 Å may be present. However, we find no trace in the spectrum of its counterpart at 3483 Å. A feature at 4051 Å remains unidentified.

4026 Å : He I 4026 is undoubtedly the principal contributor with a conspicuous violet absorption edge. C II 4017 is present weakly in the violet wing. He II is also present.

3965 Å : He II is dominant. O III 3962 Å and He I 3965 Å are present. C II 3972 Å, 3974 Å, 3977 Å, 3980 Å and O IV 3957 Å, 3975 Å add up to the intensity.

3929 Å : C IV 3929 Å is undoubtedly the dominant contributor. In the shortward wing, C II at 3919 Å and 3921 Å and He II 3923 Å produce the observed asymmetry.

3889 Å : The most conspicuous contributor is He I. C III 3884 Å, 3886 Å, 3889 Å, C II 3876 Å, He II 3887 Å and O II 3882 Å are present. The profile is not flat topped as is seen in the WN stars. Presumably the presence of C III distorts the profile from flatness. An unidentified feature may exist at 3893 Å unless one assumes that He I is flat topped from 3882 Å to 3894 Å on which lies superposed the C III contribution.

3858 Å : He II is present with additional weak contribution from C II 3857 Å, 3862 Å and 3869 Å.

O VI *at 3811 Å and 3834 Å* : Two features with mean wavelengths 3814 Å and 3834 Å exist in the spectrum. He II 3814 Å and 3834 Å can contribute to these, but considering the intensity of the next earlier line in the Pickering series 3858 Å, such contribution to these positions is apt to be small. These are identified as due to O VI. He I 3820 exists with appreciable intensity .A weak contribution from C II 3831 Å and 3836 Å is possible. An interesting feature is the absorption seen off the violet edge of 3811 Å. A weak dip can be seen corresponding to a likely absorption edge of O VI 3834 Å. A remarkable coincidence produces the ambiguity. He I 3820 Å $(2^3P^0-3^3D)$ is a member of the series $(2^3P^0-n^3D)$. Others in this series like 4471 Å and 4026 Å all have violet absorption edges. The next line in the series falls at 3705 Å with a violet absorption edge that multilates the peak of C IV 3689 Å, which is otherwise intense. A suggestion of a weak violet absorption edge exists at 3622 Å caused by He I 3634 $(2^3P^0-8^3D)$. Hence we must conclude that the major share in the violet absorption edge at 3804 Å is that of He I 3820 Å.

This does not rule out the detection of displaced absorption by the O VI doublet. We infer its presence from the spectrum of HD 165763 (WC5) in Figure 9 where the absorption does not reach the continuum level. However, the shapes of the violet wings of 3811 Å and 3834 Å are such as to show the existence of the displaced absorption. Assuming the depression at 3745 Å seen in Figure 8 as due to O III displaced absorption one finds that the wavelength interval seen in HD 165763 between the displaced absorptions of O VI and O III at 3804 Å and 3745 Å respectively is retained in the case of HD 192103. The spectrum of HD 115473 reproduced in Lindsey Smith's paper shows the violet edges for the O VI doublet; a higher dispersion spectrum should settle the issue.

Emission complex 3745 Å–3795 Å : This complex has been described earlier to justify the identification of N III. The interesting feature is the shallowness of the displaced absorption edge caused by O III 3755 Å, 3757 Å and 3760 Å.

The complex 3680 Å–3745 Å : O IV 3726 Å, 3727 Å and 3739 Å undoubtedly exist on the longward side and individual humps can be seen at these locations. Definite identifications can be made of C IV 3720 Å (7–14) and C IV 3690 (6–9). There are numerous O III lines at 3698 Å, 3703 Å, 3707 Å and 3715 Å which contribute to the profile. There is also He I 3705 Å, the displaced violet edge of which affects the peak intensity of C IV 3690 Å.

3609 Å : The main contribution arises from C III 3609 Å, with some additional intensity caused by He I 3614 Å and O VI 3622 Å. The profile is typical of one that has several contributing constituents.

A similar detailed examination of intensity tracings of blue and red spectra of the WN6 star HD 192163 cause the following comments:

6678 Å : He I 6678 Å and He II 6683 Å, contribute predominantly.

6563 Å: The most intense contributor is He II. He II 6527 Å (5–14) falls on the violet wing. Possible O V contribution at 6460 Å, 6466 Å and 6500 Å exist.

6406 Å: A broad band at 6406 Å is the combined effect of 6406 Å He II (5–15) and 6380.7 N IV. The He II line is quite intense and N IV is sufficiently well resolved.

6234 Å: He II 6234 Å (5–n) is seen. The (5–n) series can be followed to $n=21$.

5875 Å: He I 5875 Å principally is a lone contributor with a striking violet edge. The profile is almost flat-topped.

5806 Å: Principally due to C IV with likely violet edges at 5781 Å and 5792 Å as seen in Figure 10. The profile looks as though it has contamination at 5836 Å and this could be due to O V. But this is not certain since stronger O V lines are not seen.

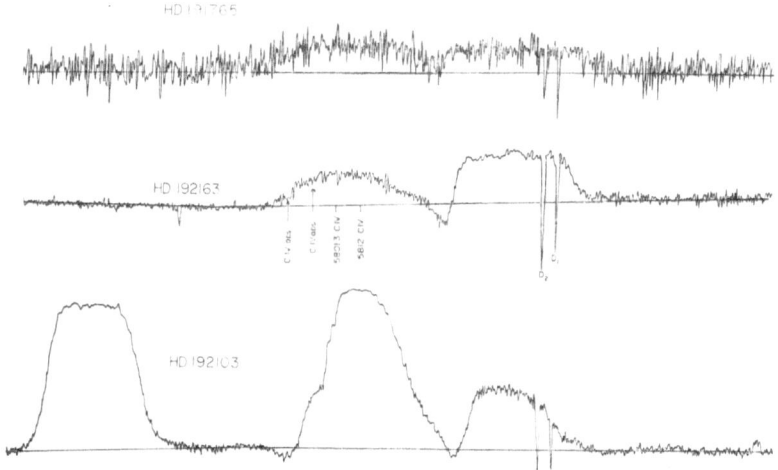

Fig. 10. The 5800 Å region in WN stars. A tracing of the WC star HD 192103 is included to establish the identification of C IV, 5801 Å, 5812 Å.

5411 Å: He II is the principle contributor. The profile of the line indicates contamination on the longward side. In particular a redward wing extension going upto 5528 Å is seen. One can identify O III 5508 Å. But the more intense line 5592 Å exists only weakly. In this group one may also include a weak contribution from O V. A large number of N II lines fall in this region and they could all be possible contributors. But here again stronger contributors between 5666 Å and 5710 Å do not exist. C IV 5470 Å could be assigned on the basis of wavelength coincidence.

4860 Å: He II is the principal contributor. The line has N III contamination and at least 10% is due to N III. The profile on the longward side indicates additional contamination which is unidentified. Very much like 5411 Å, there is an extended red wing going upto about 4965 Å. N V 4944 Å is certain in this extended feature, and the same applies to O V 4930 Å.

4686 Å: The contributions of 4713 Å He I can be inferred from the extension of the redward wing of 4686 Å. 4686 Å is predominantly caused by He II with a weak contribution, if any, of the Civ hydrogenic transitions 4680 Å (8–17), 4685 Å (6–8) and 4689 Å (7–11). The strongest of these three is likely to be 4685 Å, (6–8), in which case, one could see Civ (6–9) 3689 Å. This is not seen. Hence, it is unlikely that the hydrogenic transitions of Civ are generally seen in this star. This does not rule out the possibility of the (5d–6f) 4646 Å, (5gf–5hg) 4658 Å and (5f–6d) 4664 Å transitions being present.

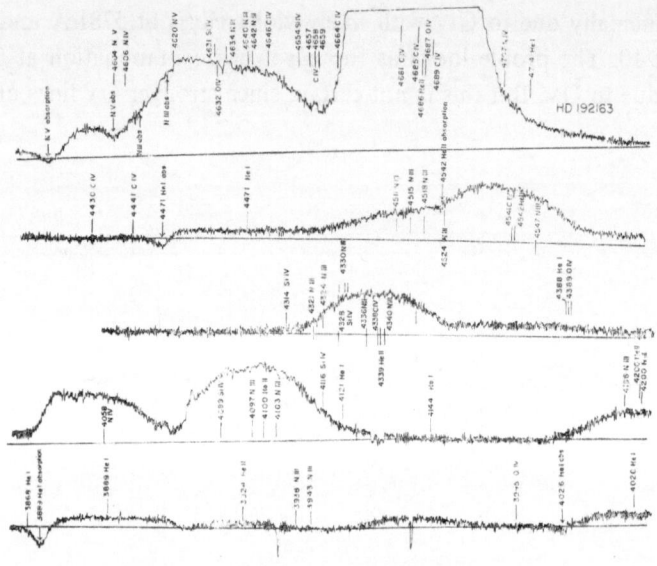

Fig. 11. The blue spectrum of HD 192163.

4640 Å complex: The principal contributors are NIII 4634.2 Å, 4640.6 Å, 4641.1 Å and Nv 4619.98 Å and 4603.7 Å. From the shape of the profile, SiIV 4654 Å is present and hence 4631.4 Å should also be existent. Civ 4646 Å is present and hence Civ 4657 Å, 4658 Å, and 4664 Å are likely contributors. Depressions in the contour at 4615 Å and 4621 Å can be tied up with NIII violet absorption edges. These are not as striking as those of Nv that are at 4590 Å and 4602 Å as can be seen in Figure 11.

4542 Å: HeII 4542 Å is undoubtedly the principal contributor. In the violet wing there is considerable contribution by NIII, going down from 4534 Å to 4511 Å. The violet wing merges with HeI 4471 Å. A violet edge for 4542 Å is seen at 4525 Å. There is also unidentified emission at 4504 Å.

4471 Å: This is a perfect example of a flat-topped profile. There seems to be no contamination. A striking violet edge is seen.

Emission at 4379 Å: In the longward wing HeI 4387.9 Å can be seen to be present. Most of the band is taken up by NIII 4379 Å. The profile of this is remarkably flat.

4340 Å: The principal contribution is from HeII. The line is contaminated appreciably by NIII both on the longward and shortward sides. The contamination is greater on the longward side as can be seen from the distension on the longward side by more intense NIII 4348 Å. On the shortward side SiIV at 4328 Å and 4314 Å are definite

contributors and can be noticed. At least 20% of the line intensity comes from contributors other than Heɪɪ.

4200 Å : While Heɪɪ is the principal contributor there is appreciable contamination
from Nɪɪɪ 4196 Å, 4200 Å and 4216 Å. Also contribution from Sɪɪv 4212 Å can be
significant. This is not a pure Heɪɪ line.

4144 Å : Weak Heɪ can be seen at this wavelength.

4100 Å : Heɪɪ 4100 Å is dominant. Sɪɪv 4116 Å, 4089 Å are substantial contributors.
Nɪɪɪ 4097 Å and 4103 Å also contribute well. In the longward wing there seems to be
some contamination from emission at 4134 Å.

4058 Å : One associates this entire feature with Nɪv 4058 Å. The profile, however,
has little relationship to that of Nɪv 3483 Å. Contamination may exist though abnormal
excitation causing a patchy surface distribution on the star with different kinematical
characteristics, is a possibility.

4026 Å : Contributors here are Heɪ 4026 Å and Heɪɪ 4025.6 Å. The line has a violet
edge.

3965 Å : Heɪɪ. No other contributor is apparent.

3940 Å : Weak Nɪɪɪ at 3938 Å and 3934 Å is present.

3923 Å Heɪɪ: No other contributor is apparent.

3888 Å : Heɪ. There is weak Heɪɪ contamination at 3887.4 and this must be about
10%, as judged from the weakness of the next line in the Pickering series. The most
striking feature is the violet absorption edge.

3858 Å : Heɪɪ very weakly present.

3834 Å : Heɪɪ weakly present.

Complex at 3762 Å : Nɪɪɪ is principal contributor.

3483 Å : This is due entirely to Nɪv. However, there must be something on the
redward side which is contributing. This could be Oɪv and Heɪ. But the identification
is not certain. 3483 Å has a violet edge.

It is impossible to be certain about the presence of Nɪɪ in the blue. The closest is a
slight contamination in the violet side of the 4058 Å Nɪv line. The presence of this ion
has to be established in the red.

3. Some Features of Bright Wolf-Rayet Stars Seen at High Dispersion

The effects of a companion, when such a star is comparable in intensity to the Wolf-
Rayet star, are easily noticeable on spectra of good resolution. Illustrated here in
Figure 12 are the 3800 Å regions of three stars HD 193576 (WN5+06); HD 193077
(WN5+OB) and HD 192641 (WC7+Be). Notice the appreciable intensity of the
Balmer series of the companion in each case even though continuum and emission
features of the Wolf-Rayet star tend to mask them. A more striking case is the detec-
tion of the companion's presence at 6560 Å in the case of HD 192641 (Bappu, 1957).
Miss Underhill later independently found this feature (1962) and ascribed it to a Be
shell companion. Figure 13 depicts the companion's presence at 6560 Å, 4860 Å and
in the near ultraviolet. At my request Olin Wilson had examined in 1957, five other

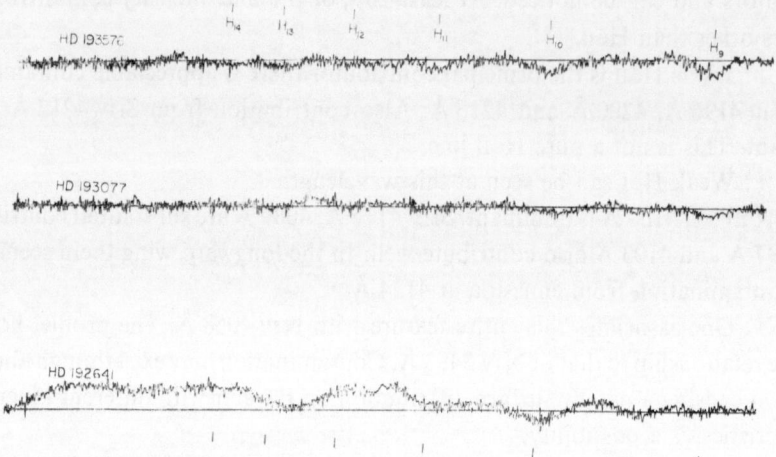

Fig. 12.   Effects of a companion. The tracing of HD 193576 shows the higher members of the Balmer series originating from the O-type companion in a well-established Wolf-Rayet binary system. Note the presence of the Balmer series members in the spectra of HD 193077 and HD 192641. No velocity shifts are measurable in these spectra but the presence of a companion can be inferred.

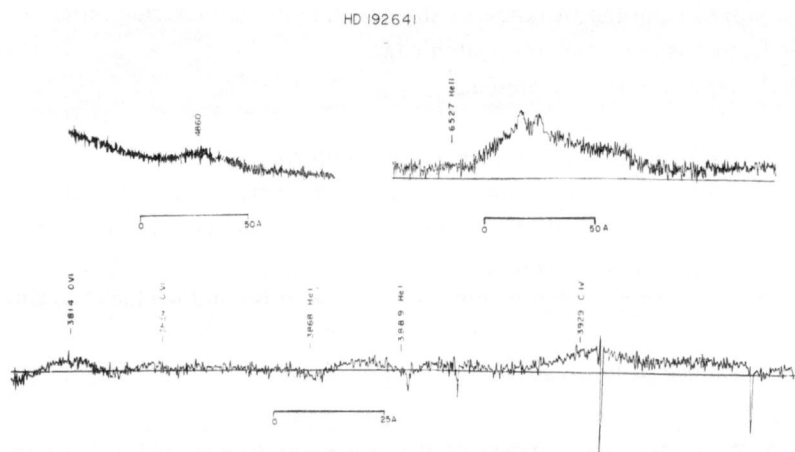

Fig. 13.   Effects of a companion in HD 192641. Note the reversals, ascribed to the companion, at 6560 Å, 4861 Å and in the near ultra-violet.

coudé spectra in his possession of this star, and measured the narrow reversal for velocity shifts. None was seen.

An interesting feature of studies similar to those made by Smith and Aller for HD 184738 and by us for the brighter Wolf-Rayet stars is the finding that violet absorption edges are more common than we had assumed initially. Smith and Aller give a sizable list of these features for a variety of ions that include transitions from metastable levels or ones connected to these to normal levels of the abundant ions. These violet edges are easily seen in the later stages of the Wolf-Rayet temperature sequence. Apart from the fact that the lower temperature enhances these features and brings out many that do not exist in the earlier types, the relatively narrower emission lines enhance the possibility of measurable detection. Table I gives my measures of violet displacement for the different ions in different stars. This table includes the Smith-Aller values for HD 164270.

The presence of violet absorption edges to some of the emission lines has been known since Beal's studies. We have been unaware of the selection which enables some lines to display the phenomenon. Underhill (1968) finds two classes of lines giving rise to the absorption edges. There are those which arise from levels over-populated by dilution effects, like the He I triplets, C III 4647 Å, 4650 Å, 4651 Å and N IV 3478 Å, 3482 Å and 3484 Å, where the lower level is either metastable or connected to one such by a strong transition. There are violet edges seen also on lines arising from relatively low-lying normal levels of ions of the abundant species like C IV 5801 Å, 5812 Å and N V 4603 Å, 4620 Å. In HD 151932 Struve has reported that the violet absorption edge of 4542 Å of He II varies.

Smith and Aller find no violet edge in HD 184738 for the common ion C II which has low lying levels while others like O III 3755 Å, C IV 5801.12 Å, He II 4542 Å that are not low lying, display them. They derive from their observations the following criterion.

A line of a given ion has a violet absorption edge if and only if the excitation potential of the lower level of the line is less than some critical value, which depends on the ion but is always greater than or equal to the ionization potential of the previous level of ionization.

The strength of the foregoing depends on the finding that no isolated line which satisfies it fails to show a violet edge.

A feature of the tabulation of the violet absorption edges is that the largest displacements are usually for He I, while the heavier ions which also have a higher excitation potential display a smaller absorption shift. The expanding shell is thus non-uniform in its kinematical characteristic. The suggestion is reasonable since an ion can exist when there is enough energy to ionize the previous state of ionization. This energy also suffices to maintain a reasonable population in any level for which the excitation potential is less than the ionization potential of the previous ion. Hence Smith and Aller rightly conclude that the violet absorption edge may be formed in regions of the atmosphere that do not have the energy required to form the corresponding emission lines.

The Smith-Aller postulate is not violated by lines observed to have violet edges. However, there are numerous anomalies. There are some ions that have violet edges despite the fact that the E.P. of the lower level is much greater than the ionization potential of the previous ion while other ions having E.P. of lower levels closer to the I.P. still do not possess violet absorption edges. Notable is the difference between He I (I.P. = 0.0; E.P. = 21 eV) and C II (I.P. = 11.2 eV and E.P. = 14 eV). Smith and Aller point out that since He is more abundant and also has only two levels of ionization, any line of He is formed over a much greater depth in the atmosphere than are other

TABLE I

Measures of violet displacement for the different ions in different stars. (The Smith-Aller values for HD 164270 are also included.)

Atom/Ion	Violet displacement								
	HD 165763 WC5	HD 193793 WC + O5	HD 192641 WC7 + Be	HD 192103 WC8 + OB?	HD 184738 Plan. neb. nucleus	HD 164270 WC9	HD 192163 WN6	HD 191765 WN6	HD 50896 WN5
				Km/s					
He I 3705				− 1125					
He I 3820				− 1154					
He I 3889	− 1889	− 2552	− 1041	− 1195	− 510	− 800	− 1380	− 1611	
He I 4026				− 1124	− 458	− 740	− 1296		
He I 4471	− 1946		− 1154	− 1127	− 444	− 670	− 1402	− 1537	
He I 5876		− 2342	− 1332	− 1148	− 490	− 840	− 1378	− 1546	− 1431
He II 4542				− 759	− 370	− 560	− 1188	− 1214	
C III 4647		− 2664	− 1638	− 1213	− 400	− 880			
C IV 5801	− 1660	− 1664	− 1204	− 997	− 415	− 670	− 1035	− 1015	
C IV 5812	− 1620	− 1578	− 1161	− 892		− 575	− 1104	− 979	− 815
O III 3755	− 1404			− 750	− 393				
O VI 3811	− 1015			− 551					
O VI 3834	− 962								
N III 4634							− 1300	− 1294	
N III 4640							− 1288	− 1238	
N V 4603							− 958	− 1243	
N V 4619							− 849	− 1194	

lines and presumably this accounts for the difference in behaviour. For He II lines they have seen absorption edges only on two of the Pickering series (4542 and 5411). This needs confirmation.

Before going on to a discussion of line profiles in these objects, I would like to briefly discuss the near infrared spectra of the Wolf-Rayet stars of the two sequences. My comments are based on low dispersion spectra of 75 Å/mm, 111 Å/mm and 250 Å//mm and serve to give a qualitative idea of the appearance of the spectrum in the photographic infrared. The C IV spectrum is dominant in HD 165763. In the nitrogen stars the N IV, N V spectrum is striking in HD 50896 and HD 192163, while N V is almost absent in the two southern stars HD 92740 and HD 93131. A feature of much interest is the apparent continuum of HD 151932 which shows even in this wavelength region a much greater infrared excess than the others do. This star needs a photoelectric scan study of the continuum urgently. The excess may be intrinsic or may be caused by a red companion.

Work in the infrared accessible to the S1 photosensitive surface has been essentially the splendid coverage by Kuhi (1966). With a 10 Å resolution he has been able to extend identification into the infrared till 1.1μ. One hopes that information in this spectral region will be available on many southern stars in the near future.

Line profiles: Most profiles in WR stars have rounded tops and steep sides. A few selected lines have flat tops like 5696 Å in the carbon sequence. And 3889 Å has in the nitrogen sequence a flat topped profile together with a violet absorption edge that falls in perfectly with the classical model of Beals. Most of the prominent lines can be fitted with a function given by exp $[-\Delta\lambda^2/\sigma^2]$. The values of σ thus obtained describe the degree of random motion inherent in the atmosphere that contributes to the width of the profile, if it is free of all blends. Such a fitting procedure must, therefore, be adopted only where one can be reasonably sure of freedom from blends. In HD 192103, the only lines that are almost free of blends are the C IV lines 5470 Å, 4229 Å, He II 5411. C IV 3933 Å and 3567 Å fall closely to the standards prescribed. In table II we have σ in km/s derived from the observed profiles. Notice the remarkable uniformity in the values derived. The superposed profiles for four lines are shown in Figure 14. Consider the alternative case of the Pickering series in HD 192163. Our identifications made earlier lead us to believe that only a few lines in the spectrum, like 5411 Å, are free of blends that distort the profile. The values for the Pickering series demonstrate this without doubt. In particular σ for 4100 Å, is as expected, unusually high. Only He II 5411 and N IV 3483 Å share a common value which we can consider as depicting the real kinematic state of the emitting atmosphere.

Coming to the flat topped profiles, we find that only a few lines display these and in many cases blending distorts the profile. C III 5696 Å sustains its flat topped nature throughout the list of stars for which we have been able to possess high dispersion spectra. The relative fluctuations of intensity over the flat top can have a two-fold origin. Either superposed blends produce them or they originate from the non-uniformity in spatial intensity distribution of a expanding shell. Such non-uniformity can be present in lines where selective processes of excitation prevail as in C III 5696 Å and

TABLE II

σ in km/s derived from the observed profiles.

HD 192103

Ion	λ	σ (km/s)
He II	5411	798
C IV (7–10)	5470	793
C IV (7–12)	4229	781
C IV (7–13)	3933	863
C IV (7–15)	3567	776

HD 192163

Ion	λ	σ (km/s)
He II	6560	1474
He II	5411	1274
He II	4860	1334
He II	4340	1461
He II	4200	1396
He II	4100	1882
Ni IV	3483	1236

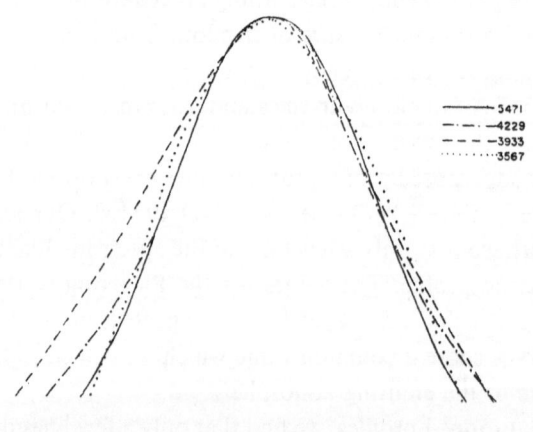

——— 5471
—·—· 4229
— — 3933
·········· 3567

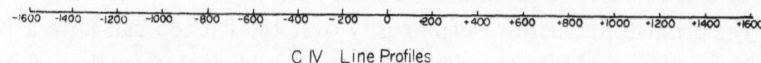

C IV Line Profiles

Fig. 14. Normalized profiles of four lines of C IV. These are the hydrogenic transitions that are relatively free of blends.

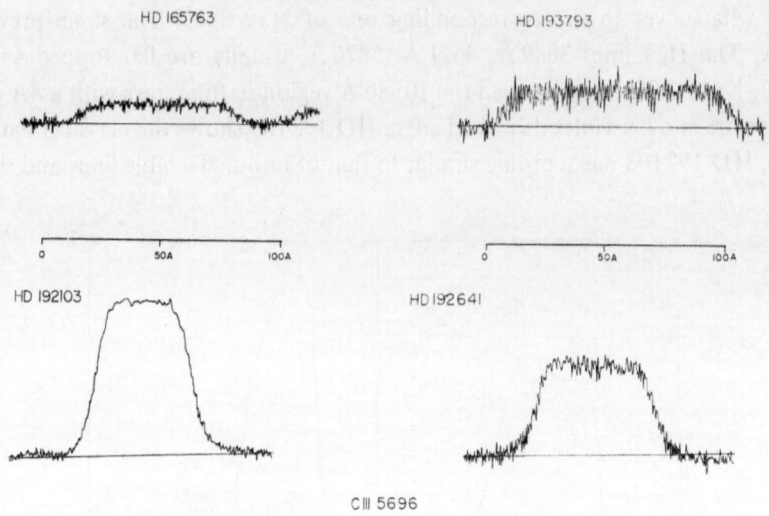

Fig. 15.   The flat-topped nature of C III 5696 Å.

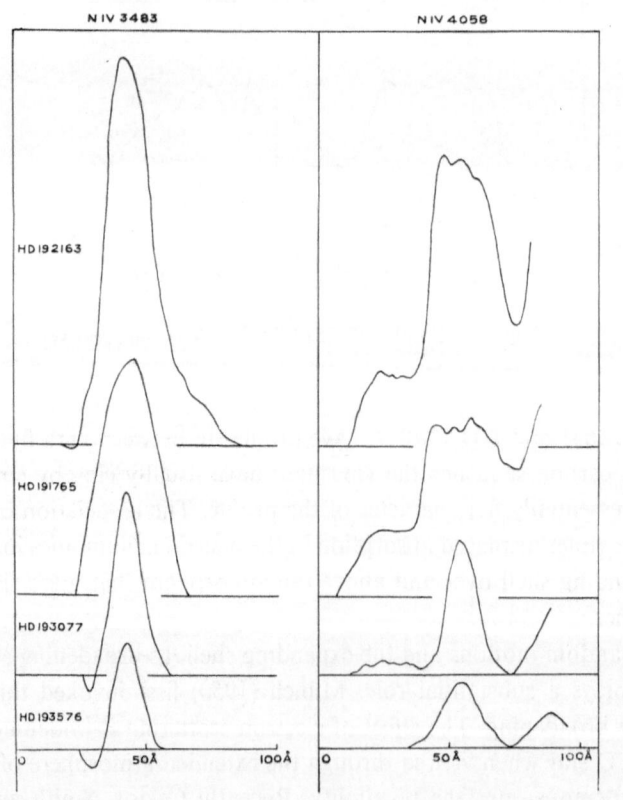

Fig. 16.   Profiles of N IV 3483 Å and 4058 Å in a few WN stars.

N IV 4058 Å. The profiles of 4058 Å in HD 192163 and HD 191765 bear no resemblance whatsoever to the corresponding one of N IV 3483. The same prevails for
5696 Å. The He I lines 3889 Å, 4471 Å, 5876 Å, usually are flat topped as in HD
192163. Kuhi (1968) has scanned the 10830 Å region in four stars with a 4A exit slit.
All the stars show a violet displaced edge. HD 192163 shows the classical flat topped
nature, HD 192103 has a profile similar to that of non-metastable lines and the other

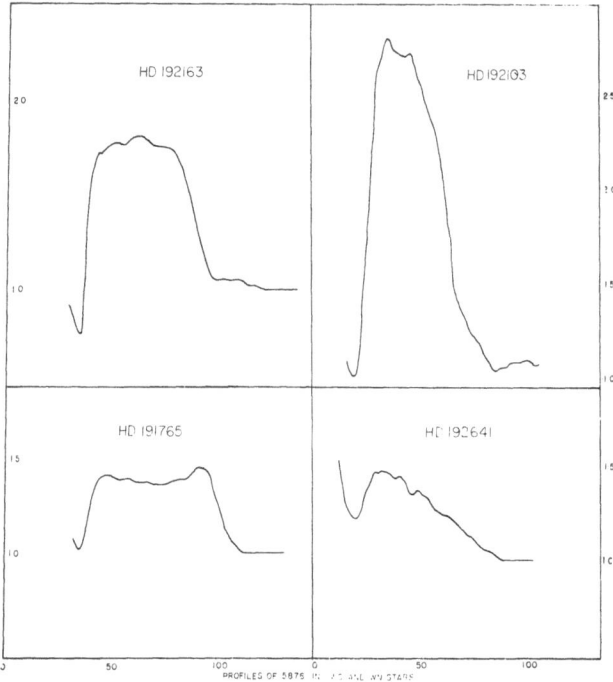

Fig. 17.   Profiles of He I 5876 Å in WC stars and WN.

two, HD 192641 and HD 193077, have profiles in between with fine structure at the
top. In the carbon sequence the He I lines have usually closeby emission lines that
distort the essentially flat character of the profile. The association of the flat topped
profile with violet displaced absorption in the neutral helium lines justify our concept
of an expanding shell over and above the atmosphere that gives rise to the normal
emission lines.

Besides random motions and the expanding shell, as broadening agencies, electron
scattering plays a substantial role. Munch (1950) first invoked this mechanism to
explain line broadening. The effect of electron scattering in widening the absorption
lines of the O star when viewed through the extended atmosphere of the Wolf-Rayet
star amply demonstrates the possibility. Recently Castor, Smith and Van Blerkom
(1970) have satisfactorily explained the broad emission wings of 3483 Å in HD 192163
by non-coherent scattering of electrons.

Smith and Aller (1971) have measured the total width at half intensity for both HD 184738 and HD 164270 for several of the ions. They have then determined a mean value of line width for each ion and plotted it as a function of ionization potential. A correlation between line width and ionization potential is clearly seen. They claim that an improvement over the past has been possible because they have taken into account that an ion will dominate the spectrum of a given element over a wide range of available excitation energies. Vertical bars in their diagram extend from the energy

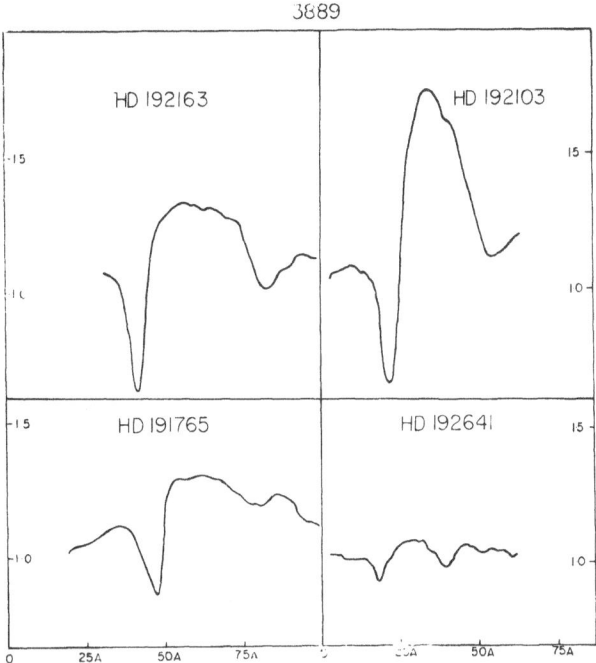

Fig. 18. Profiles of He I 3889 Å in WC and WN stars.

required to ionize the ion, to the energy needed to ionize the next stage of the atom. The good correlation observed eliminates the last outpost of opposition to Beal's model of a radially expanding symmetric atmosphere. Other objections to the model have been 'knocked out' systematically. The transit time effect is too small to be seen according to Castor (1970a) and optically thick shells in a radially expanding atmosphere can produce line shapes that simulate those caused by turbulence (Castor, 1970b; Castor and Van Blerkom, 1970). Smith and Aller favour the model where both velocity of expansion and the amount of energy available for excitation of the atoms in the atmosphere are monotonic functions of the radius. The spectrum of a given ion thus originates from a shell defined by the radius limits at which the available energy is in the range of the vertical bars in their Figure 7, and the widths of the emission lines reflects the radial velocity and range thereof in that region of the atmosphere.

Smith and Aller prefer material to leave the stellar surface at zero speeds and accele-

M. K. V. BAPPU

rate outwards. While deceleration is plausible they consider that accelerated motion is more likely. Hence, the correlation of ionization potential with line width facilitates the inference that the ionization decreases outward. The mechanism of acceleration is assumed to be radiation pressure. Many Wolf-Rayet stars are near the limit for

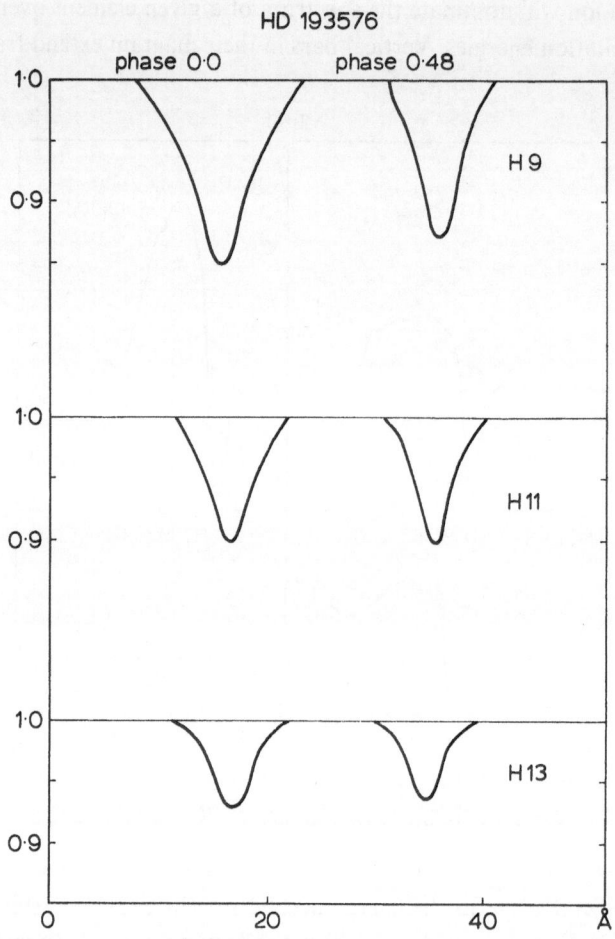

Profiles of H_9, H_{11} and H_{13} in the spectrum of HD 193576 at primary and secondary minimum.

Fig. 19.   Electron scattering effects on the higher members of the Balmer series in HD 193576
(From Ganesh, K. S.; Bappu, M. K. V.; and Natarajan, V., 1967,
Kodaikanal Obs. Bull. No. 184).

stability to radiation pressure. For a pure He atmosphere Rose (1969) states the condition for instability, due to radiation pressure on the electrons as $L/M \geqslant (L_\odot/M_\odot) \times 60000$. For $M = 10M_\odot$ the limit is $L = 600000\ L_\odot$ or $M_{bol} = -10$ mag. For a temperature of 40000 K the bolometric correction $= 4$ mag. Hence $M_V = -6$ mag., a value not inconsistent with Smith's range of values of M_V for the W's.

Line intensities: There have been several efforts made in the past to provide line intensities for the various emission features. Several limitations prevail in utilizing these values effectively for physical interpretation. If a line is not freed of blends its intensity value is hardly representative of the observed transition. Work in the future will need to be very specific on this point and use only those lines for astrophyscal calculations wherein we can be sure of no contamination. Once this criterion

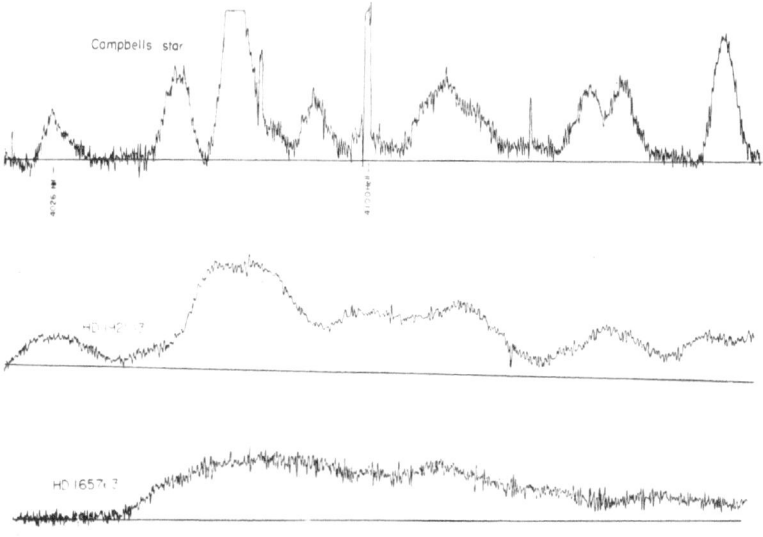

Fig. 20.   Blending effects in Wolf-Rayet spectra.

is accepted there are few lines that cater to such measurement and one can be methoddical and precise in deriving the intensity. Practical difficulties in intensity estimation are enormous for several reasons. Firstly the latitude of the emulsion restricts the chances of accurate estimates of the peaks of the very strong lines. Here obviously the photo-cell has the last say in accuracy. The photo-cell coupled with good resolution can really provide a good line profile. The problem of a continuum exists and specially that of drawing in the wings of the line. Some of the C IV lines in the carbon sequence and particularly in the narrow lined objects are quite amenable to all these procedures of care. They also have by virtue of their similarity to hydrogen fewer limitations imposed from theory. They are, therefore, ideal for use in acquiring information on the WC sequence. The WN sequence in the domain 3400–6700 Å has only 5411 Å and N IV 3483 Å that come close to these criteria. An extension into the infrared may be helpful.

In determination of continua of these stars one has few guide lines which could be applied rigorously. Obviously the narrower lined objects are the most helpful in this regard. What appears as continuum at a specific wavelength for one spectral class range does not often serve the same purpose in another. The problem is severe in the blue. There are often stretches of continuum in the red which can be utilized with ease. This limitation applies not only for measurement of line intensities but more so for flux determination and magnitude estimates through narrow band photometry.

References

Aller, L. H.: 1943, *Astrophys. J.* **97**, 135.

Bappu, M. K. V.: 1957, *Bull. Nat. Inst. Sci. India* No. 9, 155.

Bappu, M. K. V. and Ganesh, K. S.: 1968, *Monthly Notices Roy. Astron. Soc.* **140**, 71.

Castor, J. I.: 1970a, *Astrophys. J.* **160**, 1187.

Castor, J. I.: 1970b, *Monthly Notices Roy. Astron. Soc.* **149**, 111.

Castor, J. I. and Van Blerkom, D.: 1970, *Astrophys. J.* **161**, 485.

Castor, J. I., Smith, L. F., and Van Blerkom, D.: 1970, *Astrophys. J.* **159**, 1119.

Edlen, B.: 1933, *Z. Astrophys.* **7**, 378.

Edlen, B.: 1956, in A. Beer (ed.), *Vistas in Astronomy* **2**, Pergamon Press, London, p. 1456.

Kuhi, L. V.: 1966, *Astrophys. J.* **145**, 715.

Kuhi, L. V.: 1968, in K. B. Gebbie and R. N. Thomas (eds.), *Wolf-Rayet Stars*, Natl. Bur. Stds. Special Publ. No. 307, p. 21.

Munch, G.: 1950, *Astrophys. J.* **112**, 266.

Payne, C. H.: 1933, *Z. Astrophys.* **7**, 1.

Rose, W. K.: 1969, *Astrophys. J.* **155**, 498.

Smith, L. F. and Kuhi, L. V.: 1970, *Astrophys. J.* **162**, 535.

Smith, L. F. and Aller, L. H.: 1971, *Astrophys. J.* **164**, 275.

Smith, L. F. and Aller, L. H.: 1971, *Astrophys. J.* **164**, 293.

Swings, P.: 1942, *Astrophys. J.* **95**, 112.

Swings, P. and Jose, P. D.: 1950, *Astrophys. J.* **111**, 513.

Underhill, A. B.: 1959, *Publ. Dominion Astrophys. Obs., Victoria* **11**, 209.

Underhill, A. B.: 1967, *Bull. Astron. Inst. Neth.* **19**, 173.

Underhill, A. B.: 1968, *Ann. Rev. Astron. Astrophys.* **6**, 39.

DISCUSSION

Johnson: Did I hear you say that you thought there could be a red companion to HD 151932?

Bappu: The redness, of the companion I visualize, comes from the appearance of the slope of the energy curve in the near infra-red. I have no other basis. In fact when you take a spectrum you really have to overexpose the 8000 Å region in order to pick up features at 7000 Å and that I thought was rather anomalous considering some of the other Wolf-Rayet stars. It is not necessary that we should have a red companion for this purpose, but I have been biased for many years towards the possibilities of Wolf-Rayet stars looking red in one aspect and Wolf-Rayet-like in other. This has influenced my statement.

Kuhi: Could you give us some more details on your line identification procedure; for example, were you able to use more or less one value of ionization and excitation potential as a criterion?

Bappu: Yes, I have adopted procedures in line identification that follow earlier work. And I believe that these have been possible because of the very thorough work which has come out of the laboratory in Sweden of Edlen. It then reduces to identifying the right bumps with the correct wavelength assignments and intensity values and see if these fall in place. And most of them do. But I think it is really the laboratory analyses that furnish the key to the entire problem.

Kuhi: Could you not also say that the atmosphere is terribly complicated, that there are stratification effects?

Underhill: Not necessarily. The laboratory people excite these in hot plasmas. And they would get very different intensities for the multiplets in different ionization stages. You cannot take their relative intensities and expect to find the same. But within one multiplet the intensity ratios can be duplicated. I do think your study of the O VI lines is particularly beautiful. I have struggled with that region but could not satisfy myself whether it was there or not there.

Kuhi: As regards the O VI absorption, could you indicate how you establish its presence; I am not quite convinced about it.

Bappu: It is to some extent inferred indirectly. There is no clear cut indication as you see in the case of C IV or N V. The argument runs as follows. To add to the difficulty we have superposed on the same region a violet edge of the He I 3820 line originating from the transition $^3P-n^3D$. So we establish first the presence of the O III violet edge a hundred angstroms or so earlier and measure the spacing of the O VI violet edges with respect to this in HD 192103. The same spacing is maintained in HD 165763 also, which is hotter and has less of the harmful effects of the He I violet edge. Hence the violet edges must exist in HD 192103 for O VI. Also you would have seen that I proved that the O VI lines were free of contamination by He II which falls in this position, since earlier members of the He II series were quite weak. Hence the emission features are uncontaminated. And when such a situation prevails you can only explain the steep slopes on the violet side as due to the violet edges only, since there can be no other obvious causes for the asymmetry in the profiles.

Underhill: You showed very clearly that C III 5696 is flat topped and no other C III emission lines are so flat topped. This, I think, is good evidence that the line is excited in emission in some particular way. Recently Nussbaumer has rediscussed the subject and although I have not studied this paper seriously yet, I believe that the tenure of his conclusions was that there was nothing particularly unexpected with λ 5696. If you were going to get a C III emission line, you ought to get λ 5696. I think the fact that this line has a distinctive shape in WC stars is evidence that something unusual is going on.

Bappu: I believe that in the four thousand five hundred angstroms of spectral region I have covered the aspects shown by λ 5696 are not shown by the other C III lines.

Smith: So, the absence of λ 5696 from the spectrum of HD 50896 does not surprise you?

Bappu: No. What bothers me is that when you have C IV established in the visual region and in the ultra-violet and then you have a comparable excitation of C III ions, where does λ 5696 vanish to?

Underhill: If there is no C III 5696 you do not see C III in any WN star; you do see weak C IV. I rather suspect that this can be accounted for by restrictions on the electron temperature. I shall expand this idea later when I give my talk.

Kuhi: The behaviour of C III 5696 is not definitely unique in WC stars since other C III lines tend to show the same effect. However, as Bappu has pointed out very clearly, it is almost impossible to get such lines free of blends, and, hence, the C III 5696 line (which is free of blends) remans the outstanding example of a flat-topped profile. The C III line at λ9710 also shows a somewhat similar behaviour, becoming flat topped at an earlier spectral class than C III 5696.

Westerlund: You started your paper saying that 19 Å mm⁻¹ is a suitable dispersion for studying line identifications and line profiles in Wolf-Rayet stars, but towards the end of it you said that for defining the continuum you would need as high a dispersion as possible. Could you possibly specify for all of us here your thoughts on what dispersions should be useful for various types of problems particularly the one dealing with violet absorption edges. We have coude spectrographs available in more places now than before and that is why it interests us.

Bappu: I am happy that the Southern Hemisphere is getting so many coude spectrographs in operation. It is particularly gratifying that opportunities for looking at some of these very exotic objects are becoming increasingly available. I have a feeling that the large widths of these lines have tended to catch the investigators off their guard and tended to make them not so strict in the procedures that they need to adopt regarding the kind of dispersions, the slit widths required and so forth. The main idea has been to just get a spectrum, and when you have a broadlined object, it was supposed that opening the slit a trifle and admitting more light into the spectrograph would cause little harm to the information yielded. Hence, apart from the most striking details, faint features have been overlooked because of the lack of adequate resolution. My feeling is that dispersions comparable to say 20 to 10 Å mm⁻¹ are the figures that one would require to see these undulations and so forth on the contours and to establish the identifications with some degree of confidence. It is not possible with dispersions of the order of 60 to 70 Å mm⁻¹ to do anything better than just say that here is a large complex and the contributing ions are the following. As far as the continuum is concerned it is better to be able to get as much of the spectrum as possible with some resolution, before one undertakes to examine which region of the spectrum is really free of the emission lines. And, I think, it is particularly important to do so specially when one is going to use the scanner on these objects. One tends to do this in the case of the absorption line spectra where blanketing corrections for the absorption lines are made in the spectrum admitted by the scanner onto the photomultiplier. I do not believe it is going to be as simple as all that when you have the emission features coming through and, therefore, I would suggest that you really look for those few gaps between complexes and change the investigation accordingly to suit these continuum regions rather than decide *a priori* the wavelengths you would like to work at and then make appropriate corrections. When it comes to a question of intensities my feeling is that it is not as simple as it looks. First of all let us assume that you have the best of photometric procedures. Even then we must be prepared to admit the possibility of a large error coming in because of the well known fact known to every investigator of the WR stars that it is not easy to draw the continuum, and to extend the wings of these emission features to intersect the continuum. It is here that you can have errors of the order of 5 to 10 %, easily coming in, no matter what the photometry initially was to start with. I suppose you could overcome this problem by drawing in the wings many times and taking a straight mean by improving on your continuum many times and again taking a mean value, and thus reducing the overall scatter of your final measurements considerably. But this is a feature you should be well aware of before utilizing the intensities subsequently for hair-splitting theoretical conjectures. That is in the nature of things and I am sure that is something we are all going to have to live with and it is better that we realize it soon enough and be aware of the dangers that lie ahead. For line contours and general purpose photometry, wavelength identifications, I would settle for a dispersion in the neighbourhood of 10 to 15 Å mm^{-1}. Anything higher than that is probably welcome if you can afford it, and I do not think that you can really afford it for many objects besides γ_2 Velorum. So that is probably the limitation I should think you would have. On the other hand, using a 250 Å mm^{-1} spectrum and trying to go through what we did yesterday on the 10 Å mm^{-1} plates is probably something that you should not easily accept.

Johnson: When you observe absorption lines, do you always detect motions? Are you sure the absorption lines never belong to the Wolf-Rayet stars?

Bappu: I can be more specific in this matter, as regards HD 192641. When on the high dispersion spectra I first saw these absorption features, I had no access to plates other than what I had, which were just about three or four. And, therefore, I requested Olin Wilson to go through his collection of plates on HD 192641 and make measurements of these absorption features. And I think he had about 6 to 7 plates on which he made his measurements and he found absolutely no changes in the velocities as given by these lines. So presumably we are seeing the system face on.

Johnson: Do you know for sure that the absorption lines do not belong to a single star, namely, the Wolf-Rayet star itself.

Bappu: We have indirect means of saying that it possibly may not. Let us go back to the case of V444 Cygni where you have a definite case of an O companion and you have the superposed Wolf-Rayet star spectrum. You see the hydrogen lines shifting with respect to the emission lines. Now supposing the hydrogen lines did not shift with respect to the emission lines as they would if the orbit were face on, then I think another aspect that we must consider is the relative lack of contrast between the emission features and the continuum. You can easily determine the spectral type of the Wolf-Rayet star and when you have such a spectral type you assign mentally a certain intensity pattern for the emission lines. And when it gets washed out considerably, because of the presence of the other continuum which would be necessary to show up your absorption lines, then I believe the inference becomes simpler that you are really running into a situation of having the spectrum of another object rather than of having absorption lines intrinsic to the same star itself. The argument is that we pick out cases where we know we have O plus W or B plus W and then extrapolate to this situation wherein you have no relative movement and yet you have a duplication of the same features. The only thing lacking now is the relative velocities and, therefore, the arguments that lead to such a conclusion should be right.

Seggewiss: You mentioned that HD 151932 perhaps has a red companion which is not to be seen in the spectrum. Could it be possible to detect orbital motion from the emission lines of the Wolf-Rayet star?

Bappu: You should be able to detect orbital motion if the geometry of the system is favourable.

Seggewiss: I have taken a lot of spectra of this star at 12 Å mm^{-1} at La Silla.

Bappu: Such spectra on this narrow-lined object would be most useful. I do feel it is worth looking into as far as binary characteristics are concerned. I would say that you have some chance of success in this matter essentially because of the fact that I remember Struve considering the possibility of a 3.4 day period in spectral variations. Not only did he suspect periodicity, but he did find violent changes in intensity patterns of $\lambda 4542$ and others which are typical of happenings under orbital motion and, therefore, you have every chance of bagging a good binary in this case.

Underhill: You may know that I have been assigned the job of summarizing the problems of Wolf-Rayet stars and their nature. Problems I have found in abundance and I have been getting a few more ideas about the nature. I thought it might be helpful if I put a short summary of yesterday's scussions and in particular some of the things that Bappu said at the end, with the hope that we might be able to sort one or two of these things out so that my list of problems of Wolf-Rayet stars and their nature might fit into one hour on the last day of the Symposium. We started with an abstract when Thomas discussed several problems of extended atmospheres. This comes down to the question of what regimes we should divide the atmosphere into in order to obtain a theoretical insight. Next I have to put down a very brief summary of a paper which we have not taken up yet and which will be expanded upon by Van Blerkom. These first attempts at theory by Van Blerkom and Castor in 1970 are important because this is the first paper that departs from the truly classical ways of handling a Wolf-Rayet atmosphere, yet the authors, by the difficulty of the subject, are forced to make quite restrictive assumptions. I mention this because I think you ought to know what these are. Castor and Van Blerkom are trying to interpret the He II lines plus or minus hydrogen in the case of a Wolf-Rayet atmosphere. You have a spherical expanding atmosphere of uniform characteristics, the model comes out with a photosphere of 13 $R_\odot$, the outside edge at 70 $R_\odot$ and they take a representative point at about 40 $R_\odot$. This gives you the dimensions. They assume an 'on the spot approximation', that is, anything that happens to the radiation is only concerned with the local temperature, pressure and so on. They use a thirty-level atom as the first way of dealing with non-LTE transfer effects, uniform expansion and the escape probability method, to take into account the effects of motion.

The results are only for WN6 stars, that hydrogen relative to He is 0.2 by number, N_e is about 10^{11}, T_e is about 10^5. This is the electron temperature. Here are some remarks which occur to me as points worthy of discussion; something about these has been implicity assumed in some of the things we are saying. Are we right or are we wrong? Castor and Van Blerkom showed that the He II lines with $n \geqslant 14$ were optically thin. They assumed the same to be true for Hε and then derived the relative H to He abundance. Under these conditions with $N_H/N_{He} = x$, is the optical depth, small in H and in He, at $N_e = 10^{11}$ and $T_e = 10^5$? I know the arguments but I do not recall that Castor and Van Blerkom actually calculated an optical depth, or a pathlength. Now do you assume on the above model that your real pathlength is something like 40 $R_\odot$ or is it a short distance and all the rest of the atmosphere does not count because it has the wrong velocity? How thin is that layer? The other thing I worry about is, given N_e and T_e, what is the free-free emission? Free-free emission gets very strong in the optical wavelengths at high temperatures. You have got a certain number of hydrogen atoms present, depending on the value of x which is 10, or considerably less. You also get free-free emission from He III and you get electron scattering. These last two are not really affected by the velocity field because they cover wide ranges. I am very much worried about not having calculated the optical depths of the hydrogen lines and for the free-free emission. However, that is our first theoretical attempt and it is quite good, offering considerable new insight. The next paper we had was that of Lindsey Smith. I will just pick out of this the assertions made here. They are not yet published, I believe, so I cannot give you a reference. Here a study of the evolutionary state of WN spectra was presented. Very interesting remarks that I extracted from Lindsey Smith's presentation are the suggestion that WN stars are pure helium stars at a helium-burning stage, the masses are from 6 to about 15 solar masses, the stars are Population I, there are ring nebulae around some WN but not WC stars, and the bolometric absolute magnitude is about -8.0 to -9.2 Following this we had further discussion by Hugh Johnson about the properties of some of the ring nebulae. We saw the uneveness of the intensity distribution in the nebula and considered the question of the radio radiation and what it told about the effective temperature. According to Morton we get effective temperatures running from 50000°–25000°. One remark that occurred to me on reflection and which, perhaps, you might discuss further or clarify now, is how do you get X equals 0, Y presumably greater than 0.99, Z an order of 0.01, for a pure helium, helium burning star. This I take to be the meaning of pure helium. One starts with the original Population I composition, which is X of the order of 0.7, Y of the order of 0.27, and Z around 0.03. Kippenhahn and colleagues have calculated evolutionary tracks for masses of 3 to 15 solar masses, but I do not remember them ending up with a pure helium core of the mass required for a Wolf-Rayet star.

Bappu gave us a beautiful summary of some detailed and careful work on the identifications in Wolf-Rayet spectra. O V is present in WN stars. The presence of O VI in WC stars but not in WN stars indicates that some part of a WC atmosphere must be hotter than any part of a WN atmosphere, although on the whole one gains the impression that the electron temperature is higher in WN atmospheres than in WC atmospheres. One wonders if a model containing 'cool' condensations moving in a 'hot' medium (cf. spicules and the interspicule medium of the Sun) is relevant for understanding Wolf-Rayet spectra.

Thomas: Those T_{eff} for the WC stars range enormously. Do you mean all WN stars have T_{eff} greater than all WC stars.

Underhill: There is a wide range in nominal effective temperature but you have to have something more than just that, to explain the observed spectra.

Thomas: If I adopt what people put on the board, specially in Morton's table, I see a range from 22 000° to 55 000°.

Underhill: That has little to do for an understanding of the line spectra. I would start from the ground rule that the effective temperature has nothing to do with the spectrum. From the spectrum we find out the energy in the electrons. Then we know what energy has to be put into the atmosphere. Then we can go back and ask whether we can get this energy from the radiation field. If we cannot, we have to get it from elsewhere.

Thomas: Let me be specific about what you are talking about. What you have is an effective temperature that means something somewhere, but you are not sure what. And then you have a rise in temperature or may be a fall in temperature somewhere in the outer atmosphere, and the effective temperature is about the same for all WC stars and about the same for all the WN stars and it may differ between these two classes. Is that a good caricature of what you say?

Underhill: Your model is much simpler than any model I dare to make.

Thomas: So you are talking about T_e of about 10^5 for the line-forming medium which in some way increases from the $3–5 \times 10^4$ of the visual continuum. If I understand well the effective temperatures essentially refer to the flux, total bolometric magnitude.

Underhill: It refers to a total radiation flux.

Morton: It is essentially a color temperature – a ratio of short wavelength to long wavelength flux – with sufficient separation to give a useful comparison between the models and the observations.

Thomas: So in a very caricatured way again what Anne Underhill is saying, is that I have an effective continuum flux in the star and then, above that, I have a layer which does not contribute much to the continuum and it has an electron temperature much exceeding that corresponding to the continuum.

Underhill: That is the sort of model I am playing with in this discussion.

Thomas: If I ask Castor and Van Blerkom, what they would say is that really the effective temperature by itself fixes conditions in the stellar atmosphere. If I take what other people are talking about now, the electron temperature is fixed by something else, most likely a mechanical energy flux.

Van Blerkom: No. In the case of the WN6 stars we analyzed, the electron temperature in the envelope was higher than the effective temperature we assumed, implying that mechanical effects determine the electron temperature.

Thomas: From what Castor told me before I left, if you judge from the evolution of his thinking, his current position is the opposite of this.

Van Blerkom: His current position is based on an analysis of γ_2 Velorum, where he found that he did not require a higher envelope temperature than the core temperature, as it was necessary in the previous case.

Thomas: I think you need one general picture of Wolf-Rayet stars. I do not believe that in one case we can have mechanical heating and in another case no mechanical heating. Whatever we do it is going to have to apply to the class of stars as a whole. I think that is what we would like to agree on this week, to try to get some overall physical picture even if we cannot get the details.

Van Blerkom: That is what our simple model can accomplish.

Thomas: A simple model had better come out with correct general physical details. Otherwise it is not going to be simple.

Underhill: Let us wait till we have had the general presentation of theory before we go further into these details. I prefer Thomas' arguments that a model, however simple must be able to embrace the major number of things you have got here. I am trying to isolate what appears to me as significant factors of all that is given to embrace.

Thomas: That is basically what you are doing when you tell me that the effective temperature of the star is one thing whose meaning you do not know, and the electron temperature in the envelope in another thing. You are really focussing on what is the basic point. Does the effective temperature control the electron temperature or is it a perturbation, or is it the whole works. That overall picture is what one wants to get into in some detail.

Underhill: That is the great worry. For I am not sure. I have chosen one type of analysis, the paper by Castor and Van Blerkom which attempted for the first time to give us an electron temperature from the observations. It is a simplification which is literally necessary in order to make any type of progress. We have given you the answer. Now we are looking at some of the observations. I cannot help having intuitive feelings about physics and exhibit them on occasions like this. I have used these ideas to suggest that you might want to consider a model of the type used for the Sun. Now, you are a solar physicist, Thomas, and models for the Sun and all the things that are in them are familiar to you. I have associated with solar physicists over the last eight years and I have got used to hearing their words and to understanding them. Before that I could not care less about the Sun. It was a lot of words that I did not understand and the models meant nothing. I see that most of these people deal with stars and I assume that they are not familiar with solar terminology and detail.

Thomas: Now, remember one thing. For every solar physicist, there is another solar model and the situation is just as bad as for Wolf-Rayet stars. It does not look as though there is a consensus of opinion.

Niemela: I have a spectrum of one Wolf-Rayet star, HD 104994, a WN 3 star, with O vi emission lines.

Smith: Presence of O vi lines in high excitation WN and WC spectra was demonstrated at the last Conference, e.g. see Kuhi's review.

Paczyński: I was going to present some of the best established results of the theory of stellar evolution next Friday. However, the discussion we have heard here today convinced me that it may be worthwhile to spare about 10 or 15 minutes now for such a presentation. Let us consider a star with a typical Population I composition: the initial hydrogen content of 0.7, and metal content of 0.03. The evolution is not changed very much if the abundances are varied to some extent. The pre-main-sequence contraction is not likely to be relevant for the Wolf-Rayet stars. Let us consider the main sequence first. From the theoretical point of view this is the sequence of stars of various masses that burn hydrogen in their cores. The sequence has a width of about one magnitude in stellar luminosity. This phase of stellar evolution is terminated by hydrogen exhaustion in the core. It is followed by a brief phase of hydrogen burning in a thick shell, while the models remain close to the main sequence. They have no hydrogen left within their isothermal helium cores. As soon as the helium core mass exceeds about 10 % of the total stellar mass (the so called Schonberg-Chandrasekhar limit) theoretical models depart from the main sequence and cross the Hertzsprung gap on a thermal (i.e. Kelvin-Helmholtz) time scale.

Subsequent evolution depends on the total mass of a star. If this is below 2.5 $M_\odot$ then the contracting helium core becomes degenerate, and the star becomes a red giant with the hydrogen burning shell as the main energy source. As soon as the helium core mass grows to about 0.4 $M_\odot$ the helium flash takes place. If the total stellar mass exceeds 2.5 $M_\odot$ the contracting helium core does not become degenerate and helium is ignited soon after the departure from the main sequence. In all these cases the star enters the evolutionary phase of helium burning in the core and hydrogen burning in the shell. Helium ignition takes places while the star is a red supergiant. It is so atleast with those stars that have masses below 15 $M_\odot$. The situation with the more massive objects is uncertain. These massive stars have semiconvective regions in their interiors, and then nobody knows to what extent matter is mixed in such regions. Theoretical models of massive stars ignite helium either as blue or as red supergiants, depending on the assumptions applied to the semiconvection.

Stellar models evolve along complicated loops on the H-R diagram during the phase of core helium burning. Unfortunately, the size of those loops depends very strongly on the input physics and the details of the numerical technique used for the model computations. However, all the computations indicate that stars enter the red giant or supergiant region as soon as helium is exhausted in their cores. At that time we have a carbon-oxygen core surrounded by the helium burning shell source, the hydrogen burning shell source, and finally, the extremely extended hydrogen rich envelope. The carbon-oxygen core becomes degenerate in the stars with a total mass below 8 $M_\odot$. In a more massive object carbon is ignited soon after helium exhaustion, and the core never becomes degenerate. It is very unlikely that such a star may lose enough mass to become a white dwarf. It is more likely to explode as a super-nova at the end of its nuclear evolution. A star below 8 $M_\odot$ may ignite carbon explosively if the degenerate carbon-oxygen core will increase in mass to 1.37 $M_\odot$. However, if enough mass will get lost from the red supergiant envelope of such a star the core may never reach the 1.37 $M_\odot$ limit, the carbon will never get ignited, and the core will become a white dwarf.

Massive stars that are relevant for the Population I Wolf-Rayet stars spend 90 % of their lifetime on the main sequence. If we believe that Wolf-Rayet stars are post-main-sequence objects, and if we think that two different subgroups of W-R stars have different ages then it means that they originate from main sequence stars of different mass or different chemical compositions, and that these two subgroups do not make an evolutionary sequence. This statement applies to the single and binary stars as well. I think this statement is correct as I believe we cannot notice a 10 % difference in age by studying the distribution of stars. This 10 % is the age difference between a star that has just exhausted hydrogen in the core, and a star that explodes as a supernova, provided the two stars had the same mass on the main sequence.

Westerlund: Would you comment on the question of where the evolution of massive stars would end up according to your ideas?

Paczyński: It depends on what you assume about the evolutionary status of a Wolf-Rayet star. I hope to show convincingly in my talk later that if you have a Wolf-Rayet star in a binary, that Wolf-Rayet star must be essentially a helium star. There may be a small hydrogen envelope left on top of the helium core. The luminosity of such a star is about the same as the initial luminosity of the hydrogen star from which our helium star has been formed. The temperature of the star depends on its radius, and the radius depends on the amount of hydrogen you have in the envelope, and also on the possible instabilities. You cannot meaningfully compare the radii of theoretical models, which are assumed to be in hydrostatic equilibrium, with the radii of Wolf-Rayet stars which are not in hydrostatic equilibrium: we do see a rapid mass outflow from these stars. However, you may meaningfully compare bolometric magnitudes. It is very important to have reliable bolometric magnitudes for Wolf-Rayet stars. One magnitude is sufficient for a meaningful comparison.

Thomas: You say there is no hydrostatic equilibrium in Wolf-Rayet stars. I say that is the real problem; it is not obvious where the departure from it sets in. The reason is that, in some of your interior calculations that you are talking about, you do not have hydrostatic equilibrium either.

Paczyński: If you do not have hydrostatic equilibrium in the centre, it is likely that the whole star will be blown out or it will collapse on a dynamical time scale, i.e. one hour or so. But we know that Wolf-Rayet stars live for at least 100 yrs!

Thomas: It seems that there are three points here which are getting mixed up very successfully. One point is what the bolometric magnitude of the star is; another point is whether I do have or I do not have any hydrogen in the shell; and the third one is simply defining a Wolf-Rayet star. After all, definition is based on the spectrum, and has nothing to do with the absolute luminosity, nothing to do with the size of the star. It is just a very characteristic spectrum. If you take that last point very seriously, then you must remember that we have Wolf-Rayet spectra characterizing these so-called massive stars which should be derived from some other consideration; we have Wolf-Rayet spectra characterizing the central stars of the planetary nebulae, which have their own mass; and you do have Mrs. Gaposchkin's characterization of the solar rocket UV spectrum. Now I think this is a very difficult point, because, if they talk about the absolute luminosities and bolometric magnitude of the stars, that is just something characterizing the mass of the star. And if from this I come up to the effective temperature of the star, again, that is something characterizing the internal structure of the star. Then if I say I want to decide whether or not this star has hydrogen in the envelope, the fact that the star may be evolved or whether I have a pure helium core is almost irrelevant because I do not see how you get rid of the hydrogen in the envelope, unless some real catastrophe leads you down to that point where you have only He burning. It is these three considerations we must place in perspective. If I say I must have a He star *because of these internal characteristics*, that has nothing to do with the spectrum which you observe. It may or it may not have any hydrogen in the envelope. If you embroider to make the star with no hydrogen in that part of the atmosphere producing the observed spectrum, that requires too some real catastrophe. So now we have to say what are the kinds of catastrophes that get rid of all the hydrogen of the envelope. Then I have to answer the question, and I have to reckon always the fact that there are three types of stars. Those, according to some people's analysis, which have no hydrogen. Those, like the central stars of the planetary nebulae, for which nobody claims I do not have a lot of hydrogen. And thirdly, the solar atmosphere where there is no attempt to say there is no hydrogen. I would be awfully suspicious if I am going to say the Wolf-Rayet spectrum primarily reflects lack of hydrogen. If you are going to say there are stars which have a Wolf-Rayet spectrum, which do not have any hydrogen at all, that simply means that the Wolf-Rayet spectrum is a phase phenomenon, and it really covers a wider area of objects.

Paczyński: Let me make one thing perfectly clear. We should make a distinction between the interior structure of stars and their spectral characteristics. If we accept observational evidence according to which Population I Wolf-Rayet star have bolometric magnitudes of -8 or -9, and masses close to $10 \, M_{\odot}$, then we put rather strong restrictions on the possible interior structure of those stars. In fact we may get such luminosity and mass combination for a helium star only. I mean there must be predominantly helium in the stellar interior. If you want to add (or to leave) some hydrogen at the top of the star, that is all right with me. As long as the mass of the hydrogen rich layer is small, it is not going to affect the mass-luminosity relation for those objects. In fact, Kippenhahn's models with mass exchange have some hydrogen left at the surface. If you consider Wolf-Rayet nuclei of planetary nebulae you find that their masses are believed to be close to $1 \, M_{\odot}$, and their luminosities are of the order of $10^4 \, L_{\odot}$. Unless the present day theory of stellar interiors is entirely wrong, such objects must be in a helium shell burning phase of evolution, and they must have degenerate carbon-oxygen cores. I believe that the available information about masses and luminosities of Wolf-Rayet stars is sufficient to establish with a fair degree of certainty the interior structure of those stars. The reliability of the available information is a different thing. We have to realize that the interior models have very little to do with the appearance of stellar spectra. In particular they cannot explain what is the mechanism for the observed mass loss, estimated to be in the range of 10^{-6}–10^{-4} solar masses per year. I think this mass loss is the most critical phenomenon one has to understand in order to explain the observed spectra. There is no theory known that could predict this kind of mass loss. This must be a different phenomenon from that discovered in early type supergiants by Morton, as there is a 2 or 3 order of magnitude difference in the estimated rate of mass loss. The Morton-type mass loss may be understood in terms of radiation pressure in resonance lines being the driving force, as demonstrated by Lucy and Solomon. This mechanism is not efficient enough to operate at the mass loss rate increased by a factor of 100 or 1000. This is a difficult problem and you cannot expect that simple interior models may solve it. These models may account just for the observed masses and luminosities. These are two different things.

Thomas: That is precisely the point I am trying to reach. Separate things carefully. What you mean when you talk about a Wolf-Rayet star, and when you talk about a Wolf-Rayet phenomenon. It is the Wolf-Rayet atmosphere spectral phenomenon which embraces a wide variety of stars. Then, separately, you can talk about a class of objects which have a helium core if you like, a class of objects which have a carbon core, a class of objects like the Sun; all of them betray in some sense a Wolf-Rayet spectrum. But do not confuse the two concepts. I think we are saying the same thing

Paczyński: We agree on that statement.

Underhill: That is the clearest thing we have heard this week!

Conti: That was a very good argument about Population I Wolf-Rayet stars. It is predominantly helium in the interior; hydrogen is not seen in these objects. Two independent arguments say the star is mostly helium.

Thomas: But neither of them implies the other.

Conti: No, neither of them implies the other, but I want to emphasize that there is no observational evidence for the presence of hydrogen in Wolf-Rayet stars.

Thomas: I disagree with your second point. Agreed we may have a central structure which does not have any hydrogen in it; but it has hydrogen in the outer layer, unless some way the star succeeds in blowing it off. So you have to reconcile the two sets of observations.

Conti: We have Population II Wolf-Rayet stars, which are nuclei of planetary nebulae. Now we believe that one solar mass stars can very often get to the state where they loose all their hydrogen envelope and you are left with a helium core. The interior evidence says it is a helium-burning star. And you look at them spectroscopically and it looks like a Wolf-Rayet star. Again you do not see any hydrogen.

Underhill: You do not see some hydrogen in some of the stars.

Conti: It appears to me that if you take this star and you peel off most of the hydrogen, for reasons which are perhaps mechanical or something else, either a large mass or a low mass star looks like a Wolf-Rayet star.

Thomas: I disagree with this and I say you should better consider what I quoted about the Sun.

Conti: The Sun may look like a Wolf-Rayet star in the UV spectral regions, but in the rest of the spectrum you see lots of absorption lines which you never see in Wolf-Rayet stars.

Thomas: Because I am looking in a different part of the atmosphere. You see, this is my argument. If there is a kind of an atmosphere region where I produce what looks like a Wolf-Rayet star, then I should better understand how I produce that kind of atmospheric region.

Conti: I agree about the phenomenological argument you are making. The Wolf-Rayet phenomenon in a sense means a certain spectral region or atmosphere that gives you emission lines. But there are lots of stars that have their entire atmosphere in this way.

Thomas: There are stars whose internal characteristics or whose blast characteristics produce an interior configuration, and associated with that interior configuration, an atmospheric configuration which comes we do not know how yet, which has a Wolf-Rayet spectrum.

Niemela: I want to make a comment on a possible mechanism to have interior models that could produce Wolf-Rayet stars. If we have complete mixing in a star, then, when it enters the main sequence it has high abundance of C and O. Then, if it burns H through the CNO cycle the luminosity of the star rises and when it leaves the main sequence it has high abundances of N. May be we can explain in this way the two sequences and the overluminosity, because in complete mixing the luminosity is proportional to the mean molecular weight. This may also explain why WC stars are less luminous than the WN sequence stars.

Morton: I think that the references to the solar UV spectrum really confuse the issue. The solar UV spectrum does not look very much like the UV spectra we have of Wolf-Rayet stars.

Thomas: Not really, if you are naive enough.

Underhill: There is a fourth type of objects with a very Wolf-Rayet-like optical spectrum. There are two of them known, namely, Sco X-1 and WX Centauri, both of which are X-ray sources. There may be a third object belonging to this group.

Paczyński: I object against calling the Sun a Wolf-Rayet star. For me the main characteristic of a Wolf-Rayet star is the tremendous rate of mass loss. This is the common feature of all the WR stars. This phenomenon produces high density expanding envelopes, and within those envelopes the prominent emission lines are produced.

Thomas: Let us be very clear now. Never have I called the Sun a Wolf-Rayet star. I only say it looks like a WR spectrum in the UV. Mrs. Gaposchkin, an old time spectral classification person, says, at the immediate first appearance, the rocket UV solar spectrum is like a WC6 spectrum. Let us be very specific. First, there is a Wolf-Rayet object, as which you classify some stars. Then there is what you may call a Wolf-Rayet phenomenon which may be exhibited by an atmosphere in producing a spectrum which looks like this spectrum, or some of whose spectral characteristics resemble those spectral characteristics of WR stars. Be very careful. I am pushing you, Morton, because to me the WR spectrum is characterized by an atmosphere with a high rate of energy input and momentum input. A Wolf-Rayet star has such an atmosphere. So may something else. May be it is mass input that causes this, may be something else. That is why I tried to make these three categories of transfer effects: energy, momentum, and mass.

Morton: Are you not putting too much weight on Mrs. Gaposchkin's comment? Because really, what particular feature in the UV spectrum of the Sun reminds you of a WC6 star?

Thomas: The lack of nitrogen compared to carbon, the presence of high excitation phenomena. Go back fifteen years ago, when we were talking about what were the features of the solar chromosphere and corona and things like this, before you had a UV spectrum. Then we got one, and it displays those excitational ionization effects that one had in only one class of stellar spectra: the WR. Remember we discuss a conceptual evolution in understanding how to produce spectra. What is the big characteristic of the outer solar atmosphere? Twenty years ago one said the solar corona began at 50000 km outside the surface. Now we say it begins at 1500 km, a high density region, where we have lots of things happening we would not have thought of before. We are trying to understand the mechanism of production of that solar structure.

Walborn: Do you know the equivalent widths of these emission lines in the Sun and the wavelength region roughly? Because for WC6 one of the criteria is that the lines are extremely broad and I wonder if $\Delta\lambda/\lambda$ is compatible with that kind of classification for the Sun.

Thomas: You know, one of the most difficult things in the solar spectrum is how do you get the observed widths for ordinary things like Ca II and Mg II. You cannot do it with thermal motions only, you have to have some kind, of motion other than local thermal motions; the broadening mechanism for spectral lines profiles is one of our major problems. There is obviously a big difference in degree between these and WR problems, but that is what we are trying to understand; what is the physical mechanism and what size does it take in various situations?

Walborn: Yes, but I think it is important that a criterion for calling something WC6 is the line width.

Thomas: No, Mrs. Gaposchkin emphasized that *presence* of ions, not how wide the lines are, is the major point. The whole phenomenon of macroscopic motions, macroscopic turbulence in stellar atmospheres, is not understood at all.

Walborn: Morphologically, I object to placing something in a category if it does not satisfy an important criterion defining that category.

Thomas: Well, criteria either characterize something or they do not.

Kuhi: If Mrs. Gaposchkin were given the spectrum of γ_2 Velorum now, taken in the rocket UV, and were to compare it with the spectrum of the Sun, I am sure she would say that there is no resemblance.

Thomas: Remember, caricatures are what I am talking about. Let me return again to these non-classical effects which I assert are necessary to produce a stellar spectrum: population and transfer effects. They take you from the deepest atmospheric layer to the interstellar medium. Think back on what was the point of spectral classification when people first set it up. They thought that if they gave you all the characteristics of a stellar spectrum, that any two stars whose spectra were identical, were exactly the same object. Probably that is true; the only trick lies in specifying all the spectral features you need to know and what you mean by 'identical objects'. Some theoreticians in the field of stellar atmospheres, of which I am *not* that kind at least, assert that if you give somebody the effective temperature and the gravity, they will tell you all about the stellar atmosphere. I deny this possibility. I myself do not know yet what information you need to know in order to predict the spectrum of everything from photosphere to the interstellar medium. I do not know either, in terms of what observation I have, enough to be able to infer atmospheric structure. I think the Wolf-Rayet spectrum is a good example. In spectral classification, one first talked about an O star, a B star, a Wolf-Rayet star. Then one started to talk about a B star with five luminosity classes. Now we start to talk about a B star with five luminosity classes, an infrared excess, a UV excess, and with certain characteristics of a velocity field, interstellar wind, etc. *If* we really know all these characteristics, we can distinguish whether two objects may be identical. But in the Wolf-Rayet situation, there is a *Wolf-Rayet spectrum* but not a unique one. We talk about different kinds of Wolf-Rayet spectra; the central star of a planetary nebula almost looks like an ordinary Wolf-Rayet star which we think has 10 solar masses because some other star, with a similar looking spectrum was measured as a component in a binary. So I have a Wolf-Rayet spectrum. For some stars we talk about Population I, Population II, and something else. All these stars have Wolf-Rayet spectra, they also have certain mass characteristics. I had a big lecture from Lindsey Smith to the effect that we should not confuse these things by saying the Sun belongs to them because anybody knows the Sun does not resemble much in any way a 10 solar mass Wolf-Rayet star and that is true. But the Sun certainly resembles a one solar mass central star in a planetary nebulae in mass characteristics, may be only in mass characteristics. It may be it has no other resemblance to a planetary central star which has possibly a He core burning, the Sun not being that at all. So you have to define what it is with respect to which you say two things are similar. If it is our aim to be able to look at a star and say two stars, with the same set of observational parameters must be identical, then I assert that we do not yet know what are those sets of observational parameters with respect to which they must be identical. Until we know that, if there are some features of the stars which are similar, then I say those stars are similar with respect to the conditions fixing those features. When I talk about the effective temperature, then I can be talking about the model in which the star is built: the effective temperature for a given radius. When I talk about the observations in a particular region of the star where I produce O vɪ, C ɪv, whatever the lines are, then I assert that is a very interesting spectral region, atmospheric region, and what we want to do is to ask how do you produce that? If you make the assertion that if you tell me what the effective temperature is, and the gravity, then you will make a star, then remember we have four kinds of non-classical effects in the atmosphere that disprove this assertion: population effects, and three kinds of transfer effects, energy, momentum and mass transfers. Now if I look at the current stage of evolution so far as model atmosphere calculation go, then, all of these non-classical things at one time were thought to be irrelevant; today one accepts population effects, that is, non-LTE effects, still preserving radiative equilibrium, hydrostatic equilibrium and the like. That is the basis on which Auer and Mihalas computed their B star and O star models. But they are models. If in these models I introduce a mechanical energy transfer, I introduce another effect. If I introduce a momentum transfer, it is another effect. What do we need to introduce the WR model? Kuhi asserts that the main characteristic is a mass transfer or mass loss. That may be the main characteristic of several classes of those objects defining Wolf-Rayet spectra. May be it is true for only the purest kind of Wolf-Rayet spectra. May be in those I only need this mass input. Do I need in addition a mass flow through the star? Is that sufficient? Or must I also produce a momentum flow through the star? Must I also produce an energy flow through the star? Or is it sufficient, from the assertion that Castor makes with respect to one star, that I need only to have a radiative energy input? I am trying not to be dogmatic in saying this. I am only saying be very clear what you say semantically; be very clear what you are saying by implication when you say a spectral class implies a particular stellar model.

Underhill: You are being semantic. All these things were not considered to be irrelevant in the early studies. First, the physical problem was reasonably well analysed by Milne and by Eddington in 1928, but the problem was not susceptible to analytical analysis. So the decision was made to leave out much detail and to do the simplest case.

Thomas: Oh, yes, I agree. Eddington, if you read all he wrote, was aware of all these things in there. But people have offered models which they claim to represent stars, that ignored all these effects. But now we have better data and better knowledge of non-equilibrium physics. So now we can begin to handle them. But we are nowhere near other than scratching the surface. Let us just be honest and admit it. Either you believe your simplified models represent reality or you are doing computing exercises.

SECTION IV

CHAIRMAN: D. C. MORTON

O AND Of STARS

PETER S. CONTI

Lick Observatory, University of California, Santa Cruz, Calif., U.S.A.

and

Joint Institute for Laboratory Astrophysics, University of Colorado, Boulder, Colo., U.S.A. *

1. Introduction

My intention here is to discuss the 'high temperature' portion of this symposium and call attention to those stars that are called Of. There are some similarities in spectral appearance to WR stars, e.g. emission lines. I should first like to define what I think are the essential differences among four groups of hot stars;

O stars: Stars that have *only* absorption lines in the visible spectrum. Type O is distinguished from type B by the presence of He II 4541 at MK dispersion. It may be that some (supergiants) O stars will have emission lines in the rocket UV region but this description will be primarily concerned with ground based observations.

Of stars: These are O type stars that also have $\lambda\lambda$ 4634,40 N III in emission above the continuum. In addition to normal O star absorption lines and N III emission, they may also have other lines in emission. I will discuss this further below.

Oe stars: These are O type stars that have emission in the hydrogen lines (or at least at Hα), but with no emission in N III or in other lines. I personally think that this small class of objects is related to the Be stars in their evolutionary status and in their emission mechanism.

WR stars: These stars are primarily characterized by emission lines. The *only* absorption lines seen are *violet shifted* (P Cyg type). Although in some cases emission lines appear which are similar to those found in some Of stars, the latter types *always* have some *unshifted* absorption lines present. Several Of stars have P Cyg profiles in some lines.

I think these definitions will clearly differentiate between Of and WR type stars. My personal belief is that the evolutionary status of Of and WR stars is completely different, although for some lines the emission mechanisms may be similar.

Some of the work I will describe today will be published soon in the *Astrophysical Journal* (Conti and Alschuler, 1971) but other material concerning these spectra is newly presented here. I will first briefly review the results described there and then go on to more recent work.

2. Relation Between O and Of Stars

Conti and Alschuler (1971) discussed 16 Å/mm blue spectra of 130 O type stars, obtained at the Lick Observatory 120″ coudé. About fifty of these stars had absolute

* Visiting Fellow, 1971–72.

M. K. V. Bappu and J. Sahade (eds.), Wolf-Rayet and High-Temperature Stars, 95–107. All Rights Reserved.
Copyright © 1973 by the IAU.

magnitudes which were obtainable from memberships in clusters and associations with 'well established distances.' From direct intensity tracings made of these spectra, it was possible to measure equivalent widths of certain lines normally used in spectral classification. The spectral classification adopted was entirely based on the measured ratio $\lambda\lambda$ 4471 He I/4541 He II.

This relatively high dispersion and availability of direct intensity tracings also enabled us to make a careful decision as to whether or not a star was 'Of' using the definition above. Perhaps the most important result from that paper is shown in Figure 1, where the O and Of stars are plotted with separate symbols. There is a relatively clean separation between the O and Of stars in that the latter are more luminous and hotter than normal O stars. Earlier than O6, *all* O stars are Of; for later types all the brighter O stars are Of, up to O8.5. The O9 and O9.5 supergiants do not show N III in emission, but as we will discuss below do show λ 5696 C III in emission.

Also shown in Figure 1 is an evolutionary track of 30 solar masses, adapted from Simpson (1971), using the temperature scale and bolometric corrections of Morton (1969). This figure shows there is nothing anomalous about an Of star, it is merely a more massive O star that is hot enough and/or luminous enough, to have N III in emission. There can be age-zero Of stars, and evolved Of stars, just as for normal O stars. The dividing line for O and Of stars is not very sharp but is near 30 solar masses.

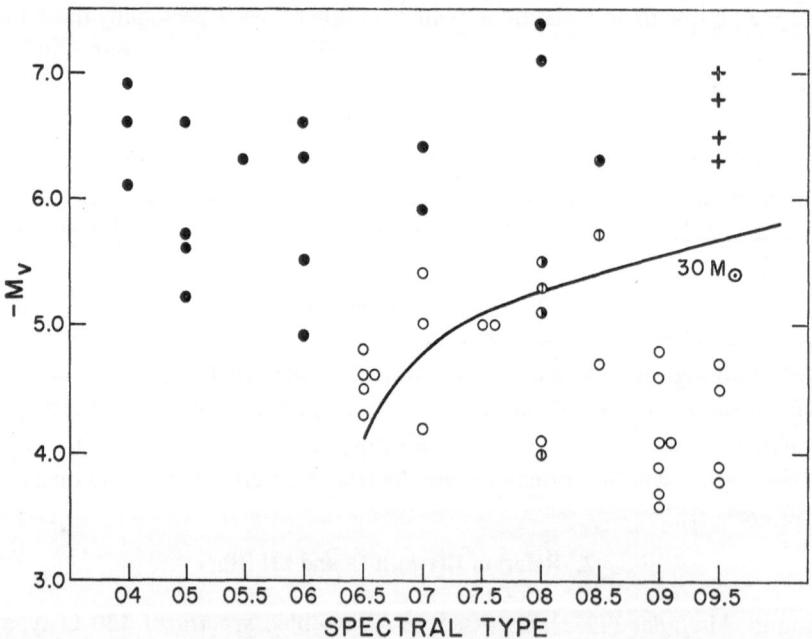

Fig. 1.   HR diagram of the O stars of Conti and Alschuler (1971). The O stars are *open circles*, the Of stars, *filled* or *half filled circles*. *Circles with vertical lines* are luminosity Type III and *crosses* are O9 supergiants. Also shown is an evolutionary track of 30 $M_\odot$ from Simpson (1971). The Of stars are concentrated in the brightest and hottest part of the HR diagram.

The fact that Of stars are generally brighter than O stars can be used as a luminosity indicator. Conti and Alschuler (1971) also discussed an absorption line luminosity indicator, namely 4089 Si IV/4143 He I. Using the same O stars with known distances, they were able to calibrate this ratio and demonstrate that for stars of type O6.5 and later, an estimate of M_v could be made. In earlier type O stars these lines disappear but for such hot stars there is not too much of a spread in M_v. Almost invariably, those O stars with a large ratio for Si IV/He I also had N III in emission, and conversely.

Conti and Alschuler (1971) also discussed the appearance of $\lambda 4686$ He II. In most O stars the line is strongly present in absorption. In some Of stars this line appears in emission or is not present, and in some Of stars the line is strongly present in absorption. There does not seem to be any certain relation between the absolute magnitude of an O star and the strength of He II. The tendency to emission of $\lambda 4686$ does not appear to be related in a simple manner to T_e or L. It is likely that this line is excited by a 'selective' mechanism via emission at $L\alpha$ and a ground state line of He II.

3. Emission Lines in Of Stars

Figure 2 is a schematic version of Figure 1. The O and Of stars are all found within the region of the HR diagram enclosed by the outside solid lines; the O and Of stars are separately found above and below a line near the 30 solar mass evolutionary track. The hotter, brighter regime is where N III 4634,40 is found in emission. Mihalas (1971)

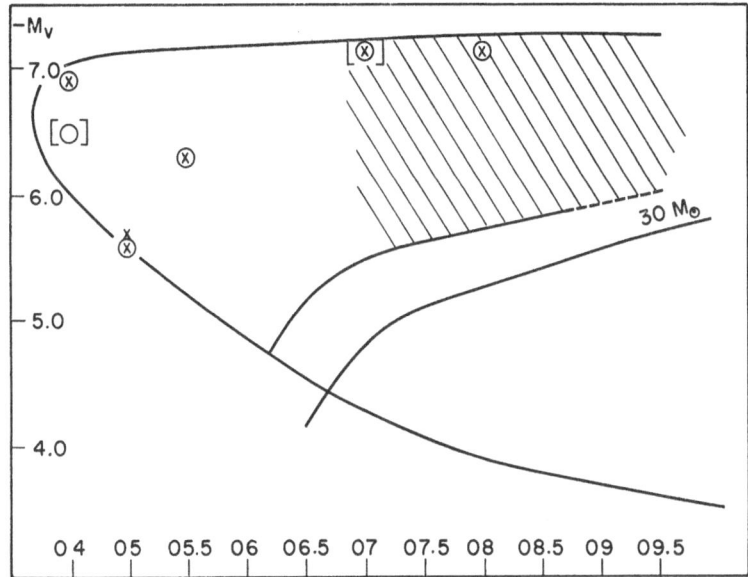

Fig. 2. Schematic version of Figure 1. The lower line is a 'ZAMS' for O stars, the upper boundary the bright M_v limit. The line, partly dotted, near the 30 $M_\odot$ track is the rough dividing line for Of and O stars. The cross hatched area is where all stars showing λ 5696 C III are found; also the regime for the lines λ 4485, 4503. The X denotes stars with C III λ 4647, 50 in emission, *open circles*, Si IV λ 4089, 4116 is emission. The two stars with brackets have M_v estimated from their spectral types only.

has suggested that N III emission is due to a recombination from an autoionizing level just above the N III continuum. The absence of N III in O9 and later type supergiants may indicate that there is not yet enough N IV population. There must be also another ingredient besides sufficient N IV population necessary for the formation of N III emission else *all* O stars hotter than O9 would show these lines in emission. It is tempting to suggest that this other condition is an extended atmosphere, but other physical conditions might be imagined.

Singlet C III 5696 appears in emission in those stars inside the cross hatched portion of Figure 2. I am not yet certain that all stars in this region have this C III emission (and conversely) since I do not yet have complete spectral data but it looks as if it will be so. The C III emission extends into the region of O9 and O9.5 supergiants, which do not show N III emission. C III emission lines are not found earlier than about type O7. A C III λ 5696 emission mechanism has been discussed by Castor and Nussbaumer (1972) for WC stars. It may be the same mechanism for these Of stars but this is not yet certain.

There are two emission lines at λ 4485 and 4503 in some O stars which have long defied identification (Wolff, 1963). These two lines are roughly of equal strength and do not show any absorption components, nor do any absorption lines appear at these wavelengths in other stars. It was a surprise to me to find that the appearance of these lines correlates very nicely with the appearance of emission at λ 5696 C III. As far as my data go, and they are nearly complete, when the former lines are present, the latter are, and conversely. This strongly suggests that *these unidentified lines are* C III. An interpretation in which the lines behave similarily to C III without the ion being C III is not completely ruled out but appears less likely. It is important to note that the $\lambda\lambda$ 4485, 4503 lines are found in O9 and later supergiants and are not found earlier than about O7 If. This behavior is rather dissimilar to that of N III emission suggesting that that ion is unlikely to be the source of these lines.

The remainder of the emission lines I will discuss here are only found in a few Of stars; apparently the mechanism of emission needs a rather unusual stellar atmosphere and/or a very extended envelope.

Triplet C III 4647,50 are found in those stars shown in Figure 2 with the 'X' sign. Four Of stars with these lines in emission are considerably hotter than any Of stars showing the C III singlet in emission, but two later type Of stars show both. These latter stars have P Cyg profiles in hydrogen lines and some other lines and are probably an extreme example of the Of phenomenon.

Si IV 4089,4116 are found in those stars with the open circles in Figure 2. In the two later type Of stars these lines show P Cyg structure. Generally when a star shows these lines in emission it also shows the C III triplet, but there is one counter example for each emission feature. Unfortunately, there is no obvious luminosity effect for either of these emission features as two Of stars on the age-zero main sequence show both ions in emission. The emission mechanisms for S IV and C III (triplet) are not well understood.

Si III 4552,4567 is found in emission in only one Of star, HD 108, which is also shown in Figure 2 as the star at spectral type O7 If with C III and Si IV in emission. N IV 4057 is seen in emission in the two hottest stars in Figure 2, both of which also show Si IV emission. In both of these stars, and only these stars, N V absorption at λ 4603,4618 is also seen. This is also the case for two other very early Of stars discussed by Walborn (1971). These data may suggest a relation between emission at λ 4057 and N V 4603,4618 being present.

4. Emission Lines and Classification in Early Of Stars

As I have pointed out, all early O stars are Of. Figure 3 shows a montage of spectra of six early type Of stars. The absorption line widths are not very dissimilar for Hγ, He I and λ 4541 He II, and are similar to the emission line widths in N III. The spectral classification is given by the ratio 4471/4541. Although HD 46223 has been called O4 by Abt *et al.* (1968), the observed ratio of these lines is more nearly O5 by comparison with other stars classified by Morgan and his associates. However, there are stars which have decidedly weaker λ 4471 He I and these stars I have classified O4 by analogy with Abt *et al.* (1968).

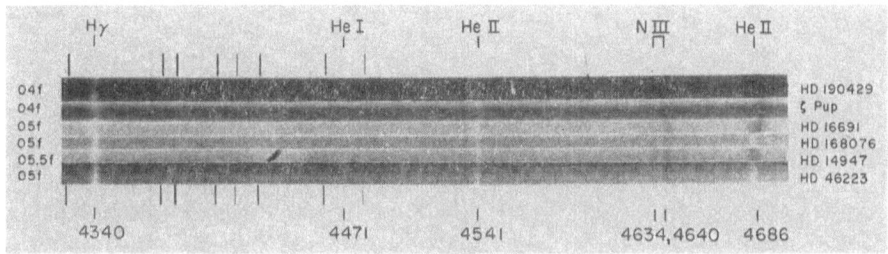

Fig. 3.   Montage of six early Of stars. Spectral types are given by the ratio 4471 He I/4541 He II. Although the emission strengths at N III are not too dissimilar, the He II line at λ 4686 differs drastically from star to star. The emission line widths are similar to the absorption line widths, except for the two O4f stars in which λ 4686 is broader.

We see in Figure 3 that the strengths and breadths of N III emission are a little different from star to star but not markedly so. There is a drastic difference in appearance in λ 4686 He II among these stars. It changes from a strong absorption line in HD 46223 to a strong emission line in HD 190429. This behaviour does not depend strongly on either T_e or L as all these stars have similar values for these parameters. It presumably does give us some useful information as Auer and Mihalas (1972) have suggested this line can be strongly in emission *only* in the presence of an extended envelope.

5. Relation Between N III and He II 4686 Emission

Figure 4 shows the measured line strengths of $\lambda\lambda$ 4634, 40 N III and λ 4686 He II for the Of stars. We are unable to measure with confidence equivalent widths of He II smaller than about 100 mÅ so that the ordinate scale is 'squeezed up' in this region.

The open circles are Of stars of type O6.5 and later; the filled circles are earlier types. We note immediately that there is no difference in behavior of these lines between these classes.

For λ 4686 He II in absorption, but of various strengths, there is little correlation with the strength of emission of N III. Auer and Mihalas (1972) suggest this behavior

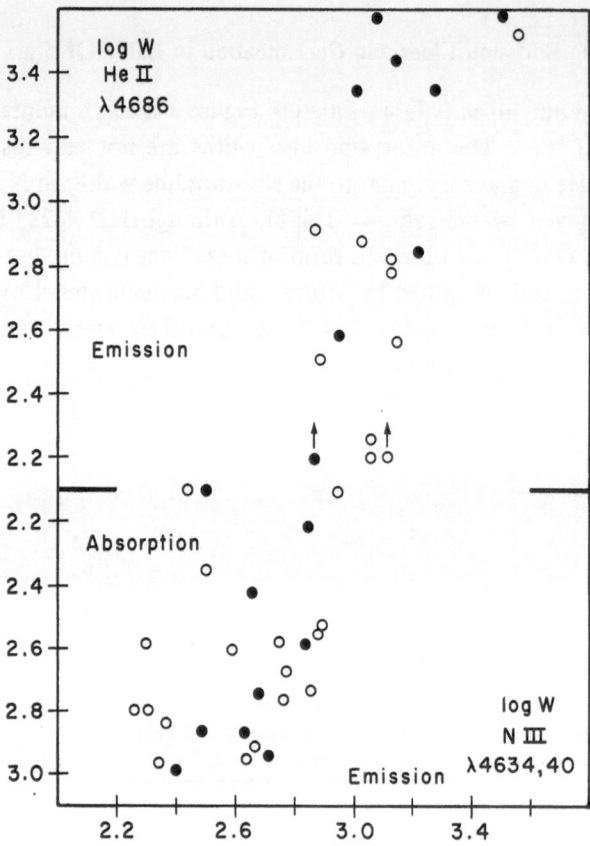

Fig. 4.   Measured line strength (log W) of N III 4634, 40 and He II 4686 in Of stars. *Filled circles,* stars O6 and earlier, *open circles,* type O6.5 and later. There is little correlation between these lines except when He II is in emission.

of He II can be understood in terms of non-LTE and plane parallel atmospheres alone. For λ 4686 filled in, or strongly in emission, they suggest an extended envelope must be present. N III emission is also generally stronger in these stars. There does not seem to be a predictable relation between the strength of N III emission and He II emission except in these most general terms. This suggests the emission mechanisms are dissimilar although both may require certain atmospheric conditions, e.g., an extended atmosphere. Roughly half of the Of stars have λ 4686 absorption, and half have this line in emission.

6. Discussion

I would like to stress what I think is an essential difference between Of stars and WR stars; the former have a more or less 'normal' O star photosphere plus a few emission lines; the latter have a completely different atmospheric structure in that a 'normal' photosphere and its associated absorption lines are not present. There is also an apparent composition difference since Of stars have definite evidence of the presence of hydrogen in their atmospheres, but as far as I am aware, this element has *not* been demonstrated to be present in WR atmospheres. My feeling is that all Of stars have extended atmospheres. For some Of stars we have direct evidence of this, e.g. P Cygni profiles in the rocket UV region, or the visible region. Extended atmospheres probably come about because these stars are hot enough, and/or luminous enough so that radiation pressure in resonance lines (or continuum?) is sufficient to drive out the atmosphere.

It appears that the microscopic mechanism causing the emission lines of several ions, e.g. N III 4634, 40, C III 5696, can be understood as peculiarities in the term structure of the ions. Probably these mechanisms are the same in both Of and WR stars. Other emission lines are not well understood with our present knowledge of the microscopic physics and much work will be necessary in this regard. The identification of the lines at 4485, and 4503 would probably greatly further our understanding of such physics.

An important question is the mass loss rate for Of stars, given the P Cygni profiles we observe in some of them. It seems that this rate is not significant on an evolutionary time scale. I am not certain this is beyond all doubt as the non-LTE line formation problem in an expanding, extended atmosphere has not yet been solved. There is still lots of room for theoretical insight concerning these very interesting Of stars.

References

Abt, H., Meinel, A. B., Morgan, W. W., and Tapscott, J. W.: 1968, *Atlas of Low Dispersion Grating Stellar Spectra*, Kitt Peak National Observatory, Tucson.
Auer, L. H. and Mihalas, D.: 1972, *Astrophys. J. Suppl.* **24**, 193.
Castor, J. I. and Nussbaumer, H.: 1972, *Monthly Notices Roy. Astron. Soc.* **155**, 293.
Conti, P. S. and Alschuler, W. R.: 1971, *Astrophys. J.* **170**, 325.
Mihalas, D.: 1971, *Astrophys. J.* **170**, 541.
Morton, D. C.: 1969, *Astrophys. J.* **158**, 629.
Simpson, E.: 1971, *Astrophys. J.* **165**, 295.
Walborn, N.: 1971, *Astrophys. J. Letters* **167**, L31.
Wolff, R. J.: 1963, *Publ. Astron. Soc. Pacific* **74**, 485.

DISCUSSION

Alcaino: How many O stars are found as members of very young open clusters such as those analyzed by Walker?

Conti: NGC 2264 has one O star, 15 Mon; and NGC 2244 has five O stars. 15 Mon is one of these stars in the first diagram that I classify as an O III star that has emission lines; NGC 6530 has two O stars, 9 Sgr being one of them. I did not include NGC 6611 because I did not have a good distance, which is unfortunate, for there are four O stars in it. It has moduli given in the literature ranging from

12.6 to 11.2 depending on who did it. I think you could get the distance from the B stars, but people have not yet tried this.

Underhill: I know, I tried to reach it from Victoria, but it was too far south and the stars were too faint.

Paczyński: Some of those stars do show P Cygni type profiles for some lines. What is the typical velocity shift? What is the expansion velocity?

Conti: Up to about a few hundred kilometers per second in the visible. Hutchings has made some measures for two stars in the upper part of the HR diagram, one of them HD 152408, and I really cannot improve on those numbers. Velocities measured in the rocket UV come out about 1–2 thousand km s^{-1}.

Sahade: From your description of the Of spectra it was not clear to me whether you found any lines with Wolf-Rayet like characteristics similar to those that Robert Wilson found?

Conti: I want to look into that, but I have not had a chance to do so. I am not convinced of the presence or absence of really extended wings. There are a few stars that have rather broad lines; ζ Puppis is a good example. But I do not think that is what Wilson was talking about. I think the emission lines in ζ Puppis are rather broad but other Of stars also have lines this broad.

Sahade: What is your opinion?

Conti: I am going to go and look very hard, but there is nothing obvious to me. I think my spectra are at least as good as Wilson's, but he may have been looking harder.

Morton: To see the Wilson effect one really wants to make the measurement with a photoelectric scanner in order to get a better signal-to-noise ratio than possible with the photographic plate.

Underhill: I did not believe in it at first, but I found that I had to put it in. The continua are just not flat in the 4600 Å region but rise up by 3 to 5 %. It is very hard to draw the continuum in this region because of the changing photographic sensitivity.

Conti: I have used the same plates to measure interstellar λ 4430, which is a somewhat similar measurement. At least down to the 5 % level I was getting the same as the photoelectric measures Wampler reported, so if these emission lines are less than 5 % over the continuum they may still be there and I have not seen them yet. I have generally only one plate per star. (*Added 21 September, 1971*): I looked 'real hard' at my tracings of 9 Sge and there is *no* extended emission near λ 4640.

Underhill: If it is less than 5 %, you have to average several plates to see it.

Bappu: I remember that Henry Smith looked at ζ Puppis both in the infrared and in the blue violet and seemed to suspect that there was a Wolf-Rayet companion to ζ Puppis. In other words, he did find certain spectral details that were not purely Of and this led him to suggest a possible Wolf-Rayet companion.

Conti: All I can say is that the width of the emission lines that I have shown you in ζ Puppis is a little wide for a 'normal' Of. But then HD 190429 needs some attention too; it has absorption lines of the same width as the N III emission and the He II is much broader. The point is that if ζ Puppis is a Wolf-Rayet plus O one expects velocity variations. So far as I know there are none. That is an interesting question. I will confess that the emission width in ζ Puppis is a little large; however, you do see absorption lines which are also very wide.

Morton: The measurements of ζ Puppis with the Narrabi intensity interferometer indicate that any companion must be considerably fainter by at least 1 mag. and probably by as much as 3 mag.

Bappu: If it is two magnitudes fainter, you would not see it in the spectra either.

Wood: Would you comment about the energy budget of these stars? Yesterday, at lunch, you and Lindsey Smith said you thought radiation pressure was sufficient to support an extended atmosphere and get a Wolf-Rayet-type spectrum, but now you point out that in the instability part of the HR diagram, the mechanical energy flux may be important.

Conti: Lucy and Solomon calculated for two O-type supergiants, that the radiation pressure in resonance lines is sufficient to provide an expanding envelope. As you proceed hotter it is going to still be sufficient to remain extended. These supergiant stars have been calculated to have mass loss. I am not sure mechanical instability has anything to do with extended atmospheres in these stars. I would like to point out the possibility of radiation pressure in the continuum to form an extended atmosphere. The calculations of Lucy and Solomon, by the way, have not treated in detail the line formation problem, which is not a trivial calculation.

Morton: I think in fairness to Lucy and Solomon, they did more, the full details are not there, but they did go one extra step. They tried to put the transfer effects in photons being absorbed in lines, transferring momentum. It is not all quite there, but at least they have gone quite a bit further than these back-of-the-envelope arguments.

Thomas: They did treat the aerodynamic interaction of this material with the rest of the atmosphere?

Morton: No.

Paczyński: They were unable to construct static models above a certain line on the HR diagram, because of the radiation pressure in resonance lines.

Thomas: Did they solve the differential flow problem?

Paczyński: They found that the mean free path is sufficiently small to produce the hydrodynamic flow. Recently Lucy wrote me that the mean free path of the heavy ions with respect to hydrogen is small but not negligible. This results in the heating of the medium, because the ions of carbon or nitrogen are accelerated significantly before they lose their energy by means of collisions. This may be a heating agent for the expanding envelope. Still the mean free path is sufficiently small to justify hydrodynamic description.

Thomas: Once you justify the hydrodynamic description, then you have to solve the hydrodynamic flow problem.

Paczyński: They have done it. This is in a recent letter I received from Lucy. In the paper which is published they obtained a steady flow in the expanding envelope. If you cannot find a static solution then the next simple assumption you may make is that you have a stationary outflow, like in the solar wind. You may get a critical point of the flow, and you may calculate the rate of mass loss.

Thomas: What about the stability of the solution? If one introduces mechanical heating, then that is the critical thing so far as asking what efficiency of conversion do I have. I have then a mass flow, a momentum flow and an energy flow, and I want to ask, is this assumed steady flow pattern really stable against breakup?

Paczyński: This has not been done. However, the results obtained for the steady flow have been compared with the observations, and the agreement was quite reasonable. We can compare line profiles and equivalent widths, and the agreement is fairly good.

Conti: In the rocket UV lines.

Morton: There I think I disagree. For example, the calculations predict that only one strong line should appear, say N v, whereas the rocket observations show that at the same time, C ɪv, O vɪ, and Si ɪv are also strong. Moreover, the predicted profiles do not have zero intensity over a large part of the line as suggested by some of the observations.

Paczyński: Another small question to discuss is where are the Oe stars to be placed on the diagram.

Conti: There is one star in NGC 2244, with M_v near the main sequence.

Paczyński: If I understand you properly you find Of stars all the way down to the main sequence. Of supergiants have been found to loose mass rapidly according to the observations in the ultraviolet. If we exptrapolate this finding to all the Of stars, then this phenomenon should operate on the main sequence too. So, there is no disagreement between Lucy and Solomon and the observations at this point.

Conti and Morton: They predicted mass loss for B-type main sequence stars.

Conti: By the way, there are three very luminous Of stars that show strong P Cygni profiles in the visible. None of them have been looked into in the rocket UV region, presumably the P Cygni profiles will be stronger there.

Sahade: At the beginning of your talk, Conti, you said that you were going to explain the difference in widths of certain lines in Of stars.

Conti: What I said and I thought it was important, was that the widths of the emission lines were not the same for N ɪɪɪ and for He ɪɪ. There were cases where He ɪɪ was considerably broader than N ɪɪɪ. There were cases, for example ζ Puppis, where the widths were both very large. There were cases where these lines were of about the same width. That suggests to me that at least the region of the stellar atmospheres in which these lines are formed is not the same. The only additional comment that I can also make is that when one looks at the very broad emission line profiles, in some stars they are Gaussian while in other stars, as in ζ Puppis, they look suspiciously like scattering profiles.

Thomas: You mean you are dealing with flat-topped profiles?

Conti: No, they are not flat-topped like Kuhi's description of λ 5696 in WC stars. In some cases the profile is narrow, in others it gets very broad.

Underhill: You cannot exclude that ζ Puppis, with its double line spectra, could be double, because you could have a Wolf-Rayet with a continuum two magnitudes fainter and the strong lines, such as λ 4686, would still stand out against the combined two continua.

Conti: It is possible, but I do not believe that ζ Puppis has a WN companion.

Underhill: I would like to make one other remark. I find it fascinating that the Hα emission is so weak in Of stars. There is no suggestion ever made that O stars are short in hydrogen. They have almost all their hydrogen ionized in the outer atmosphere and the result is very little emission. This is a point that I wanted to make about the apparent absence of hydrogen lines in Wolf-Rayet stars. We can have hydrogen present and fully ionized and yet not see the Balmer lines strong in the emission.

Morton: Do you ever see the O vi lines in absorption in any of the Of stars?

Conti: None of these O stars have shown it. Lots of them show O iii in absorption and probably a few of them show O iv, but I have not really played around with O vi. I have not observed O vi.

Walborn: Since this is the day for quasi-Wolf-Rayet phenomena, I would like to start out with some slides of certain OB stars, and work my way back through some remarks about Of stars and comments on Conti's beautiful results, and then end up with a couple of comments about a Wolf-Rayet star. (For illustrations see *Astrophys. J. Letters* **161**, L149; **164**, L67; **167**, L31; *Astrophys. J. Suppl.* **23**, 257.)

These are 63 Å mm^{-1} classification spectrograms 1.2 mm wide, taken at Kitt Peak National and Cerro Tololo Inter-American Observatories, in connection with a program to attempt refinement of the MK spectroscopic parallax system for OB stars. The classification O9.7 is an interpolation between the previous types O9.5 and B0; the primary criterion is comparable strengths in the two lines He ii 4541 and Si iii λ 4552. In O9.5 supergiant standards the He ii line is much stronger than the Si iii line; conversely, in the B0 supergiant standards the Si line is by far the stronger. This top star is a very luminous star, HD 195592 O9.7Ia. I say that because not only is the Si iv, which is a primary luminosity criterion, extremely strong at $\lambda\lambda$ 4089 and 4116, but also there is no feature at λ 4686 of He ii. The negative luminosity effect at λ 4686 is one of the primary, well-established luminosity criteria at late O and B0, that is, it is very strong in the dwarfs, it declines smoothly with increasing luminosity, and it is completely absent in the most luminous stars at O9–B0. You will note also in HD 195592 the N iii blends at $\lambda\lambda$ 4634–40–42, which are the ones that come into emission in Of stars. We see absorption there in this very luminous star, which is very interesting because at O9.5, these lines are not present or they may be very weakly in emission in the most luminous stars. The second star, HD 191781 (ON9.7 Iab), is of similar spectral type although slightly less luminous, and I think you can see the very striking behaviour in the region of $\lambda\lambda$ 4640–50, in which the N iii blends are quite strong and the C iii blend is almost absent. This is very unusual, of course, and this is the reason that I have introduced the classification 'ON' for such spectra. For contrast here is HD 194280 (OC9.7 Iab) with the same spectral type and luminosity class and yet the N iii line λ 4097, which is very strong in the other two stars, comparable almost with λ 4089, is extremely weak, although it is there. Also, the C iii blends at λ 4070 and 4650 are stronger even than in the more luminous star.

Considering next some B2 supergiant classification standards, the expected behaviour of N ii λ 3995 is greater strength than λ 4009 of He i in the Ia star and about equal strengths in the Ib. At B2 Ia the maximum of the N ii spectrum occurs within the two-dimensional MK reference frame. Then we have the star HD 14443 (BC2 Ib) which has an interesting, very peculiar spectrum; if you just looked at it out of context, you would say it is not peculiar as are other spectra that show strange configurations, which are however, partly accidental results of the normal two-dimensional variation in the criteria. This star is peculiar because there is a very strong contradiction between the behaviour of the helium and silicon lines, which are the primary basis of my classification with this higher dispersion, and the carbon and nitrogen lines. (The classification is explained in detail in *Astrophys. J. Suppl.*, No. 198). The star cannot be earlier than B2, it cannot be B1.5, because of the absence of Si iv. It cannot be B2.5 because of the weakness of Si ii relative to He i, so it has to be B2, on the basis of the criteria. But, for a B2 spectrum, the CNO line intensities are completely abnormal; λ 3995, which should be about equal to λ 4009, is just barely visible. I classified it B2 Ib, by the way, and there is no deficiency of silicon, because Si iii λ 4552 is normal in intensity relative to He i, for the luminosity class. The normal behaviour in a B2 supergiant in the region of λ 4630–50 is, comparable to intensities in N ii λ 4631 and the oxygen blends at λ 4639–42 and 4649–51. In HD 14443, however, λ 4650 is stronger than λ 4640, indicating a contribution from C iii at the former wavelength, and λ 4631 is essentially invisible. Finally, consider HDE 235679 (BN2.5 Ib:); oxygen appears to be completely absent and the very unusual green region of the spectrum is dominated by the N ii lines at 4601, 4607, 4614, 4621, 4631, and 4643 Å. This star also has abnormally large He/H line intensity ratios, which is the reason for the noted uncertainty in the luminosity classification. Next consider the spectrum of HD 201345 (ON9 V), which is the prototype of the nitrogen-enhanced, late-O dwarfs. All helium-silicon criteria indicate that it is in fact a dwarf, but the N iii lines are comparable in strength to those in a supergiant. Moreover, the great strength of N iii $\lambda\lambda$ 4634–40–42 (greater than that of C iii λ 4650) also provides evidence that the star is a dwarf, since these lines are never seen in absorption in high-luminosity stars at spectral type O9 (e.g.,

HD 210809, O9 Ib), although they may be present weakly in normal dwarfs (e.g., 10 Lac, O9 V), but with less intensity than C III λ 4650.

The fact that these particular N III lines have a selective negative luminosity effect at O9–B0, as does He II 4686, suggests that the selective emission effects in these same lines in the O stars may be a related phenomenon, and that the latter may, therefore, also be a luminosity indicator. These effects have formed the primary basis of a proposed luminosity classification for the earlier O stars. The following extension of the f designation is introduced: ((f)), strong λ 4686 absorption and weak N III emission; (f), λ 4686 absorption filled in or 'neutralized' and N III emission; f, λ 4686 and N III emission. Main sequence stars are those with strong λ 4686 absorption (e.g., HD 46149, O8.5 V; 15 Mon, O7 V((f)); HD 46223, O4 V((f))), intermediates have weaker absorption or no feature at λ 4686 (λ Ori, O8 III((f)); HD 225160, O8 Ib(f); HD 15558, O5 III(f)), and the most luminous stars are those classified Of (HD 151804, O8 Iaf; λ Cep, O6 If; HD 15570, O4 If). When the present two-dimensional classifications are compared with published equivalent widths of Hγ in O stars, a remarkably tight correlation is found to exist as early as O4, in the sense that the Hγ strength decrease with increasing luminosity.

The present conclusions concerning the O stars are in substantial agreement with those of Conti. It will be noted that his definition of 'Of' includes all three categories used here, so that some 'contradictory' statement are only apparently so. The principal disagreement concerns the luminosity effect in λ 4686 (and hence the luminosity classification earlier than O6.5); Conti concluded that it was not a reliable indicator. However, a star-by-star consideration of his figure shows that the number of stars in serious disaccord is actually rather small and includes two peculiar objects (weak P Cyg profiles at λ 4686), as well as a few (in the region of the Perseus Arm) for which there may be some uncertainty in the absolute magnitudes assumed. Further investigation of these particular cases is needed. Finally, I would like to drawn your attention to the remarkable high-excitation spectrum of HD 93129, which is so far unique and in several respects appears intermediate between those of a very early Of star (e.g., HD 190429, O4 If) and certain Wolf-Rayet stars. This star is one of four in the region of η Carinae classified O3 because they are earlier than the earliest MK standards. HD 93129 is additionally unusual in that it has N IV λ 4058 emission of greater intensity than that at N III $\lambda\lambda$ 4634–40–42, and N V absorption at $\lambda\lambda$ 4604 and λ 4620 of intensity comparable with that of He II λ 4541. These characteristics are also possessed by the Wolf-Rayet stars HD 92740 (WN7-A) and 93131 (WN6-A) of the same region, but in addition they show other N III lines in emission; P Cyg profiles at hydrogen, helium, and N V lines; and much broader and stronger λ 4686 emission. I would like to suggest that further study of HD 93129 may contribute to an understanding of the late-type WN stars of the narrow-line sequence (Hiltner and Schild, *Astrophys. J.* **143**, 770). Also, it is of considerable importance to determine directly whether HD 92740 and 93131, which show high Balmer lines in absorption, are in fact spectroscopic binaries or not.

Niemela: I have some spectrograms of HD 93131 in which absorption lines of the Balmer series upto H 11 are visible; it looks very much like a WN 7 + O spectrum. I have also measured radial velocities of some lines and they are variable.

Walborn: I might mention that among the old Harvard observations of HD 93129 variable P Cygni emission at the hydrogen lines is reported, so it is an extremely interesting star. I have never seen another one like it, with N IV emission stronger than N III.

De Groot: I would like to point out that in HDE 235679 there is a series of N II lines, that you find also in P Cygni.

Walborn: I should mention that the N II lines between λ 4601 and λ 4643 are there also in χ^2 Orionis, the normal B2 Ia standard; you can see them, but they are much weaker relative to the carbon and oxygen feature, than is the case in HDE 235679.

De Groot: P Cygni is believed to be a star of quite normal composition. There may be a little over abundance of hydrogen.

Underhill: The N II lines in the middle B supergiants are queer lines in the sense that they get very, very strong. If you compute a middle B supergiant model atmosphere with log g = 2, which you can do, you can compute from that the wings of the Hγ line following Strom and Peterson using LTE, and fit the model to the star satisfactorily. Then compute the N II lines using LTE as the only available method – a sort of zero order approximation – and you get lines very much too weak. This is one more indication that the lines of the second and third spectra of the metals in supergiants and stars with extended atmosphere are very sensitive to local conditions. The minute you get to an atmosphere with an electron density between 10^{11} and 10^{12}, if the model of the hydrogen lines means anything, then your simple line theory is well off and you have got to do the theory properly.

PETER S. CONTI

I just wonder what the empirically assigned small difference in luminosity class and spectral type mean. Have you really got the correct empirically selected type lines to be sure that all your stars have the same basic flux coming through them, that is, effective temperature? You then ask, if I did the solution for the hydrogen spectrum correctly, will the rest give me something that I can relate to luminosity? Is it just a difference in the extent of the atmosphere? The part that bothers me – and I am curious – is that when you assign these various letters (and empirically you have probably got a very consistent system) with those luminosity classes, have you any outside confirmation of difference of luminosity classes that are significant?

Walborn: Yes, this brings up a very basic point about the MK approach to the problem of spectral classification. What Miss Underhill says is right, the MK approach is empirical and is essentially hypothetical at first. It is an empirical arrangement of the spectra according to certain arbitrarily defined criteria which are a function of what is available in the spectrum and which also may be a function of the resolution and of other things. Of course these are not blind hypotheses; there was earlier work and there are physical reasons behind the choices of criteria, but the question of the validity and the usefulness of this kind of approach can only be answered in terms of the calibration in terms of physical parameters. One must derive the absolute magnitudes and I can state that for some number like 90% of the OB stars, the calibration work that I have done indicates a smooth relationship between the two-dimensional classification and the physical parameters. Of course we are primarily concerned with getting spectroscopic parallaxes in order to do galactic structure and the relevance of this work to stellar evolution is incidental. The sort of thing I have been showing is a by-product and not illustrative of the main purpose of this work. I think it illustrates, however, that the MK system provides a very useful method for discovering and describing unusual objects with respect to a very tightly defined two-dimensional reference frame.

Underhill: I am still wondering about the meaning of this redefinition.

Walborn: I have in my paper a comparison with previous MK classifications and one can see no systematic differences. In the case of the O stars my experience indicates, in comparison with the Morgan and Hiltner lists, essentially no disagreement greater than one tenth of a spectral class.

Conti: Walborn may be right about this business of the classification, but I would like to add a few cautions. He is using a luminosity classification based on He II. I personally feel that the luminosity dependence of He II has not been well established in the early O stars independently of the absolute magnitude. Now, what he did show was a very nice correlation between Hγ equivalent width and the He II 4686 equivalent width. If a star had λ 4686 in emission he called it a Type I, if it had strong absorption he called it a Type V, and anything intermediate as Type III. If λ 4686 was in emission the Hγ equivalent width was weak and conversely. That may or may not be related to absolute magnitude. It may be related, as I said, to some non-LTE population effect.

Walborn: That is right The illustration does not say that there is an absolute magnitude effect in Hγ for the O stars. All it shows is a relationship between two spectroscopic features. However, if the proposed luminosity classification is correct, it follows that there is a luminosity effect in Hγ.

Conti: The second point I wanted to talk about is the MK classification which I have also redone. My spectral classification actually agrees well with Walborn's. An interesting sidelight is θ Orionis; I called this star class O7, the MK class was O6. Peimbert has studied the Orion Nebula and he has very good reasons to believe from the size of the nebula that it is excited by a 38000° star which is exactly the temperature of an O7 star.

Walborn: There is nebular emission in the helium lines in the spectrum of θ Orionis.

Conti: I want to emphasize again that the N IV emission is seen in the two stars you showed with N v in absorption and I want to call people's attention to the fact that of the four or five stars that have N IV emission, all have N v absorption and this may be telling something about how the lines are formed in emission. I also want to say by my definitions that I noted earlier that HD 93131 would be called an Of because it has absorption and it would be very important to see if it is one star or two stars.

Walborn: I think the absorptions are blue shifted.

Conti: The N v are also blue shifted?

Underhill: They are definitely blue shifted. The emissions on tracings made of a Radcliffe plate are always broad. You get the idea of a drowned Wolf-Rayet. The N v lines are sharp.

Conti: You do not see N v absorption in any Wolf-Rayet star.

Underhill: N v absorption does come in many Wolf-Rayet WN stars. It is very strong.

Smith: Anne Underhill is correct. Violet absorption edges on the N v lines 4603, 4620 are usual in

WN spectra. However, in the spectrum of HD 9974 (WN3), N IV 4057 is absent and there is no absorption on the N V lines. So the correlation noted by Conti appears to apply to WR stars as well.

Paczyński: What is the fraction of binaries among the Of type stars?

Conti: I know of three Of binaries and I am now working on their spectra. Two of these systems are double-lined and have been studied in the past. One of them is HD 228766 which was called a WR star but is now called on Of, and the other is BD + 40°4220. Both these are of considerable interest because their mass ratios derived 15 to 20 yrs ago show that the Of star has a mass lower than the O star. I personally do not believe that result. If it were correct it would suggest that the Of star is going to become a Wolf-Rayet star.

Underhill: And there is one other case. Sally Heap showed me a tracing of an O sub-dwarf with a WN8-like spectrum. It is like that picture of HD 93131. I showed her my tracings of a Radcliffe plate of this star; almost identical in all respects.

Paczyński: Does it mean that out of 130 Of stars only two are members of binaries?

Conti: Two Of's out of about 50 Of stars are known double line binaries. There are a number of O-type binaries.

Morton: For how many of these O stars have you measured radial velocities to see whether they are variable?

Conti: There are only two that are known Of binaries with double lines. There may be some more. I think, I may have found double helium lines in another Of star.

Morton: For how many of these stars do you have more than one plate?

Conti: Mostly I only have one spectrogram. I would say that two thirds of the stars are listed in the radial velocity catalogue. It may be possible to detect velocity variations by measuring my plates and comparing to the catalogue value. I am working on this now. I think my statistics are premature at present.

Smith: We seem to be having a little session here of putting odd things on record. There is one I would like to contribute which I have noticed amongst the WN spectra; it concerns the absorption components of the helium Pickering lines. You can see from Table 4 of my review paper that absorption edges are definitely observed on the Pickering lines in 3 of the WN stars studied, the WN4, WN7 and WN8 stars. Let me describe first the WN7 star, HD 151932 (it is a member of Scorpius OBI). The derived H^+/He^{++} ratio is 1.0, i.e. there is hydrogen present, and its contribution to the strength of the emission lines is equal to that of helium. However, a plot of the strength of the absorption lines is smooth, apparently indicating that hydrogen does not contribute to the absorption at all. That strikes me as exceedingly odd. I would also note that it rules out the possibility that the lines arise in the spectrum of a companion star, since I believe that in O absorption spectra the hydrogen always makes some contribution to the absorption.

An even more extreme situation occurs in the WN4 spectrum of HD 187282: The derived H^+/He^{++} ratio is 0.4, so again hydrogen is contributing to the emission lines. The odd-n lines show an absorption component, narrow, double and violet shifted; the even-n lines, to which hydrogen contributes, show no absorption components. Similarly the spectrum of the WN8 star, MR 119, shows strong violet shifted absorption lines on the odd-n Pickering lines, but not on the even-n lines. Apparently the hydrogen somehow masks or fills in the helium absorption; this strikes me as even more peculiar than the previous case. I note again that it is obvious that such absorption cannot be due to a companion.

Underhill: In the very early O type stars, where you run down the equivalent widths of $H\beta$, 4542, $H\gamma$, etc, the contribution due to hydrogen absorption is quite small.

Smith: Is that correct? Is that corroborated?

Walborn: $H\beta$ measurements have been done and compared with Mihalas models and the hydrogen lines hardly decline in intensity at all, to the earliest types, a non-LTE effect. If you look at the grating Atlas, in which the sequence goes all the way to O4, you can see that the hydrogen lines seem to have almost constant intensities all the way from O8 to O4.

Conti: If you can see what you are calling Balmer lines down to H16, then you are sure it is hydrogen.

Underhill: We still have to solve that problem. If you get a very hot plasma you are going to ionize all the hydrogen, and there will be very little chance of recombination with the consequent emission of the Balmer lines. In the Of stars $H\alpha$ emission is very weak. You know there is plenty of hydrogen in these stars yet you would never say that the emission is strong. We cannot determine the abundance of hydrogen from the emission-line strength without setting up and doing the calculations for a hot extended plasma.

SOME RELATIONS BETWEEN WOLF-RAYET AND
P CYGNI-TYPE STARS

MART DE GROOT

European Southern Observatory, Casilla 16317, Correo 9, Santiago, Chile

1. Introduction

In taking up the subject of the relation between different types of early type stars one inevitably comes on ground set foot upon before by C. S. Beals. Already in 1940 he published an extensive paper entitled 'On the Physical Characteristics of the Wolf-Rayet Stars and Their Relation to Other Objects of Early Type' (Beals, 1940). The purpose of the present contribution is mainly to make the discussion more complete by including observations of later years.

One of Beals' biggest problems in 1940 was the lack of sufficiently detailed observations on the subject. Now, more than 30 years later, the situation has improved only a little. Hence, in discussing the characteristics of the P Cygni-type stars, one often can only refer to observations of P Cygni itself, because this is the only well-studied star among the P Cygni-type stars. However, it should be remembered that P Cygni, although the denominator of the group, probably is not really representative of what normally is called the group of P Cygni-type stars. This is because P Cygni has a rather early spectral type when compared to the other P Cygni-type stars and because none of the other stars in the group shows the P Cygni phenomenon as completely as P Cygni itself.

Beals (1940) remarked that "it is very difficult to set up any arbitrary division line between P Cygni and Be stars". However, this paper does not include a discussion of Be stars proper, the separation being made on the ground of the difference in spectrum. Be stars are all high-rotational-velocity stars, while P Cygni-type stars normally are normal-rotational-velocity supergiant-like stars. Beals (1940) also pointed out the close relationship between the α Cygni-type stars and the P Cygni-type stars; this close relation is retained in the present discussion.

2. Definition

The fundamental work of P Cygni-type stars is Beals' study (1950) in which he discusses the spectra of 69 stars showing in at least one line of their spectrum one (or more) of four typical spectral-line profiles. All these profiles have in common a nearly undisplaced emission component, flanked on its short-wave-length side by an absorption component. Beals' group of P Cygni-type stars is in fact a very heterogeneous group, ranging from spectral type O5feq to F2eαq. From his 69 stars 15 are α Cygni type (i.e. supergiants, normally showing only a P Cygni-type line profile in Hα), 3 are

M. K. V. Bappu and J. Sahade (eds.), Wolf-Rayet and High-Temperature Stars, 108–125. All Rights Reserved.
Copyright © 1973 by the IAU.

Of stars, 7 are close binaries of β Lyrae type, 3 are symbiotic objects, one is a planetary nebula of low surface brightness and there are several unique stars (η Car, XX Oph).

For a discussion of the similarities of P Cygni-type stars and Wolf-Rayet stars one should try to limit the class of P Cygni-type stars to a more homogeneous group of stars. Since, in the context of WR stars it seems worthwhile to distinguish between WR objects, quasi-WR objects and the WR phenomenon (Thomas, 1968), I suggest that we define similar terms for P Cygni-type stars, although on a different basis as Thomas' classification.

The P Cygni phenomenon, then, consists of the fact that there are stars showing in their spectra one or more of the four P Cygni-type line profiles defined by Beals (1950). Such stars will be called quasi-P Cygni-type stars if the number of their spectral lines showing the P Cygni phenomenon is very limited, e.g. if it is only shown in one spectrum, or, as even more often is the case, only in the first few Balmer lines. A pure P Cygni-type star is then a star showing a spectrum in which the P Cygni phenomenon is shown by a substantial number of spectral lines of at least two different elements. Such stars will conveniently be called P Cygni-type stars. Most supergiants of early spectral type are showing the P Cygni phenomenon in a rather limited sense (often only in Hα) and so present only an extreme case of the occurrence of the P Cygni phenomenon. They will be referred to as α Cygni-type stars and are forming part of the group of quasi-P Cygni-type stars.

In this context the close binaries of β Lyrae type showing the P Cygni phenomenon will not be discussed. In those stars the P Cygni phenomenon is produced by the duplicity of the star and the presence of a gaseous ring or envelope encompassing the two stars. This selection yields a group of 34 pure P Cygni type stars out of Beals' original 69. There are, however, some more pure P Cygni-type stars not included in Beals' list. Nevertheless, the pure P Cygni-type stars still form a rather small group including quite different objects. Furthermore, good observations have only been made of a relatively small number of these stars. Theoreticians concerned with WR stars often feel a trifle embarrassed because, owing to the limited number of stars, there are only few observations available to compare their theoretical results with. In the case of the P Cygni type stars the situation is even more difficult, and so is a comparison of these two groups of stars.

3. Galactic Distribution

Because of the just-mentioned circumstance little can be said about the galactic distribution of the P Cygni-type stars. It should be remarked, however, that all known P Cygni-type stars are lying close to the galactic equator. Of Beals' P Cygni-type stars, apart from those in the Large Magellanic Cloud, only one star is found at a distance of more than ten degrees from the galactic equator. Also, since the P Cygni-type stars are closely related to the O and B-type supergiants, they, like the classical WR stars, belong to Population I.

4. The Line Spectrum

As has been stated earlier (cf. Section 2) the line spectrum of a P Cygni-type star is characterized by spectral lines showing a nearly undisplaced emission component flanked on its short-wave-length side by a violet-displaced absorption component. Due to the heterogeneity of the group a general description of the main spectral characteristics of the P Cygni-type stars cannot easily be given. The spectral types of the pure P Cygni-type stars range from O6 to A4 and this reflects well how different one P Cygni-type star can be from any other.

The most prominent spectra normally are those of H and He I, except for the extreme spectral types. In the early A stars the role of He I as a prominent spectrum is taken over by Fe II, and Ti II. In the earlier spectral types, besides H and He I, lines of N II, N III, O II and Fe III often are present as medium strong lines. Especially in the B-type stars the N II multiplet between $\lambda 4601$ Å and $\lambda 4643$ Å appears to be stronger than in normal stars of the same spectral type. Abundance analyses along classical ways have been done for P Cygni itself by Ghobros (1962) and by Luud (1967a, b). The latter finds that the composition is essentially the same as for the early B-type supergiants with N probably a bit overabundant in P Cygni.

An abundance analysis by Caputo and Viotti (1970) of the spectrum-variable star AG Car suggests that this star, which has shown a variable number of P Cygni-type lines since 1950 (Thackeray, 1950, 1956; Bond and Landolt, 1970), is underabundant in C and O.

Absorption-line radial velocities in P Cygni-type stars are found to vary with the excitation potential of the spectral lines from which they are determined. The normal interpretation is that this is produced by a stratified atmosphere showing a velocity gradient. Hutchings (1968a, 1968b, 1969) studied this effect for several stars and gave a method of determining the gradient of the excitation temperature in the atmosphere. Similar procedures have been applied by other authors and the general conclusion is that the radial velocities increase with distance from the stellar surface from a few tens of km s^{-1} to a few hundreds of km s^{-1} at 3 to 10 stellar radii. Rocket observations, e.g. by Morton (1967), have even indicated velocities up to 2000 km s^{-1} in the far-UV resonance lines of abundant elements such as C, N and Si.

The excitation temperature is highest near the stellar surface and decreases when going outward. This agrees with what Bappu said yesterday about the behaviour of the excitation temperature in WR stars.

It may be useful to compare some of Conti's information of this morning with observations of P Cygni. The latter star (spectral type B 1.5 according to the absorption lines) is showing the C III $\lambda 5696$ line in emission and very weak absorption, but there is emission at N III $\lambda\lambda 4534$–4640. The unidentified lines at $\lambda 4485$ and $\lambda 4503$ found in Of stars and which seem to behave like C III are not found in P Cygni. The He I emission lines only found in the most luminous late-type Of stars all show P Cygni-type profiles in P Cygni. C III $\lambda 4650$ which only appears in emission in certain hot Of stars shows a very weak emission and absorption in P Cygni. In Of stars the same behaviour is found for the Si IV lines $\lambda 4088$, $\lambda 4116$; in P Cygni these lines only appear in absorption.

4.1. INTERSTELLAR LINES

The interstellar lines in the spectrum of P Cygni show a remarkable behaviour. Apart from a complex interstellar line with components between -10 and -23 km s^{-1} there also are weaker components at $+17$, 0, -36 and -49 km s^{-1} (referred to the Sun) i.e. at -33, -16, $+19$ and $+33$ km s^{-1} with respect to the star. Thus there occur weak interstellar lines at higher-than-normal velocities and in pairs suggesting both material moving away from and falling towards the star. Yesterday Bappu presented a similar observation in a WC8 star. No such behaviour of the interstellar lines is known for other P Cygni-type stars, but, again, this is largely due to a lack of high-dispersion spectrophotometric studies.

5. The Continuous Spectrum

The continuous spectrum of 20 P Cygni-type stars has been studied by Arkhipova (1963). Although the stars studied do not form as homogeneous a group as one might wish it can be concluded that for pure P Cygni-type stars the colour temperature normally is lower than their excitation temperature. For three P Cygni-type stars the colour temperature varies inversely with wavelength to some appreciable extent in the sense that the colour temperature is lower at longer wavelengths. These observations allow an interpretation in terms of the presence of extended envelopes for these stars and do not differ qualitatively from the observations of WR stars (Kuhi, 1966). There is only a slight trend of these colour temperatures with spectral type. The extent of the envelopes seems to be the real determining parameter. In this respect it should be remembered that Of stars and B-type supergiants with only weak P Cygni-type characteristics show hardly any intrinsic reddening. On the other hand peculiar binaries and symbiotic stars on the average show considerable reddening like the pure P Cygni-type stars. Also, the normal colour indices of P Cygni-type stars are redder with respect to the $B-V$ index and bluer with respect to the $U-B$ index than those in ordinary B-type stars.

6. Absolute Magnitudes

Like for the WR stars the determination of absolute magnitudes of P Cygni-type stars is a difficult task. The narrow band photometric systems used by Westerlund (1966) and by Smith (1968) to determine the absolute magnitudes and intrinsic colours of WR stars would be most useful for determining the absolute magnitudes and intrinsic colours of P Cygni-type stars as well. This is especially so since the P Cygni-type stars, with their narrower emission lines, would contaminate even less the photometric system. However, these photometric systems have not yet been used for this purpose and absolute magnitudes and intrinsic colours are based upon U, B, V photometry and upon distance estimates from the interstellar K line. In this respect it should be borne in mind that distance estimates based upon K-line radial velocities and equivalent widths do not always agree nicely since P Cygni-type stars may show stellar components of the K-line, both in absorption and in emission.

Anyhow, from a study of 19 of Beals' pure P Cygni-type stars and 8 α Cygni-type supergiants Arkhipova (1964) concludes that the α Cygni-type stars, with a few exceptions, generally are more luminous than most P Cygni-type stars. She distinguishes two classes of P Cygni-type stars: A small group of bright stars with $M_v \simeq -7.5$ and a larger group of giants and bright giants of luminosity classes III to Ib and M_v between -3 and -5. On the one hand this may be taken as an illustration of the heterogeneity of the collection of P Cygni-type stars, on the other hand it must be pointed out that a study by Luud (1967) of about the same number of P Cygni-type stars shows that the visual absolute magnitudes are all between -6.5 and -8.5 for spectral types between O6 and A4. From Luud's study one gets the impression that the P Cygni-type stars form as homogenous a group of stars as can be reconciled with their variety in spectral types, all being about one magnitude brighter than the brightest supergiants of their corresponding spectral types. Both Luud and Arkhipova base their conclusions on Beals' distances, but Arkhipova is correcting in a more elaborate way for interstellar absorption. The conclusion at this moment must be that there are a number of bright P Cygni-type stars, about 1^m brighter than the B-type supergiants. In Arkhipova's study these are at the same time the stars of earliest spectral type and of highest excitation. Thus, in general the P Cygni-type stars are a few magnitudes brighter than most of the WR stars; the WR stars of subclasses WN7 and WN8 forming the only exception to this rule.

7. P Cygni-Type Stars in Binaries

Very few P Cygni-type stars are known to be members of binaries. Perhaps the best known example is AR Pav, but this is a very complex system (Thackeray, 1959) and very difficult to interpret. A more promising object would be HD 152667. This system has recently been studied by Walker (1971) who finds that the Balmer lines show weak P Cygni-type line profiles of Beals' type III. The mass function of the system is 0.487 which leads to a minimum mass of the system of 4.1 solar masses. Since the light curve shows two unequal minima the total mass of the system cannot be much larger.

Another possible candidate may be HD 98922. Bond (1970) found this star to be a spectroscopic binary, probably with a large velocity amplitude. Spectra taken by the author show that the stronger H lines show a P Cygni-type profile superposed on a quite normal B9 absorption. Although no pure P Cygni-type star seems to be involved, this system like HD 152667, is quite interesting and may yield some information about the mass of a star showing the P Cygni phenomenon at least in its H lines down to H8. A few photometric observations by the author have not yet given a positive answer to the question if the system shows eclipses as well; the observations are being continued. Clearly, the situation with respect to determining masses of

P Cygni-type stars in a more direct way is not as favourable as for the WR stars of which at least one third is known to be double. Of a large number of bright O and B-type supergiants studied by Hutchings (1970) none, except one, are known to be binaries.

8. Masses

Since in the case of the P Cygni-type stars the method of determining masses from binary systems fails because of their nearly complete absence from those systems, the only way of determing masses is through the mass-luminosity relation.

Here we are facing two problems. The first is the uncertainty in the absolute magnitudes. Only for a small number of P Cygni-type stars having some relation to galactic clusters can absolute magnitudes be obtained in another independent way. The other problem is more fundamental: do P Cygni-type stars obey the mass-luminosity relation? The WR stars do not; they are overluminous for their masses. As long as the reason for this discrepancy is not fully understood we can, *a priori*, say very little about the behaviour of the P Cygni-type stars in this respect. Because of certain similarities to the WR stars (extended expanding atmospheres, emission-line spectra) it may well be that also the P Cygni-type stars are overluminous for their masses.

If, to have some basis of discussion, we adopt the absolute magnitudes as derived by Arkhipova (1964) ($M_v \simeq -7.5$ for the early B-type stars and between -3 and -5 for later types) the masses of the stars in these two groups as derived from the mass-luminosity relation are about $100\ M_\odot$ and between 8 and 20 $M_\odot$ respectively. If on the other hand, we adopt Luud's (1967c) bolometric absolute magnitudes, the corresponding masses range from 46 to 144 $M_\odot$ with a mean value at 76 $M_\odot$. There are strong arguments in favour of these higher masses in the light of the variability and mass loss of the P Cygni-type stars. This will be discussed further in the next section. If, following the suggestion of Arkhipova (1964) the P Cygni-type stars really are to be divided into two groups, the brighter group most probably owes its characteristics to their state of instability against nuclear-energized pulsations (Luud, 1967b; De Groot, 1969). The masses of the stars in the less-bright group are more in the mass range of the WR stars. However, other characteristics are very different so that not too much weight should be attached to this similarity.

9. Variability

9.1. LIGHT VARIATIONS

P Cygni-type stars normally show light variations. Most of the variations reported have been characterized as 'irregular' and of small amplitude: normally less than a few tenths of a magnitude in at least several days. However, some stars are known to have shown larger variations. P Cygni was discovered in 1600 as a nova and has since varied rather irregularly over a range of more than three magnitudes (Schneller, 1957). In modern times AG Car has shown irregular variations with an amplitude up to two magnitudes (Greenstein, 1938; Mayall, 1969). Smaller variations have been reported for HR Car (Hoffleit, 1940; Wisse and Wisse, 1971). When discussing these observations it should be borne in mind that the photometric behaviour of most P Cygni-type stars has been studied quite erratically only, so that possible short-period variations may well have been overlooked.

The only star of which frequent photometric observations have been made is P Cygni itself. Magalashvili and Kharadze (1967a, b) have found that this star, superposed on its slow irregular light variations, shows a very regular variation with a period of $0\overset{d}{.}500656$ and an amplitude of $0\overset{m}{.}1$. The light curve resembles that of a W UMa system, but this conclusion has been disputed by several authors (Fernie, 1968; Luud, 1969; De Groot, 1969). Also HD 152667 shows small intrinsic light variations superposed on its eclipse variations, but as yet nothing can be said about amplitude and period (Walker, 1971). Apart from light variations due to duplicity quite a number of WR stars have been found to show probably irregular variations of small amplitude (Roberts, 1962; Smith, 1966; Smith, 1967a; Smith, 1967b; Ross, 1961; Kuhi, 1967; Demers and Fernie, 1964). However, different authors do not always agree upon which stars vary and how much (cf. e.g. Ross, 1961, with Demers and Fernie, 1964). This may be largely due to the infrequency of the observations as in the case of the P Cygni-type stars. For the moment we can only conclude that the variations of WR stars like those of P Cygni-type stars are irregular with small amplitude but that short-period variations may exist as well.

9.2. SPECTRUM VARIATIONS

In his study of the P Cygni-type stars Beals (1950) has indicated some stars which had shown spectral changes in the first half of this century. However, hardly any stars have been studied in any detail so that we can only rely upon detailed studies of P Cygni itself (Luud, 1966; Luud, 1967a; Luud, 1967b; De Groot, 1969). Luud studied the equivalent widths of most of the stronger lines in the spectrum of P Cygni and also the line profiles of the H lines. Both authors agree upon the fact that the equivalent widths of all major lines seem to vary irregularly over a range of about 30 per cent. Kupo (1955) and Dolidze (1958) found that in 1951 and 1952 P Cygni showed irregular spectrophotometric variations coincident with the light and colour variations of the star. P Cygni also shows variations in its line profiles (Astafjew, 1967; De Groot, 1969) due to the presence of different shells in the atmosphere. The outermost shell shows radial-velocity variations with a 114-day period.

Other investigations reveal spectral changes in other P Cygni-type stars. In 1952 HR Car had its He I lines in emission together with those of H, Fe I and [F II] (Henize, 1952) but in 1968 the He I lines were all in absorption while the rest of the spectrum remained essentially unchanged (Bond and Landolt, 1970). AG Car is a P Cygni-type star surrounded by a low-excitation, ring-shaped shell. P Cygni-type lines in its spectrum were numerous in 1950 (Thackeray, 1950; Thackeray, 1956) but had mostly disappeared in 1968, only remaining present in the stronger Balmer lines (Bond and Landolt, 1970).

HD 152667, apart from the radial-velocity variations due to its duplicity, shows small intrinsic radial-velocity variations with a period of about $0\overset{d}{.}6$ or less (Walker, 1971) resembling a β Cephei-type of curve.

Spectral variations in WR stars are not so firmly established. This may be partly due to the existence of their strong emission lines and their weak continuum which makes the precise measurement of equivalent widths and line profiles very difficult. Obvious intrinsic variations are known in the case of HD 45166 which looked like a WR star in 1922 and was intermediate between Of and WR in 1948 (Anger, 1933; Neubauer and Aller, 1948). Recent work has shown that at least some WR stars show irregular variations of their emission line profiles (cf. e.g. Smith and Kuhi, 1970). They attribute some variations to the fact that lines are formed by selective processes (Underhill, 1968) so that different lines of the same spectrum (e.g. N IV λ 4057 and λ 3480) are formed in different parts of the atmosphere. Intrinsic variations have also been observed by Bappu (1951).

Underhill (1968) remarks that "it is rather striking how constant WR spectra remain, considering the state of motion believed to exist in their atmosphere". On the other hand, the state of motion believed to exist in the atmospheres of P Cygni-type stars seems to be much more quiet but their variations are more pronounced. Therefore, the spectral variations of P Cygni-type stars probably have a different cause from those in WR stars. Both the light and spectrum variations of P Cygni-type stars probably are closely related to their evolutionary status. Since, however, they form a rather heterogeneous class of objects one should hesitate to draw general conclusions about the class as a whole.

The explanation of the spectral and light variations of P Cygni-type stars is sought in the following facts: From the work of Ledoux (1941) and of Schwarzschild and Härm (1959) it is known that above a certain critical mass a star cannot be stable against nuclear-energized pulsations. Appenzeller (1970) calculated the characteristics of a vibrationally unstable main sequence star of 130 $M_\odot$ and came to the conclusion that it describes nearly exactly the observations of P Cygni. On the other hand, Stothers and Simon (1968), from a calculation of massive core helium-burning stars with hydrogen-burning shells, find all their models highly stable against nuclear-energized pulsations and propose that any star with a mass less than the critical mass, $\sim 60\ M_\odot$, on the main sequence should evolve into the blue-supergiant region of the HR diagram without disruption. They believe with Lucy and Solomon (1967) that the mechanism causing the mass loss from the B-type supergiants with P Cygni-type characteristics (10^{-7} to $10^{-5}\ M_\odot\ yr^{-1}$) is atmospheric radiation pressure. In a later paper Simon and Stothers (1970) compute the evolution of hydrogen burning stars with moderate mass loss whose mass exceeds the critical mass for pulsational stability at the top of the main sequence. They conclude that these models give a good representation of the mass loss of the most luminous group of WR stars. To explain the behaviour of the P Cygni-type stars I prefer Appenzeller's calculations because they account not only for the observed mass loss but also explain nearly all other observed phenomena. The pulsation periods for unstable stars are around $0\overset{d}{.}5$ and may explain the $0^{d}5$-period light variations of P Cygni found by Magalashvili and Kharadze (1967a, 1967b). The special variations with a time scale of a few days as found by Astafjew (1967) can be explained in the same way. Since he took only four spectra in one week a more intensive series of observations probably would reveal the shorter period.

The same applies to the $0^{d}5$ period radial-velocity variations in HD 152667 found by Walker (1971). He remarks that of the irregular variations in the light curve it is impossible to say whether they are connected with the possible short radial-velocity variations. I would say instead that it is highly probable that upon closer investigation the irregular light variations will reveal a time scale equal to that of the short radial-velocity variations. The only remaining problem in the case of this system is that it probably is not very massive.

10. Mass Loss

Combining the radii determined by Luud (1967c) with estimates of particle density from equivalent widths of spectral lines and with measured radial-velocities of out-flow (Hutchings, 1970) one arrives at mass-loss figures for the P Cygni-type stars between 10^{-4} and 10^{-5} $M_{\odot}$ yr^{-1}. An extensive study by Hutchings (1970) has shown that mass losss in O and B-type supergiants is dependent on both temperature and luminosity; or, in other words, on the UV flux and the surface gravity. The lines connecting stars with the same amount of mass loss follow the zero mass loss evolutionary tracks, suggesting thereby that the rate of mass loss is approximately constant throughout the hydrogen and early helium-burning lifetime of each star. The stars with very high luminosity show mass losses of about 10^{-4} $M_{\odot}$ yr^{-1}. For normal supergiants this figure goes down to about 10^{-7} $M_{\odot}$ yr^{-1}. Mass loss seems to begin in those stars that have exhausted the hydrogen in their cores. These values of the mass loss for P Cygni-type stars compare with those for WR stars which are of the order of 10^{-5} to 10^{-6} $M_{\odot}$ yr^{-1} (Underhill, 1969; Smith, 1966).

For all these stars radiation pressure exerted on the gas in the atmosphere through absorption in resonance lines at wavelengths near the maximum of the stellar-continuum flux is an important driving mechanism for the observed mass loss (Lucy and Solomon, 1970). Besides that there is another circumstance for the brighter P Cygni-type stars. Since they have masses ranging from 50 to 140 $M_{\odot}$ many of them are more massive than the critical mass for hydrogen burning main-sequence stars and, therefore, are pulsationally unstable. This may be a very important mass-loss factor. The analogy with the WR stars, where Paczyński, I think, believes that mass loss is due to a similar instability for helium burning stars, is striking.

11. Evolution

This last part is only meant as a presentation of some thoughts and facts related to the evolutionary status of the P Cygni-type stars and should not be considered as a well-founded explanation of all observations.

There are a few examples of the evolution of single objects. Apart from the spectral changes mentioned in Section 8 one might mention HD 190073 which showed a Be-type spectrum before 1933 and P Cygni-type lines superposed on shallow high-rotation lines in 1937-'38. HD 87643 showed only H in emission in 1909; about 30 yrs later it was reclassified as P Cygni-type; in 1953 Munch listed it as an emission star with WR characteristics (Munch, 1953); finally, in 1968 the spectrum was rather nova-like (Hiltner *et al.*, 1968). Another interesting object is HD 89249 which in 1909 showed only $H\beta$ in emission, later was reclassified as P Cygni type and in 1968 showed a typical Be-type spectrum. (Hiltner *et al.*, 1968). These are some of the facts concerning the evolution of several individual objects showing the P Cygni phenomenon.

Let me now turn to some thoughts relating certain characteristics of the WR stars with those of P Cygni-type stars. Among the WR stars the WC8 and WC9 stars show the lowest excitation and may therefore, lend themselves best for a comparison with the P Cygni-type stars. The WC8 and WC9 stars have line widths of 10 to 15 Å, against more than 30 Å for the other WR stars and less than 4 Å for the P Cygni-type stars. Before this symposium the absolute magnitudes of the WC8 and WC9 stars were still at $-6^{m}.2$ and nicely intermediate between the other WR stars and the P Cygni-type stars. However, Lindsey Smith has felt obliged to bring this value back to $M_v = -4^{m}.8$ for the WC8 stars and $-4^{m}.4$ for the WC9 stars. Since this correction creates some problems with respect to their galactic distribution not all hope for WC8 and WC9 stars to be intermediate between WR stars and P Cygni-type stars need be lost. There is, on the other hand, another problem: If in the P Cygni-type stars the N overabundance is a real thing, then there may be a case for comparing the WN8 and WN8 stars with them. These WR stars still stand at their high luminosities of $-6^{m}.8$ and $-6^{m}.2$ respectively. Furthermore, these are the WR stars with more-pronounced UV absorption edges than the others as Lindsey Smith said yesterday.

Since P Cygni-type stars do not occur in binaries they have not been subject to mass exchange. Furthermore, they show the normal H/He abundance ratio. Thus, there does not seem to be any contradiction in concluding that the P Cygni-type stars are burning hydrogen as massive main-sequence stars subject to mass loss through radiation pressure in their resonance lines and/or through the effect of instability against nuclear-energized pulsations. This last effect also explains their short-period light and spectral variations. If anywhere, then Thomas' stars having a corona and an exosphere are to be found among the P Cygni-type stars.

References

Appenzeller, I.: 1970, *Astron. Astrophys.* **5**, 355.
Arkhipova, V. P.: 1963, *Soviet Astron.* **7**, 51.
Arkhipova, V. P.: 1964, *Soviet Astron.* **7**, 686.
Astafjew, E. R.: 1968, *Astrophysics* **4**, 191.
Bappu, M. K. V.: 1951, *Astron. J.* **56**, 120.
Beals, C. S.: 1940, *J. Roy. Astron. Soc. Can.* **34**, 169.
Beals, C. S.: 1950, *Publ. Dominion Astrophys. Obs., Victoria* **9**, 1.
Bond, H. E.: 1970, *Publ. Astron. Soc. Pacific* **82**, 1065.

Bond, H. E. and Landolt, A. U.: 1970, *Publ. Astron. Soc. Pacific* **82**, 313.

Caputo, F. and Viotti, R.: 1970, *Astron. Astrophys.* **7**, 266.

De Groot, M.: 1969, *Bull. Astron. Inst. Neth.* **20**, 225.

Demers, S. and Fernie, J. D.: 1964, *Publ. Astron. Soc. Pacific* **76**, 350.

Dolidze, M. V.: 1958, *Abastumansk. Astrofiz. Obs. Gora Kanobili Bjull.* **23**, 69.

Fernie, J. D.: 1968, *Observatory* **88**, 167.

Ghobros, R. A.: 1962, *Z. Astrophys.* **56**, 113.

Greenstein, N. K.: 1938, *Harvard Bull.*, No. 908, 25.

Hiltner, W. A., Stephenson, W. B., and Sandulcak, N.: 1968, *Astrophys. Letters* **2**, 153.

Hoffleit, D.: 1940, *Harvard Bull.* No. 913, 4.

Hutchings, J. B.: 1968a, *Monthly Notices Roy. Astron. Soc.* **141**, 219.

Hutchings, J. B.: 1968b, *Monthly Notices Roy. Astron. Soc.* **141**, 329.

Hutchings, J. B.: 1969, *Monthly Notices Roy. Astron. Soc.* **144**, 235.

Hutchings, J. B.: 1970, *Monthly Notices Roy. Astron. Soc.* **147**, 161.

Kuhi, L. V.: 1966, *Astrophys. J.* **143**, 753.

Kuhi, L. V.: 1967, *Publ. Astron. Soc. Pacific* **79**, 57.

Kupo, I. D.: 1955, *Astr. Cirk. Izdav. Bjuro Astr. Soobshch. Kazan*, No. 163, 23.

Ledoux, P.: 1941, *Astrophys. J.* **94**, 537.

Lucy, L. B. and Solomon, P. M.: 1967, *Astron. J.* **72**, 310.

Lucy, L. B. and Solomon, P. M.: 1970, *Astrophys. J.* **159**, 879.

Luud, L. S.: 1967a, *Soviet Astron.* **11**, 211.

Luud, L. S.: 1967a, *Astrophysics* **3**, 172.

Luud, L. S.: 1967c, *Izv. Akad. Nauk Est. S.S.S.R.* **16**, 319.

Magalashvili, N. L. and Kharadze, E. K.: 1967a, *Inf. Bull. Var. Stars*, No. 210.

Magalashvili, N. L. and Kharadze, E. K.: 1967b, *Observatory* **87**, 295.

Mayall, M. W.: 1969, *J. Roy. Astron. Soc. Can.* **63**, 221.

Morton, D. C.: 1967, *Astrophys. J.* **150**, 535.

Munch, L.: 1953, *Bol. Obs. Tonantzintla Tacubaya*, No. 8, 27.

Roberts, M. S.: 1962, *Astron. J.* **67**, 79.

Ross, L. W.: 1961, *Publ. Astron. Soc. Pacific* **73**, 354.

Schneller, H.: 1957, *Geschichte und Literatur des Lichtwechsels der veränderlichen Sternen*, Akademie Verlag, Berlin.

Schwarzschild, M. and Härm, R.: 1959, *Astrophys. J.* **129**, 637.

Simon, N. R. and Stothers, R.: 1970, *Astron. Astrophys.* **6**. 183.

Smith, L. F.: 1966, *Narrow-Band Photometry of Wolf-Rayet Stars*, Thesis, Australian Natl. Univ., Canberra.

Smith, L. F.: 1968a, *Monthly Notices Roy. Astron. Soc.* **138**, 109.

Smith, L. F.: 1968b, *Monthly Notices Roy. Astron. Soc.* **140**, 409.

Stothers, R. and Simon, N. R.: 1968, *Astrophys. J.* **152**, 233.

Thackeray, A. D.: 1950, *Monthly Notices Roy. Astron. Soc.* **110**, 524.

Thackeray, A. D.: 1956, *Vistas in Astronomy* **2**, 1386.

Thackeray, A. D.: 1959, *Monthly Notices Roy. Astron. Soc.* **119**, 629.

Thomas, R. N.: 1968, in K. G. Gebbie and R. N. Thomas (eds.), *Wolf-Rayet Stars*, U.S. Government Printing Office, Washington, p. 1.

Underhill, A. B.: 1968, *Ann. Rev. Astron. Astrophys.* **6**, 39.

Underhill, A. B.: 1969, *Astrophys. Space Sci.* **3**, 109.

Walker, E. N.: 1971, *Monthly Notices Roy. Astron. Soc.* **152**, 333.

Westerlund, B. E.: 1966, *Astrophys. J.* **145**, 724.

Wisse, P. N. J. and Wisse, M.: 1971, *Astron. Astrophys.* **12**, 149.

DISCUSSION

Thomas: Could you sketch how Ca II looks?

De Groot: The normal interstellar components (Figure 1) lie at -10 and -23 km s^{-1}; they are quite strong. Sometimes, on the shortward flank of the component at -10 km s^{-1}, another can be seen at -16 km s^{-1}. Then there are two much weaker components at $+17$ and $+26$ km s^{-1}. Finally there are the two weak components at -36 and -49 km s^{-1}.

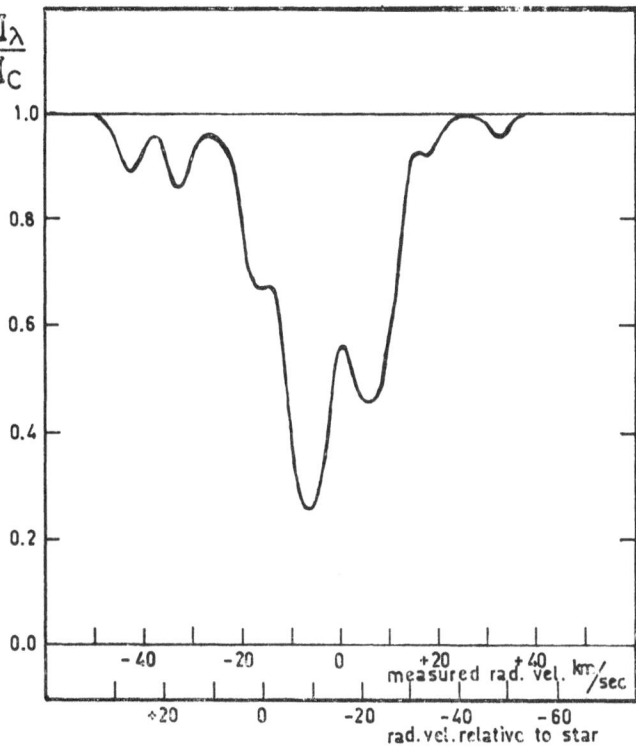

Fig. 1.

Morton: Which ones are permanent and which ones appear sporadically?

De Groot: The components at -10 and -23 km s^{-1} are always there. The components at 0, $+17$ and $+26$ km s^{-1} are quite permanent, that means visible roughly on one plate out of three. The components at -36 and at -49 km s^{-1} I have seen only once.

Van Blerkom: Do you see the same sort of behaviour with the sodium lines?

De Groot: That I did not check yet. The one at -36 km s^{-1} is sometimes showing up in the Na I D lines as well, but not at the corresponding positive velocity. However, at this wavelength the number of plates is only 5.

Kuhi: I am just wondering how you can so definitely rule out the interstellar component?

De Groot: That was based on the fact that when I found these components, I was more or less convinced that they were really there, and then the fact they just fit was such a nice coincidence that I wanted to explain it in this way.

Paczyński: I still do not understand what the coincidence could possibly mean.

De Groot: The mean velocity of two particular components is the stellar velocity. It means that you have a stream going out and one coming back at the same velocity with respect to the star.

Paczyński: So you have matter ejected from the star, and at the same time other matter falling back. I am trying to visualize the picture. It seems difficult.

Thomas: No, if you are willing to accept an outward decelerated field, you can accelerate it for a while and then decelerate it for a while and then throw it back. I am not suggesting anything.

Paczyński: You have two components always present. So you have a more or less stationary situation.

Thomas: It depends on where the excitation is. If, for example, I throw it out, its excitation changes and if it comes back in again its excitation changes. You pick the locus of where Ca is predominantly Ca II. It is a way to make a model.

Underhill: I think it is interesting that these inward components are seen against the emission because that emission gives you a background against which you see some absorption. After all, in stars like

31 Cygni you see clouds going in and out, at the same time, when you are looking at it. You only have the B star at one side of the K star but you can see plus and minus components at the same time. It is a satisfactory description.

Kuhi: Could I also ask whether you have looked at any other O and B stars in the vicinity, to see what the interstellar component out there looks like.

De Groot: I remember from the work of Adams that there is another star very close to P Cygni. It has the same four components at the same velocities.

Bappu: But one finds a temporal change. That is the striking feature. It is not a permanent feature. One of those components comes and goes.

De Groot: It has happened only once. I did not see all of them always, but quite often. The real interstellar ones are always there.

Thomas: So when you average the -49 and the $+17$, which you call interstellar, what you are seeing is really a shell associated with the star. How many shells associated are interstellar?

De Groot: No, it is what you would call a circumstellar line.

Thomas: So this is something which is partly going out and partly coming back. Is that what you suggest?

De Groot: I suggest, and it seems to be quite controversial, that you see here the sum of the interaction between the expanding envelope and the interstellar medium.

Paczyński: Is there any observational evidence that could disprove the suggestion that P Cygni behaves, as far as the light curve is concerned, like a W UMa system with a period of half a day and an amplitude of 0.1 mag.

De Groot: Fernie calculates this case. From the mass-luminosity relation, he finds a mass of about one hundred solar masses. If you take one particle and let it go round this star which has a large radius as well, then the velocity is something like 10^4 kilometers per second, and that is not possible.

Paczyński: I have asked you about the observations. According to the results of Appenzeller, a main sequence star of about 100 solar masses may become vibrationally unstable and may increase its radius by a factor of four, and decrease its effective temperature by a factor of two, and look like P Cygni itself. The pulsation period for the main sequence star is about 10 hours when the star expands by a factor of four. Therefore, there is no theoretical reason at present to deny a half-day period for the luminosity variation of P Cygni.

De Groot: I think there are photoelectric observations of Alexander and Wallerstein (*Publ. Astron. Soc. Pacific* **79**, 500, 1967), who have tried to check the half-day period, and have found no variation. I would not reject at first the half-day period, but reject at least the W Ursae Majoris system idea, and explain any variations as pulsational instability.

Paczyński: That is why I was asking for the observational evidence.

De Groot: I think the observations pointing to the half-day periods are still better than the ones denying it, since they are more numerous, but this is still a controversial point. May be we can summarize by saying that normally the variations are quite irregular.

Thomas: Could you tell us what density and radii and velocity you use, that yield a mass loss of 10^{-4} to 10^{-5} $M_\odot$ per year?

De Groot: For stars between 46 and 144 solar masses you have radii between 40 and 80 solar radii; roughly 100 solar masses give a radius of 60 solar radii.

Paczyński: What do you mean by a mass-radius relation?

De Groot: The luminosity-radius relation.

Paczyński: Luminosity-temperature-radius?

De Groot: Yes, the luminosity-temperature-radius. You have the logarithm of the electron densities which range between 12.5 to 13, and you have velocities of one hundred to two hundred kilometers per second. Now if you combine these you come to mass-loss figures in this range. The only thing is that you have to believe this. Hutchings has more extensively studied mass loss in OB supergiants and he finds that it is dependent on both temperature and luminosity. He draws a HR diagram; you have the main sequence, you have the evolution of a thirty and a fifteen solar mass star and the stars showing the highest degree of mass loss are all in the 30 $M_\odot$ region and above; we should go down to find lower masses in order to the stars of less and less mass loss.

Thomas: And what happens if you go in the other direction?

De Groot: You have more mass loss. This is only for supergiants. In OB supergiants you have mass losses of 10^{-5} to 10^{-7} $M_\odot$ per year ,so a little bit less, and the lines of constant mass loss are more or less parallel to the lines of the evolution off the main sequence of the stars of the masses of the OB supergiants.

Thomas: And how do I infer the mass loss of the OB supergiants?

De Groot: More or less as it has been done for the P Cygni stars.

Thomas: For P Cygni I have some kind of a number to give me a number.

De Groot: Because you have in the OB supergiants nearly always a velocity at Hα from the P Cygni-type profile, and you can take your velocity from there. Then, you have the observations in the rocket UV.

Paczyński: The rocket UV observations as interpreted by Lucy and Solomon indicated the mass loss rates of 10^{-8} solar masses per year. This is considerably less than the rates you quoted.

Morton: We must not forget the uncertainties in both the theoretical and observed rates of mass loss. For the theoretical estimate Lucy and Solomon put in one mechanism and found about 10^{-8} solar masses per year. This number can be taken as a lower limit, but the observations show that many resonance lines are effective rather than a single dominant line at each temperature predicted by the calculations. On the other hand, the higher rates derived from the spectra often depend on large, poorly established corrections for unseen ions.

Thomas: What makes you think that it will disagree? What is the evolutionary time-scale for these kinds of solar masses?

Paczyński: It is less than 10^7 years on the main sequence and less than 10^6 years in the supergiant phase.

De Groot: It is less interesting if it concerns a mass that comes from a mass of 100 solar masses, because then it may be only one per cent anyhow.

Conti: Theoretically we do not know. It may make a whole lot of difference.

De Groot: You could ask why we do not see surrounding P Cygni stars, anything like shells or rings or nebulae. We have to conclude that the extended atmospheres and the matter which is lost is so dense that the region cannot be very large in extent and so it escapes discovery. If you build a Strömgren sphere and you fill it and you make it very dense, the ionizing radiation cannot go very far from the star, so the nebulae you will detect are to be small ones.

Thomas: You say this is so dense that I cannot see it?

De Groot: No, the matter which leaves the star is so dense that the ionizing radiation from the star does not go very far.

Thomas: But if I take the implicit suggestion of velocities that can be a thousand kilometers per second, then you will see them in the rocket UV. I think the absence of a shell is very embarrassing.

Morton: You have the same problem with Wolf-Rayet stars.

Thomas: Oh, sure. But you still have an embarrassment. Would you think that in terms of the observations that if they are there, you should see them? Or would you accept what Anne Underhill said, that may be we have not looked enough?

De Groot: I think we did not look at enough areas. In the southern hemisphere, Thackeray has found very few, two or three objects which have real P Cygni characteristics and are associated with visible nebulae.

Thomas: Oh, he does find some!

De Groot: Yes. I did not say there were none. But the number must be impressively low.

Thomas: Do they have any particular characteristics; are they different from other kinds of shells?

De Groot: I think that most of them are circular. The star is more or less in the center.

Paczyński: Are the central stars hotter than average P Cygni stars?

De Groot: Most of Thackeray's were B's. They were not excessively hot stars.

Paczyński: The late B's could hardly ionize the nebulae.

De Groot: I think they were probably early B's, but I should have to look it up. I do not remember.

Smith: I think the WN7 and WN8 spectra are more like the P Cygni spectra than are the WC spectra.

De Groot: Yes.

Smith: The WN7's and WN8's are still estimated to have average absolute visual magnitudes − 6.8 and − 6.2 respectively. I think the WN's are the most closely related to the P Cygni's, in visual appearance.

De Groot: If you talk about line widths and about excitation, then, the answer is no, but if you take some other characteristics, especially the fact that N may be a bit overabundant in P Cygni, because at least the N lines are very marked, then you come to WN7 and WN8. The P Cygni stars do not occur in binaries, apart may be from a very few; they cannot have been subjected to mass exchange and, therefore, at the moment I do not have the problem to assume that they are really helium-burning stars at later phases of their main-sequence life and show this pulsational instability. If I had to classify the phenomena on the scheme of Thomas, I would say that they have a corona and an exosphere. Regarding mass and momentum transfer, etc., that is very difficult to say.

Morton: What is the mass of a typical P Cygni star? Do the P Cygni stars have the problem of the Wolf-Rayet star in the sense that the mass is much less than we would estimate from the mass-luminosity relation?

De Groot: At the moment it is not clear where we stand on that. But to summarize, the one bit of evidence was one star with a mass not more than 4 solar masses, that somebody just mentioned.

Walborn: It was HD 152667. This star is a B0 supergiant in Sco OB1, and it has an absolute magnitude of -6.7, which is consistent with other stars with similar spectra in the region. The mass derived, according to that analysis, is mot nuch greater than 4.1 $M_{\odot}$.

Morton: It just may be that the P Cygni stars have the same problem that the Wolf-Rayet stars have, that they are very over-luminous for their masses.

De Groot: Apart from the fact that you then do not really know their masses, you would introduce the problem that you do not understand the cause of the difference either. You have to invent something else.

Paczyński: If you feel attached to the vibrational instability of massive stars you should favour lower masses. If these stars are overluminous indeed, it means that their hydrogen content must be low. The critical mass for the vibrational instability decreases with the hydrogen content. At the limit of zero hydrogen, the critical mass is 15 solar masses. It means that you may reduce this critical mass very considerably. If there is indeed anomalous abundance of N, you may speculate along the lines you suggested yesterday: these stars may be mixed. Of course, this is just a speculation.

Underhill: What is the period of HD 152667?

De Groot: 7 days point something.

Underhill: I was wondering whether you are looking for eclipses which are not known. Is it possible in this star that the apparent supergiant character of it could be partly due to gas streams in a reasonably close pair. Is it double-lined, do you know?

De Groot: No, it is single-lined, so it gives the mass function. The star which shows the P Cygni-type line profile is the star which is in front during eclipse.

Underhill: Does it have a good light curve?

De Groot: Yes; The light curve has not been established very completely, but at least one can say that there are eclipses of 0.10 or 0.15 mag.

Kuhi: I have several questions. One is in connection with the color temperatures that you quoted earlier. What wavelengths did those refer to?

De Groot: The results I mentioned refer to the region between 3500 and 6000 Å and in several stars there is a trend of the colour temperature decreasing with increasing wavelength.

Kuhi: I raised a question about that very same point, that in OB supergiants, as well as in Wolf-Rayet stars, you do find that the colour temperature that you get is definitely a decreasing function of increasing wavelength. So it is not clear what the colour temperature means at all. And then I have another question. I am more intrigued by the presence of the circumstellar Ca II lines. Is there any way that you can even guess the size of the distance from the star at which these occur, or can you speculate on it?

De Groot: The lowest excitation stellar lines, that is the Balmer lines and, I think, Mg II, are formed at $2\frac{1}{2}$ stellar radii. That means that at $2\frac{1}{2}$ stellar radii we have the amount of excitation that produces the hydrogen lines. This would be at about 160 $R_{\odot}$. This means that you have to go a little bit farther out to find the excitation region for the calcium lines. I cannot say how much farther out but it gives you an idea; it should be between 3 and 10 stellar radii.

Paczyński: Can we observe the component of the Ca line that could show a large expansion velocity? In other words, if you extrapolate your picture going out very far from the star, you are going into steadily less excited regions of the envelope, you may approach the region where the Ca line should be visible? Is there anything at a few hundred km sec^{-1}?

De Groot: I could not say at the moment. I have a list of identifications, and at two angstroms shorter there is an unidentified weak absorption line. But it has connected with it an unidentified weak emission line, which should not be there. It has been seen only in one spectrogram, so I would hesitate to identify that as a stellar Ca II absorption line.

Morton: I can understand the Ca II resonance lines originating in some unusual clouds but I would be surprised if these lines are significant in most of the expanding shell. Close to the star the ion state is certainly going to be primarily Ca III, and to a first approximation the ionization stays constant as the material flows out. The decrease in the radiation flux is compensated by the decreasing electron density.

Paczyński: You mean it has to be really far away, in the interstellar medium.

Morton: I could imagine you might have some very particular cloud or prominence as in the Sun, where the material goes up and comes down again.

Underhill: That is about what I was just thinking. The point is, you see multiple components, atleast one up and one down. Perhaps you are seeing small condensations. The only way to bring your degree of general excitation down is to increase the pressure, and may be it just happens that this time you saw some small irregularities going out in the density. There is no real evidence one way or the other in the P Cygni stars that it is an absolutely uniform expansion. The types of observation we have in hand, can show that you have to have an eclipsing star to see that. And for that reason, this HD 152667, because it does show an eclipse, would be rather useful to be observed at rather high resolution, particularly when one star goes behind the other in order to see if you have any evidence for an irregular flow.

Paczyński: One more comment. The escape velocity from those masses and radii which are given in the table is about 600 km s^{-1}. This is considerably more than what is observed. So in fact there is no clear observational evidence for rapid mass loss.

Morton: I am willing to bet that the UV spectra will show material moving outward faster than the escape velocity, just by analogy with the OB supergiants where we find 200 km s^{-1} in the visible and 2000 km s^{-1} in the UV.

Thomas: But you do not know how much of this material is lost. These observations are all based on essentially photospheric radii. In the rocket UV this can really be extended to almost anything. So I can have a small fraction of the matter at escape velocity and a large fraction not at escape velocity.

Morton: The resonance lines in the far-ultraviolet spectra should show where the majority of the matter in the shell is.

Thomas: The majority of the matter excited to that level.

Morton: But in the far ultraviolet we can observe the absorptions from the ground state.

Thomas: Of what lines?

Morton: Almost all the abundant ions.

Thomas: Of Ca and C IV?

Morton: C IV is observable but not C III since its resonance lines lies shortward of the Lyman limit.

Thomas: So that is four resonance lines in that region of the atmosphere where I excite C IV.

Morton: But you also can get C III and C II.

Thomas: So if I only see C IV, that does not tell me how much the mass density is. It can be very much larger.

Morton: You must apply an ionization correction based on observations of C III and possibly C II resonance lines.

Thomas: So we are very uncertain until we observe these. I am sure you are right on all the possibilities you have specified. I just wish I know what they were.

Wood: This is another topic, but I wonder if the measurement of expansion velocities from a plate of P Cygni is not related to the question of the velocity variation with distance out from the photosphere of the star?

Are you not making a systematic error if you measured both emission edges or the redward edge of the emission? One never sees the maximum velocity of the expansion of the back side of the star. You do see the front side in absorption. And does the error not depend on the geometrical dilution factor, that is, how big the shell is with respect to the photosphere?

De Groot: There are several geometric factors which determine the shape of the profile, and they determine the relation in density between the emission and absorption line profiles. Normally you will have an overlap between the absorption and emission lines. And you would expect that you measure a higher-velocity absorption line for the stronger emission, just because the short part of the absorption line has been filled in. This has been corrected for as good as I could at that time. You can determine this relation from lines with various emission intensities, and then apply a correction, and then come out, at least for the hydrogen lines, with nearly the same velocities for all the Balmer lines. And then, from the densities, you can show from the Balmer decrement that they are all optically thick, that they are all formed in the same region, so they should have more or less the same velocities. Then you are more or less assured that what you did was all right.

Wood: Was it an iterative process, then? You first measure the expansion and some line strengths, calculate the geometry, then go back and correct the velocity? Is that it?

De Groot: No, I estimated the influence of the emission lines with different strengths. The other part of the reasoning is that you can plot the intensities of the hydrogen lines against the Balmer number, and then you find that they are all optically thick. Then you say they should all be formed in more or less the same part of the atmosphere and over a very large range of depth. Then, they should have the same velocity, and with your correction for emission you can have more or less the same velocity and be somewhat consistent and say the correction was probably quite right.

Wood: You do not then measure the highest point in the emission against the lowest point in the absorption?

Underhill: That is the idea you start with, I think, Then you say, with a very weak emission next to a big, strong absorption, it is probably not much distorted. You work your way out to the stronger lines. Then you start reflecting profiles, trying to get a consistent fit. It is explained in De Groot's thesis.

Kuhi: I have two comments: One, to throw the situation into complete confusion, HD 190073 has a fantastically peculiar spectrum, and I am sure you know. The Ca II H and K lines are in emission in that star, and have two, or is it three (?) violet displaced absorption component, indicating a huge amount of material streaming out in the ionized calcium resonance lines.

De Groot: You find emission components for the Na I D lines in P Cygni. They are not very strong, but they are definitely there. This is only a question of a shell or an envelope which is denser or less dense.

Kuhi: I am not objecting to anything you are saying, I am just trying to connect it to the circumstellar Ca II lines. The other thing I wanted to mention is that nobody here so far has mentioned anything about the infrared observations of the P Cygni stars. The thing is that many of them, if not all, (the observations are very spotty), are very bright in the infrared in the sense of having excess radiation at 5, 10 and 20 μ. No matter how you look at it, this suggests an awful lot of material around the star, may be in the form of dust, may be not. The appearance in the infrared is sometimes accounted for by the presence of free-free emission, and in other cases there is a very distinct bump, which looks like a 10 μ dust bump. It is all connected with material leaving the star, and in a group of a dozen or so stars surveyed spectroscopically as well, the excess seems to be positively correlated with the rate of mass loss.

Morton: Do the Wolf-Rayet stars have a similar effect or not?

Kuhi: They have been looked at, but unfortunately aside from a few, they are all a little bit on the faint side for present equipment. Consequently, nothing has been found: certainly no very large excess exists since it would have been detected easily in the WR stars which have been looked at.

Morton: What about the Of stars?

Kuhi: Some Of stars have been attempted and they do show a slight infrared excess.

Morton: What about ordinary B supergiants like ε Orionis?

Kuhi: I do not think they have been looked at. The list of things people have been observing in the infrared is a very spotty one. They just pick things that look as if they will be good candidates, i.e., have emission lines. No systematic study has been done, but I think that any star that is throwing out material as evidenced by P Cygni profiles or broad emission lines will probably have an infrared excess.

Underhill: An interesting question is how much is free-free and how much is due to a dust belt. Now with the shell stars, for instance, it turned out that a good deal of it is free-free. The intensity seems to be correlated with the intensity of Fe II emission lines. That makes sense to a spectroscopist, but if you look a little bit further in the literature they were bubbling over themselves with having rings of dust forming suddenly outside Be stars!

Kuhi: We do not know. That is why I said there were two groups: The ones that have a smooth increase in excess which will go out to some 10 or 20 μ, and these may be free-free, but there are many others that are not.

Underhill: How big are the observed rings? I think there are one or two. But it is very interesting. How do they get rings out there?

Walborn: Just one comment I wanted to make. In talking about the P Cygni stars, we have been implying they **are** a rather homogeneous group with regard to luminosity, although you indicated quite a temperature range. I seem to recall from Beals' old paper that he found this type of phenomenon in a wide range of objects, not only a wide range of spectral type, but also in several that he thought were of quite low luminosity, so I would like to suggest that perhaps there is a 'P Cygni phenomenon' also, which may occur in different types of objects, and we have been considering only one category. More specifically, I wanted to say that I think even among the high-luminosity group the term 'P Cygni stars' is a little inhomogeneous. On the one hand, we have objects like P Cygni itself which we know had an outburst and has almost a complete P Cygni spectrum, nearly all of the lines having P Cygni profiles; on the other hand, we have very-high-luminosity supergiants which, as you indicated, have a P Cygni profile at Hα only, and also it seems to be a phenomenon of varying degrees in that in many of them you see it at Hβ also, that is, the only P Cygni profile in the classification region is at Hβ. HD 190603 is an example of this. And finally, there are the two Of stars associated with NGC 6231, which, as pointed out specifically by Schild, Hiltner and Sanduleak (*Astrophys. J.* **156**, 609, 1969), have very similar spectral types, very similar absolute magnitudes (-7.4 and -7.1), yet only one of them has a complete P Cygni spectrum, very similar to P Cygni itself although it is an O8f star. The other one is actually $0^m\!.3$ brighter in absolute magnitude and it has a P Cygni profile only at Hβ. So these are two very similar yet very different stars.

Underhill: If you look at Hα from a BIa supergiant, you may get a reversed P Cygni emission. I have observed it several times in high dispersion spectrograms.

Morton: What do you mean by 'reversed' P Cygni emission?

Underhill: A profile consisting of a weak emission hump, usually less than 5 % on the violet side and the absorption definitely to the red of where you think your line should be.

Thomas: What kind of star do you see this in?

Underhill: B supergiants. η Canis Majoris had it in some plates De Groot took for me in 1970.

Thomas: So we have an implosion phenomenon.

Underhill: χ² Orionis has had it. I mention this to say that the combination of emission and absorption around Hα in the BIa stars does not clearly indicate a simple expanding atmosphere. It is quite variable. These things change. With 55 Cygni, we have gone through a whole series of observations, and sometimes there is no absorption, sometimes it is here, sometimes it is there. Sometimes it is a little faint hump of emission and sometimes nothing.

Walborn: Struve has notes about β Ori and 67 Oph, too, in which he reports variable emission in the hydrogen lines.

NUCLEI OF PLANETARY NEBULAE

LINDSEY F. SMITH

NASA Goddard Space Flight Centre, Greenbelt, Md., U.S.A.

1. Introduction

In 1969 Aller and I collected information about the spectra of emission-line nuclei of planetary nebulae. (See also Aller, 1967, for classification of both emission and absorption spectra.) We classified them and compared them to the spectra of Population I stars. Our idea was to follow this general classification by detailed observation and comparison of stars in each of the classes. Most of the detailed studies are not complete, but already the picture has changed from the way Aller and I saw it.

Let me emphasize first that the spectra of planetary nuclei frequently mimic those of massive Population I stars. This presumably results from the fact that both types of stars are hot. We are considering now only those spectra which show emission lines, and are ignoring those which do not. Note, however, that many of the latter are classified as O type and are rather like the spectra of Population I O stars.

An object is called a planetary nebula if it is in the Perek-Kohoutek *Catalogue*

TABLE I

The numbers of planetary nuclei and population I stars in the various spectral groups

Class[a]	Planetary[a] nuclei	Population I Stars with ring nebulae		Population I Stars in general	
		Definite	Possible	Galaxy	Mag. Clouds
WR	2 WC6	3 WN5		65 WN[f]	44 WN[e]
	1 WC8	2 WN6	2 WN6[b]	56 WC	16 WC
	2 WC9	1 WN8	1 WN7		
	1 WN8				
O VI	10	←?—3—?→		4[c]	
Of	1 O7f[d]	1		Many	
	(NGC 2392)	(NGC 7635)			
Of pec	5			Some[d]	
WR-Of	5			?	
	(e.g. NGC 6543)			(HD 93131)[g]	

[a] According to Smith and Aller (1969).
[b] Crampton (1971) adds one WN7 and one WN6 to the list given by Smith (1966; quoted by Underhill, 1968).
[c] Sanduleak (1971).
[d] Heap (1971).
[e] Smith (1968b).
[f] Smith (1968a).
[g] Walborn (1971).

* Present address: Max Planck Institut für Radioastronomie, 53 Bonn, Germany.

M. K. V. Bappu and J. Sahade (eds.), Wolf-Rayet and High-Temperature Stars, 126–139. All Rights Reserved.
Copyright © 1973 by the IAU.

(1967, henceforward PKC). That is not very helpful to the present discussion. (See Heap, 1970, for a suggestion of a more specific definition.) For our present purpose I suggest a simple observational distinction between planetary nebulae and the 'ring' nebulae observed around Population I WR stars – the spectrum of a planetary nebula always appears superimposed on the spectrum of the central star, while that of a ring nebula does not. This observational definition is obviously related to the relative brightness of the star and the nebula. Most of the objects considered in this paper which are in the PKC satisfy the above criterion for a planetary nebula; however, there are a few exceptions which will be noted in the section on O VI spectra.

The situation as it was in 1969 is summarized in the first two columns of Table I. I will take each class in turn and try to bring the picture up to data.

2. WR Spectra

There are six planetary nuclei whose spectra are known to be reasonable imitations of the spectra of Population I WR stars; of these, 5 have WC spectra and only one has a WN spectrum. This is in sharp contrast to the distribution of the galactic WR stars, in general, where WN and WC spectra are about equally numerous, and to that

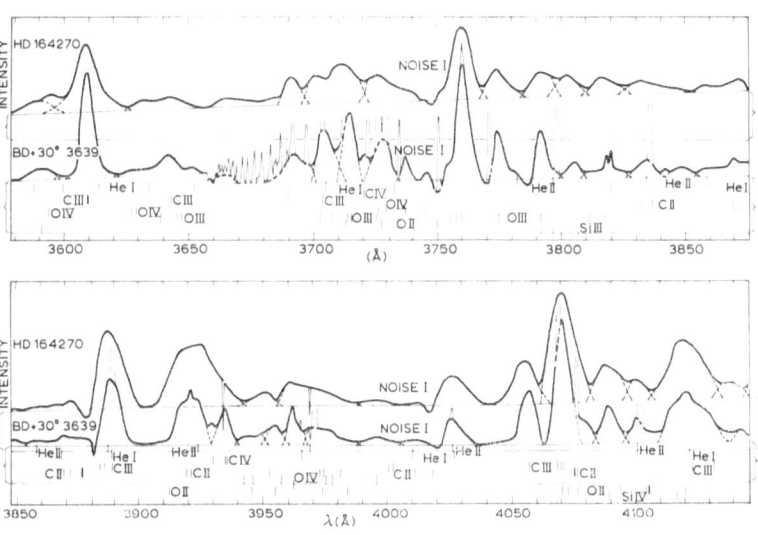

Fig. 1.

of WR stars associated with ring nebulae, in particular, which all have WN spectra. A detailed comparison of the spectrum of one of the WC9 nuclei, BD + 30°3639, with that of the Population I WR star HD 164270, shows (Smith and Aller, 1971) that the two are extremely similar, with the exception that the line widths differ systematically by a factor of 1.6. Figure 1 shows a section of the spectrum of each star.

Clearly, the spectrum of the Population I star looks like a smeared version of the spectrum of the planetary nucleus (minus the nebular spectrum). The line widths are shown plotted against ionization potential in Figure 2. The equivalent widths of lines in the two spectra are the same with the exception of those of lines of Si III, He I and C II which are 1.4 to 2.4 times larger in the spectrum of the Population I star.

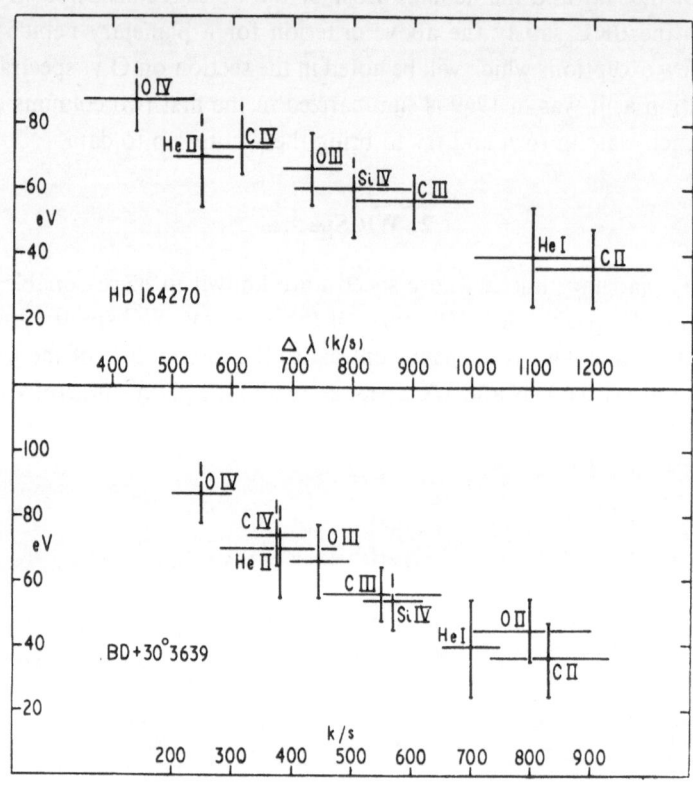

Fig. 2.

A similar situation apparently applies to the WC 9 spectrum of the nucleus of He 2–99, and probably also to the WC8 spectrum of the nucleus of NGC 40. However, comparison of the WC6 spectrum of the nucleus of NC 6751 to those of Population I WC6 stars shows a different situation; the line widths are, in this case, comparable, but the spectrum of NGC 6751 is unusual in the presence of strong N IV λ 3480 emission and possibly other nitrogen lines as well (see Aller, 1967). The spectrum should probably be classified WC6–WN5. There *are* intermediate N–C spectra among the Population I WR stars, but they are uncommon and, when they do occur, are usually of lower excitation class, e.g. WC7–N6.

The WN8 spectrum of the nucleus of M1–67 appears, to visual inspection, to be identical in every way to the WN8 spectra of Population I stars.

3. O VI Spectra

The second group of spectra Aller and I called 'O VI Spectra' because the strongest lines are O VI λ 3834, λ 3811. Some of these have absorption lines and narrow emission lines like Of spectra; others have broad emission lines and no absorption lines like WR spectra. While the O VI lines are present in all the high excitation WC spectra, Aller and I thought that their occurrence as the dominant lines in the spectrum was unique to the planetary nuclei. This is not the case. Sanduleak (1971) has pointed out five objects whose spectra fall in this class that are not in the PKC. One is in the Small Magellanic Cloud (SMC) and one in the Large Magellanic Cloud; their luminosities are comparable to those of the WR stars. Photometry and spectra of the object in the SMC are given by Smith (1968b) (the spectrum is in Plate 1 and is incorrectly labelled SMC 1 instead of SMC 2), and of one of the galactic objects by Blanco *et al.* (1968a, b).

There is a region of confusion between the planetary and non-planetary nebulae at this point. I have chosen to distinguish planetaries from Population I ring nebulae by the high relative brightness of the nebula to the star, so that the spectrum of the nebula always appears superimposed on the stellar spectrum. However, Abell (1966) has identified objects as old planetary nebulae that are characterized by large size and very low surface brightness. Consequently the nebular spectrum does not always appear superimposed on the stellar spectrum, and some of the nebulae are similar in appearance to the ring nebulae. One such object is Abell 78. A similar object is NGC 5189. Both are listed in the PKC; both nuclei show O IV spectra. The object No. 4 in Sanduleak's list is also situated in a large faint ring of nebulosity; this object is not in the PKC. I have indicated in Table I that these three objects are of uncertain population. My guess is that they are closely related to the planetary nebulae, but without some independent way of determining the luminosity of the central stars this cannot be proven.

4. Of Spectra

Among Of spectra, Aller and I considered two types: those that showed only helium and nitrogen in emission – which is the common form among Population I stars – and those that also showed carbon lines in emission with strengths comparable to those of the nitrogen lines. We believed the latter to be unique to the planetary nuclei – in the sense that, while carbon lines do sometimes occur in the spectra of Population I Of stars, they are usually rather fainter than the nitrogen lines. However, Heap (1971) has found that, for many planetary nuclei with Of pec spectra, spiral-arm stars can be found with essentially identical spectra.

In summary, it is probably safe to extend Heap's statement to all of the emission-line nuclei – the spectra of many emission-line planetary nuclei (from all defined classes) have counterparts among the spectra of Population I stars. However the relative numbers of stars showing the various types of spectra are quite different. Whereas Aller and I believed that certain types of spectra were characteristic of planetary nuclei alone, that statement now needs modification to – "certain types of spectra are much more common amongst planetary nuclei than among Population I stars".

This begins to make sense in the light of an observation made by Heap (1971). She notices that the intermediate N–C spectra are found among the *early*-Of spectra. This may imply that there is a correlation between the strength of the carbon lines and the temperature of the star, in the sense that the hotter stars have stronger carbon lines. Thus the predominance of spectra with strong carbon lines among the planetary nuclei may simply reflect the fact that planetary nuclei have higher temperatures, on the average, than do Population I stars.

The same situation may apply to the WR and O VI spectra. If we interpret the O VI spectra as a high excitation extention of the WC sequence, as suggested by Freeman *et al.* (1968), then a greater frequency of O VI spectra among planetary nuclei than among Population I stars may again reflect the greater frequency of high temperatures among planetary nuclei. (I would not invoke the same logic to explain the difference in distribution between WN and WC spectra, since I do not think the difference between WN and WC spectra is due to difference in temperature.)

5. WR-Of Spectra

I have ignored the last group in the table until now, because they are the most confused. Aller and I found there were some spectra which appeared to be intermediate between WR and Of spectra in the sense that they had narrow lines, like Of spectra, but did not appear to have an absorption spectrum. We considered these spectra also to be unique to the planetary nuclei.

The best observed star among those that we placed in this class is the nucleus of NGC 6543. Aller (1967) has found absorption lines in the spectrum and noted that at least some of them were of P Cygni type. Heap (1971) has noted that the spectrum is rather similar to that of HD 93131, a star in the Carina association which is classified WN7. She confirms that the spectrum of the nucleus of NGC 6543 shows narrow emission lines of C, N and O (see also Swings, 1940; Aller, 1943; Aller, 1956) and that the absorption lines are of P Cygni type. The spectrum of HD 93131 (Walborn, 1971) shows, of the three elements of interest, only N in emission, and P Cygni profiles on all of the H and He lines. Underhill (1968), has previously noted the presence of "an absorption dip" at Hγ and suggested that the star was composite. However, absorption edges on the Pickering series are also found in the spectrum of HD 151932, the WN7 star included in the *WR Atlas* (Kuhi and Smith, 1973); in this case the lines cannot be attributed to a companion (see elsewhere in this Symposium), and would seem to be characteristic of the class.

Mostly on the basis that the N III and C IV lines in the region 4630 to 4660 Å are much narrower in the spectrum of NGC 6543 than in WR spectra, and that the absorption lines are of the P Cygni type (rather than undisplaced as in Of stars), I am still inclined to classify this spectrum as intermediate; however, the reality and correct definition of the class are open to question.

Aller and Heap classify the emission spectrum of HD 45166 (which they believe to be a member of the disk population) as WR-Of. This spectrum has narrow emission lines (of greater number and strength than is characteristic of Of spectra) together with broad undisplaced absorption lines. In view of the intermediate H/He ratio, 5/1, that they derive, the assigned classification appears to be physically meaningful. It would be of great interest to know whether the same applies to some of the planetary nuclei that are presently classified WR-Of.

Acknowledgements

I thank Miss S.R. Heap for communicating her data to me prior to publication, and for helpful criticism of the manuscript. The final manuscript was prepared while the author was at the Institut d'Astrophysique, Liège, under the support of an ESRO Fellowship.

References

Abell, G. O.: 1966, *Astrophys. J.* **144**, 259.
Aller, L. H.: 1943, *Astrophys. J.* **97**, 135.
Aller, L. H.: 1956, *Gaseous Nebulae*, Chapman & Hall, London, p. 211.
Aller, L. H.: 1967, in D. E. Osterbrock and R. C. O'Dell (eds.), 'Planetary Nebulae', *IAU Symposium* No. 34, 339.
Aller, L. H. and Heap, S. R.: 1971, in preparation; Abstract in *Bull. American Astron. Soc.* **3**, 400.
Blanco, V., Kunkel, W., and Hiltner, W. A.: 1968a, *Astrophys. J. Letters* **152**, L137.
Blanco, V., Kunkel, W., Hiltner, W. A., Lynġa, G., Clark, G., Naranan, S., Rappaport, S., and Spada, G.: 1968b, *Astrophys. J.* **152**, 1015.
Crampton, D.: 1971, *Monthly Notices Roy. Astron. Soc.* **153**, 303.
Freemen, K. C., Rodgers, A. W., and Lynġa, G.: 1968, *Nature* **219**, 251.
Heap, S. R.: 1970, Dissertation, University of California (University Microfilms), p. 42.
Heap, S. R.: 1971, Private Communication.
Kuhi, L. V. and Smith, L. F.: 1973, *An Atlas of Wolf-Rayet Line Profiles*, Government Printing Office, Washington, D.C.; in preparation.
Perek, L. and Kohoutek, L.: 1967, *Catalogue of Galactic Planetary Nebulae*, Czechoslovak Academy of Sciences, Prague.
Sanduleak, N.: 1971, *Astrophys. J. Letters* **164**, L71.
Smith, L. F.: 1966, Dissertation, Australian National University, Canberra.
Smith, L. F.: 1968a, *Monthly Notices Roy. Astron. Soc.* **138**, 109.
Smith, L. F.: 1968b, *Monthly Notices Roy. Astron. Soc.* **140**, 409.
Smith, L. F. and Aller, L. H.: 1969, *Astrophys. J.* **157**, 1245.
Smith, L. F. and Aller, L. H.: 1971, *Astrophys. J.* **164**, 275.
Swing, J. P.: 1940, *Astrophys. J.* **92**, 289.
Underhill, A. B.: 1968, *Ann. Rev. Astron. Astrophys.* **6**, 39.
Walborn, N.: 1971, *Astrophys. J. Letters* **167**, L31.

DISCUSSION

Conti: We all know that, generally speaking, planetary nebulae are Population II objects. Could you review the evidence that shows that the nuclei of planetary nebulae must also be Population II objects? The statement is made that planetary nebulae are of Population II and, therefore, the nuclei are Population II also. Can you be sure that these objects are really Population II by evidence other than that they have planetary nebulae. In other words, is there kinematic evidence for this group or not?

Smith: I do not know off hand what the velocities of these objects are. You have only a very small sample, by the time you have selected only known emission line nuclei. I should emphasize in all discussions of the distinction between Population I and Population II nebulae, that if you take one nebula and say "prove to me that this is Population II", that is pretty difficult to do.

Conti: What I want to know is, do we have to say these are one solar mass objects because they have planetary nebulae, or could they be massive objects that have formed planetary nebulae which looks like any other planetary nebula?

Smith: There are many planetary nuclei whose spectra mimic those of Population I objects; why should there not be some that mimic WR and Of stars?

Underhill: Is there any information on proper motions and radial velocities of these particular objects? If not, then I think it is essential to get such data because these objects do have apparent nebulae around them. On Tuesday we spent a long time talking about ring nebulae around certain Wolf-Rayet stars, and one can quite easily change that idea into planetary nebula-type extensions. I think that it is very important to tighten the argument up for these objects, because you can have planetary nebulae that are barely detectable. And you can have of course the classical ones which were seen and gave the name to that kind of object. We begin from the ring nebulae in Wolf-Rayet-like stars and possibly some ring nebula O-stars, if there are any, which overlap in concept.

Smith: I do not believe that the ring nebulae form a continuous sequence with the planetary nebulae. Their properties are quite distinct from planetary nebulae: their morphological appearance is different and their electron densities are much lower – a few hundred per cc. There are, incidentally, no 'ring' nebulae around O stars.

Underhill: What do you mean by their 'morphological appearance'? I have seen pictures of planetary nebulae that just almost are not there. And they have condensations and irregularities. It is a little bit difficult to decide what is meant by morphological properties.

Smith: All the ring nebulae have a somewhat filamentary, shell-line structure to the extent that the region in the immediate vicinity of the star is obviously of very low density.

Underhill: You mean you do not see any Hα radiation from it?

Smith: Correct. This is a usual structure for a planetary nebula (although some of Abell's very old nebulae have a similar form). So far as I am aware, the morphology of planetary nebulae with emission line nuclei is not systematically different from that of other planetaries. In particular, they do not show the filamentary, ring-like structure that is characteristic of the Population I nebulae. Planetary nebulae are generally 'filled in', often with extensive structure in the inner regions.

Underhill: I do not understand what you mean by them being filled. I am thinking of the very common picture of the Ring Nebula in Lyra, the center of which in a usual photograph, is not filled.

Smith: That is right. But the nucleus is not an emission line star. My statement stands that the nebulae around emission line nuclei do not differ in any systematic way from all other planetaries.

Morton: Could you describe the difference in the following way? The planetary nebula is fairly uniform over the whole diameter, or at least uniform over finite width, like the classical Ring Nebula, whereas nebulae around WR stars all have a very sharp outer shell but no inner boundary?

Smith: Yes, that is a fair statement. The ratio on the thickness of the shell to the radius is very large in all the Wolf-Rayet nebulae, larger than is characteristic of planetary nebulae.

Morton: You mentioned the problem of differing excitation in the nebulae. Is there a notable difference between the two kinds?

Smith: In part. The nebulae around the Population I WR stars have spectra like ordinary, low excitation, diffuse nebulae. Among the planetary nebulae, the excitation varies; the nebulae with 'O VI' nuclei are of very high excitation, NGC 6751 (WC6 nucleus) is also of high excitation; such high excitation is, I believe, characteristic of planetary nebulae and is not found among diffuse nebulae. However, the planetary nebulae with WC8 and WC9 nuclei show low excitation; I could not say, off hand, whether there are obvious differences from the Population I nebulae or not.

Johnson: I think Anne Underhill has a point about the morphological difficulties, because Minkowski, perhaps the most expert of all recognizers of nebulae, thought the shell around HD 50896 was a planetary nebula and also Pikelner has the classification of NGC 6888 as a supernova remnant. So among these shells around Population I Wolf-Rayet stars there is an ambiguity of morphological classification by the experts.

Smith: I cannot emphasize enough, though, that you take one object, and it is impossible to tell the difference. But, statistically, I think the differences between the two is impressive.

Johnson: I do want to mention one other point, and that is we do not know too much about the motions in the nebulae in these Population I shells, but in the case of NGC 6888 at least two people agree that the expansion velocity is around 50–60 km s^{-1}, which would be rather high for a planetary and rather low for a supernova remnant. Whereas in the case of NGC 2359 and 3199 (ring nebulae around Wolf-Rayet Population I stars), the Courtes group find internal velocity dispersions of just a few km s^{-1}, which might be comparable with planetary nebula dispersions of velocity. So there, I think, the velocity criterion does not tell you whether the shells are like planetary nebula shells or not.

Smith: Again stressing the statistical aspect, if you divide the nebulae according to whether it is called a planetary nebula or not you find many statistical differences between the stars in the middle of the nebulae. Including, I might say, the fact that all the stars in the middle of the ring nebulae are WN's whereas most of the ones in the middle of planetaries are WC's.

Walborn: Following along that line, the arguments are statistical, but I am struck by the fact that there are three categories there on the board, each of which has only one object in it, and, therefore, statistical reasoning is very dangerous. In each one of those cases the most distinguishing characteristic is that it is different from the majority and it is most similar to Population I. That is the O7f, the WN8, and the one that looks like HD 93 131. Also, there is always the possibility of runaway stars too, even if they are out of the plane, in the case of an individual object.

Kuhi: Coming back to the Ring Nebula in Lyra, if you photograph it in the lines of Ne v, is it not true that it is completely filled in? Has anybody ever looked at any of these nebulae around your WR stars in emission lines from the higher stages of ionization to see if anything similar happens?

Smith: No, there are only ordinary red photographs, to my knowledge.

Kuhi: How many WR stars are you talking about? You say the statistics are very good, but I do not know what the number is.

Smith: Well, there are 6 Wolf-Rayet nuclei and 6 to 9 'ring' nebulae, depending on how many of the 'possible' ones you count.

Van Blerkom: Several years ago, Johnson and Hogg looked at NGC 6888 and calculated how much interstellar material would be swept up. The nebula around that star was pretty massive compared to planetary nebulae, around 4 solar masses.

Morton: Are there very many nebulae with masses less than 0.1 solar mass?

Paczyński: The total number of objects is about 30 or so. You need the electron density to estimate the mass of a planetary nebula without knowing the distance. The estimates of electron density are very uncertain and contribute to the observed scatter of nebular masses. Vauclair (1968, *Ann. Astrophys.* **31**, 199) derived nebular masses from the radioemission observed in about 20 planetary nebulae. He got masses anywhere between 10^{-3} and 10 solar masses. I think there is no observational reason to believe that all planetary nebulae have masses of 0.2 or 0.6 solar masses. This mean value was originally derived to get the distances to a large number of nebulae, and the distance estimate is not very sensitive to the assumed value of the nebular mass.

Smith: It would probably be safe to say that the planetary nebulae are less than 10 solar masses, whereas the nebulae around Population I WR stars have masses as high as 100 $M_{\odot}$ (see Johnson in this symposium). Many of them are much more massive than planetary nebulae.

Thomas: Is there any reason to speculate now that maybe the upper or the lower limit is the more reasonable value and that everything else is just observational inadequacy? If I take what Lindsey Smith says here, could I say everything should be at the upper limit of the scatter and everything else is the spread. Or everything should be at the lower limit of the scatter and everything else is the spread.

Paczyński: As long as you do not have reliable estimates of the observational errors, you can just say that the true value is somewhere in the middle and that the intrinsic scatter is unknown. I tried to derive the mass of the nebula around BD $+ 30° 3639$. Depending on whose value I choose for the electron density and for the surface brightness of the nebula, I can get anything between a fraction of a solar mass and almost 100 solar masses.

Morton: Do I understand you are saying that the masses of the rings around some Wolf-Rayet stars may not be too different from planetary nebula masses? Do you really want to suggest that a typical planetary nebula could be easily within the range of the masses of these ring nebulae?

Paczyński: I suspect that the definition of a planetary nebula is sufficiently vague to make any definite statement difficult. Depending on the definition, you may either include those objects in a list of planetary nebulae or you may consider them to be different from planetary nebulae.

Morton: Do we agree then, that the ring nebulae around Wolf-Rayet stars are in the upper mass range of all planetary nebulae? They are definitely not typical of the average.

Thomas: Let us come back to pushing the argument farther. Suppose I take what Kuhi, Conti and

Paczyński have been talking about, that the real index of the Wolf-Rayet kind of spectrum is the mass loss, in some sort of a vague way. Then, if I take the three characteristics of the Wolf-Rayet-type planetary nebula spectrum that Lindsey Smith pointed out and consider the fact that the lines have about the same equivalent widths but with a factor of 1.6 in the actual width between the two cases, that means that the central intensity of the planetary nebula-type Wolf-Rayet star is higher relative to the continuum than is the central intensity of a Population I type Wolf-Rayet star.

Smith: Aller and I attempted to account for the observed similarities and differences between BD + 30° 3639 and HO 164270 by suggesting that they were exact scale models of each other. This works quite nicely. If you believe that you have two stars which differ in mass by a factor of 10, then, in order to obtain similar surface conditions, you have to change the radius by a factor of about 3. So I suggest that the planetary nucleus is 3 times smaller than the Population I star, and, in other respects, they are exactly the same.

Thomas: It must be more active!

Smith: This immediately gives you the difference in the line width, because presumably the acceleration process is the same, and in the bigger star the accelerating force operates over about 3 times as great a distance. You have to pick the dependence of the accelerating force on the radius such that the ratio of observed velocities is the square root of the ratio of the radii, but since a $1/r^2$ accelerating force satisfies the requirement, that would not appear to be a problem. You predict that the equivalent widths of the optically thin lines will be $\sqrt{3}$ times greater in the spectrum of the larger star, in reasonable agreement with the observations. However, you also predict that the equivalent widths of the optically thick resonance lines will be $\sqrt{3}$ greater in the larger star. Resonance lines are not observed, but the optically thick subordinate lines have equal equivalent widths in the two stars. Thus, there may be a disagreement with the model at this point. Van Blerkom did some calculations showing that the properties of the nebula are completely consistent with the properties we required for the nucleus, if it is a scaled version of the Population I star. So, we did not run into any serious conflicts with the assumption that they differed by a factor of 10, except for the central intensity.

Thomas: Then I suggest that you can only reconcile the question of the intensity by increasing the activity, in which case the presence of the nebula measures it.

Smith: That may go in the direction of explaining one thing we could not cope with, so I am very happy. However, greater activity would probably also affect the optically thin lines which appear to behave as predicted.

Morton: Would Lindsey Smith like to comment on the galactic distribution of planetary nuclei?

Smith: The galactic latitudes of the planetary nuclei range up to very high numbers, so that there is obviously a strong galactic distribution difference between these emission line nuclei and the Population I Wolf-Rayet stars. For the Wolf-Rayet nuclei alone, the galactic latitudes go up to 9° and the mean (absolute) value is about 5°, whereas the galactic latitudes of the Population I WR stars rarely exceed 5° and the mean (absolute) value is about 2°. If you include the O VI planetary nuclei in the sample, you include objects up to 20° from the plane.

Morton: What is the latitude dependence of ordinary planetary nebulae?

Smith: They are up to about 20° from the plane; they are 'disk' population.

Alcaino: I understand that one planetary nebula has been found in a globular cluster; does anyone know which cluster is this?

Kuhi: M 15.

Paczyński: What is the galactic latitude of Wolf-Rayet stars with ring nebulae?

Johnson: All ten degrees, or less.

Paczyński: So they are just like those planetary nebulae with WC spectra. Another thing is, we have statistics of about 5 or 7 objects in each class. We may expect random scatter to be large. The main morphological difference between these two classes is such that ring nebulae have very low surface brightness and shell-like structure, while those few planetary nebulae are much smaller and have much larger density.

Smith: Yes, the planetary nebulae have higher surface brightness.

Paczyński: Let us forget about our prejudice that these are two different kinds of objects, and let us assume that this is one kind of object at different evolutionary stages. Perhaps we see the evolution of those nebulae proceeding together with the evolution of the central stars. When a nebula is young, the central star appears like a WC object. Later on the nebula expands and the spectrum of the central star is changed into WN. This would be much more comfortable: at present it is difficult to understand why single WC stars do not produce ring nebulae. I would not like to push this new suggestion too hard, it just came to my mind a few minutes ago – but I think it is not less justified than the sub-division of those objects into two different populations.

Thomas: So you are arguing for all these stars having the same masses, whatever the mass?

Paczyński: Yes.

Johnson: There is the interesting case of the object called GX 3 + 1, longitude 3°, latitude + 1°, which was earlier identified at Cerro Tololo with the X-ray source and later found that the position of the X-ray source is different. However, this is one of those objects which has O VI $\lambda\lambda$ 3811-34 and so forth strongly in emission, and it also has a ring nebula around it. And the ring nebula is something like 5′ in size; it is not very small. So there you have an ambiguous case which could be a WC planetary nebula with a ring, a weak ring, but nonetheless quite visible. And yet, one cannot be sure that it is a planetary nebula either, at the present time.

Paczyński: Are the expansion velocities derived from the violet displaced absorption edges known for planetary nuclei with the Wolf-Rayet spectra? BD + 30° 3639 was shown here, but are the expansion velocities known for other nuclei too?

Smith: Spectra have been obtained, but have not been measured yet.

Morton: Qualitatively, do all of the Wolf-Rayet nuclei have violet displaced absorption lines?

Smith: Yes.

Morton: And what about the O VI nuclei?

Smith: Not obviously. However, the Of nuclei often show P Cygni profiles. Heap and Aller note P Cygni profiles in NGC 6543. I am not familiar enough with other objects to be able to say.

Morton: So there is a possible problem with the O VI nuclei, that they do not all have evidence of mass loss?

Smith: From Aller's preliminary tracings of the O VI spectra, there do not appear to be strong violet absorption edges.

Underhill: Bappu attempted to see the displaced absorptions and he got results on O VI in a couple of stars. There is great difficulty in detecting those displaced absorptions which occur at O VI in very hot objects. It would indeed take a very careful spectrophotometric study, which has not been done. The nebular spectra fall on top of what you are looking and that is why the detection of expanding shells is spectrophotometrically an unproved thing at the present time.

Van Blerkom: The presence of very broad emission lines indicates that mass outflow takes place. Whether or not violet displaced absorption components appear depends on the structure of the atmosphere.

Paczyński: The violet displaced absorption component is the most clear evidence of mass outflow and it gives directly the velocity of outflow independently of any theoretical model. This should be measured and published every time when the measurement is possible. I had great difficulty in finding those data in the literature.

Sahade: I suppose that one of the conclusions of this Conference is that from now on, we will not say any more that all central stars of planetary nebulae have one solar mass. Very often in the literature we find some very definite statement which people repeat over and over again and later when we look more thoroughly into the problem we find that the statement is not true. I am afraid that the question of the mass of planetary nuclei will be one such case.

Smith: Obviously, if you are going to assert that WR nuclei are more massive than one solar mass you are saying that those nuclei are a great deal more luminous than we have assumed. Somebody must check what that means in terms of their distances and masses, etc., and see if you come up with any contradictions. It seems to me that you have no good reason for separating the Wolf-Rayet nuclei from the rest and declaring that they have the same luminosity and mass as the Population I stars. Why not assert the same for the O VI and the Of and the O absorption spectra and all the others? Because these spectra also mimic all of the Population I types.

Underhill: We come back to my original statement, that to settle this we do need the kinematic information about these stars because the population type is basically defined on kinematics. This is a very difficult observational program, but I think we can here underwrite it.

Paczyński: I believe that the radial velocities were measured by Minkowski for all the possible planetary nebulae some ten or twenty years ago. Unfortunately, those measurements were never published. Is it possible to do anything about it?

Morton: In fairness, whether or not we cross out one solar mass for planetary nuclei, we have not really had any evidence against that for the traditional planetary nebulae in this discussion. It raises just the question of the Wolf-Rayet planetary nebulae, which is, I understand, a very small sample of the total number of planetary nebulae, 1% or something like that.

Sahade: We have quite a variety of objects, and to assign to all of them the same mass does not seem very reasonable.

Westerlund: There are more than a 1000 planetary nebulae known. It would be about 10 or 20 here, which means that we are talking of about 2% of them.

Smith: Unfortunately we have not been able to observe the nuclei of most of the planetaries, so we have some nasty selection effects on this.

Thomas: How many nuclei have you observed you can be sure of?

Westerlund: Would emission lines from the nuclei not show up fairly well also in spectra taken of the nebulae only and not aimed particularly for the nuclei?

Mendez: There are more or less 50 or perhaps 60 nuclei of which spectra with some details have been taken.

Thomas: So, the percentage we are concerned with might be 30%.

Conti: Lindsey Smith has spoken of the differences in line widths of the WR nuclei with respect to normal WR; what about the Of stars? If you just look at the spectrum and did not know about the nebula, would there be any difference?

Smith: I think the line widths are similar but I have not studied the problem. I suspect you can find a Population I Of star which has the same line width as any given planetary nucleus Of star.

Conti: And the only reason for thinking these might be different, in addition to having the nebula around them, is, in fact, the galactic latitude.

Smith: Yes, and the fact that they tend to have a slightly different spectral appearance.

Bappu: I have one question to ask on the widths of the lines in a planetary nucleus as compared to the normal Wolf-Rayet star. It is just a matter of making such a comparison with a typical type star, but amongst the Wolf-Rayets of that particular spectral class, what would be the range in widths?

Smith: Much less than the difference between the two groups; there are two WC9 stars that have nebulae associated with them, and both have lines which are obviously much narrower (even with casual inspection) than lines in the spectra of seven of the other WC9 stars.

Johnson: I recall that in the papers by the Russian and French groups who did the NGC 6888 velocities getting the expansion of about 60 km s^{-1}, that when they took the mean of the expansion to and from in the line of sight, the mean of the nebular velocities came out comparable in absolute size with the expansion velocity. This nebula happens to be in the direction of Cygnus, of course, so you do not expect much differential galactic rotation and, therefore, it seems that, if you take the straight means, the nebula is moving with a Population II velocity although it is surrounding HD 192163, which is certainly a Population I Wolf-Rayet star. I do not mean to draw any conclusion from this, but I think it is an interesting point.

Morton: Do you have a measurement also of the radial velocity of the particular star?

Johnson: No. Unfortunately not, because it has such wide lines. I suppose most Wolf-Rayet stars do not have a radial velocity measurement in the usual sense.

Morton: Can you not even estimate the velocity to 60 km s^{-1}.

Smith: You get velocities differing by a hundred km from different lines.

Bappu: I think that it could be pinned down to about 10 or 15 km.

Smith: In the Wolf-Rayet spectrum, we do not know which lines to measure. The γ-velocities of binary systems derived from different lines can differ by a hundred kilometers per second from one another. So, how do you know which emission line is giving you the right velocity?

Underhill: Bappu, from his high dispersion work, is now in a position to do this job.

Morton: I presume that there are no known binaries in any of these.

Smith: In the ring nebulae, no. In the planetary nuclei, no. The O VI Wolf-Rayet star, which does not have a nebula appears to be a composite spectrum. But apart from that no other object on the board has a composite spectrum (unless you consider the peculiar Of Wolf-Rayet spectra to be composite). All the others show just emission, no absorptions.

Seggewiss: May I come back for a moment to the distribution in the galactic latitude? If these nuclei are much fainter than the Wolf-Rayet stars, as supposed by Lindsey Smith, then the distribution in latitude must be higher than that of the more luminous Wolf-Rayet stars.

Smith: That is correct; both types have galactic latitudes as high as $+10°$. However, the planetary nuclei are 10–15th magnitude objects and the WR stars are 7–12th magnitude objects. So that if you assume that they are the same luminosity, the distances of the 'planetary nuclei' from the Sun and from the galactic plane are much greater than those of the Population I stars. Temperature does not enter. I am just determining the distance from the apparent visual magnitude and the assumed absolute visual magnitude.

Paczyński: Visual magnitude depends on the temperature, and even if these objects have the same bolometric luminosity...

Smith: You are trying to tell me that these are ordinary Population I objects. Do not change your hypothesis.

Paczyński: I do not know. I just cannot see the population difference between the two kinds of objects.

Smith: It is suggested that the 'planetary nuclei' are Population I Wolf-Rayet stars. So I may assert that, therefore, I know their luminosities, etc., and may determine their distances, the same way that I have determined the distances for other Population I WR. And the result is the one I have mentioned a moment ago.

Paczyński: Let me make one suggestion. Let us forget about all the planetary nebulae and all about the Wolf-Rayet stars. Let us forget about Population I and Population II objects. Let us concentrate on those 12 or so objects which do show Wolf-Rayet type spectra and some nebulosities around them. You get about five or so objects which are classified as planetary nebulae and seven objects which are classified as ring nebulae. Now, let us forget about any absolute magnitude estimates for either classical Wolf-Rayet type stars or classical planetary nebulae, as we are not sure that there are no systematic differences between our ten objects of either classes. Let us just try to consider those twelve objects divided into two sub-classes, and see in which way these objects differ, and without referring to any knowledge of genuine Population I objects and genuine planetary nebulae. This will save us a lot of confusion, I believe.

Morton: And, what do we conclude if we do make this?

Paczyński: Let us see; is there any difference, any population difference? And is there any reason from the status of these objects, that they belong to a different population or that they have different luminosities?

Morton: Do you think we have enough information to draw any conclusions?

Paczyński: No, I believe we do not. I think that there is no reason to believe that these two sub-classes are really different, if we do not refer to other Wolf-Rayet stars which do not show nebulosities, and those planetary nebulae which do not show Wolf-Rayet-type spectra. And up to now, the procedure was the following: we assume that we know two distinct classes of objects, Population I, Wolf-Rayet type stars, which are frequently binaries, and planetary nebulae which belong to the disc population. Now, it happens that some objects were classified originally as Population I Wolf-Rayet stars, and some others were classified originally as planetary nebulae, and we had a large bias because of this original classification. So, let us forget about this. Let us just concentrate on those twelve objects that you have here, and do not assume that some of them have luminosities or masses like planetary nebulae or like Population I Wolf-Rayet stars. I have no idea what is the population assignment of those objects; that is a thing to be done. But we cannot assign those objects to Population I or II, just because there are similar objects known among Population I or Population II stars.

Morton: Can we not derive at least one conclusion from the totally alternative hypothesis that they are all at roughly the same distance? Then we conclude that the absolute visual magnitudes of the two groups must be quite a bit different, because the apparent magnitudes are different.

Paczyński: The absolute bolometric magnitudes, not necessarily.

Morton: We do not know about that, of course. But then, there is at least one natural separation by apparent magnitude.

Underhill: I have a question to throw out. It concerns the definition of population type. Do you have to assume that all Population I objects are younger than 10^7 yrs, and, therefore, that the objects appearing to have high effective temperatures must be in an early stage of evolution? Some of the Wolf-Rayet stars with nebulae may have been formed 10^7 yrs ago, yet they have raced through to a rather advanced stage of evolution where they are in a stage of producing a planetary nebula, but the development of that nebula and of the central star itself has not yet been completed. The character of these seven or eight stars that Lindsey Smith mentioned may be like the planetary nebula class.

Paczyński: I believe there is nothing wrong with this conclusion as far as stellar interior theory is concerned.

Thomas: What is the problem with the distances?

Smith: I suspect that if you assume that the nuclei of the planetary nebulae have absolute magnitudes equal to those of Population I WR stars, their distances from the Sun will be unreasonably large and their distances from the plane of the Galaxy will be too great for Population I objects.

Underhill: My question was whether there is anything against supposing that some of the hot objects are not − 4.4, recognized only from the spectra, but that they have reached the stage where they are considerably fainter. Then the nebula star is not at 6–8 kiloparsecs, because, you could not see the nebula at 6 to 8 kiloparsecs. A star within a ring nebula has, then, to be reasonably close. The only way to make these apparently faint stars close is to assume that their intrinsic luminosities are low. Consider the stars in question which are attributed to Population I. You have no kinematical knowledge of their absolute magnitude. The assumed absolute magnitude is based upon the correlation, a unique correlation it is presumed, between spectral type and visual magnitude. I question the uniqueness of that correlation.

Morton: Has any one looked at any of these planetary nebulae with a radio telescope, as it has been done with ring nebulae?

Johnson: Many planetary nebulae have been observed but I do not know specifically the answer to these things of interest.

Bappu: As far as kinematical aspects that concern the first WN8 object, let me point out that what prompted its first publication by Merrill was that the WN8 object had a real high velocity. That was the principal theme of his paper. And then it turned out to be a planetary and the high velocity corroborated this nature.

Paczyński: Do you expect to discover ring nebulae at large distances?

Smith: Ring nebulae might be difficult to detect at large distances in the galactic plane because of high absorption and their low surface brightness.

Morton: If ring nebulae have the same distribution as the planetaries, you would want to see some at high galactic latitude.

Underhill: It is a very attractive point of view to consider that by taking planetaries of different appearance, you take objects at different evolutionary stages. Then you could say that the only stars you see high up are indeed very old ones. If they have a nebula, they have got one for a very special purpose.

Smith: We have tabulated, for the ring nebulae, the Morton Roberts number of the central star, its spectral class, apparent visual magnitude and galactic latitude. The distances are derived from the absolute magnitudes tabulated in my review and the intrinsic colours derived by me in 1968. For the planetary nebulae are tabulated an identifying name, the spectral type of the central star, its visual apparent magnitude and galactic latitude (see Table Ia and Ib). The absolute magnitudes are assumed equal to those of Population I WR stars of the same sub-class, the distance modulus is derived assuming an absorption of one magnitude (on the basis that the galactic latitudes are all greater than 3°, so after about 1 kpc the line of sight is out of the galactic plane).

The derived distances of the planetary nuclei are absurdly large, 5 to 44 kpc, the derived distances from the galactic plane are correspondingly absurd. Obviously the assumption of high luminosity is incorrect.

I expect that any other set of specific assumptions along the lines being discussed here can similarly be shown to be inconsistent with the observations.

Paczyński: Is the spectrum of NGC 6543 different from that of a Population I object, as far as absorption lines are concerned?

Smith: I do not know.

Underhill: The absorption spectrum is practically identical with the blended absorption and emission line spectrum of HD 93 131.

Paczyński: Then that Population I object can also show an absorption spectrum?

Underhill: Yes, it does.

Paczyński: In addition to the violet-displaced absorption edges?

Underhill: Yes. Can you show the spectrum of HD 93131? It has beautiful O-type spectral lines.

Paczyński: This Population I object is interpreted as a binary, but the nucleus of a planetary nebula is believed to be single?

Smith: WN7 spectra do show violet-shifted absorption components to the Pickering lines. However, if HD 93131 shows undisplaced absorption and is not a binary, then it is not a WR star under the current definition of the class.

Underhill: We have no adequate radial velocity study.

Smith: There are some very big question marks in this area of the classification, and these objects deserve very careful attention.

TABLE Ia
Ring Nebulae

MR	Sp. Type	v	b	D (kpc)	z (pc)
6	WN5	6.9	− 10.1	1.6	284
7	WN5	11.7	− 0.1	6.9	12
21	WN5	11.2	− 1.0	3.6	63
34	WN8	7.8	− 4.8	4.0	336
49:	WN6–C	10.9	+ 0.2	4.8	17
60	WN6	11.4	− 1.5	4.0	105
97[a]	WN7	12.3	+ 1.7	1.0	1.0
100[a]	WN6	8.3	+ 1.6	1.9	53
102	WN6	7.7	+ 2.4	1.4	59

[a] Crampton, D.: 1971, *Monthly Notices Roy. Astron. Soc.* **153**, 303.

TABLE Ib
Planetary Nebulae

Name	Sp. Type	v	b	M_v	$v − M_v − 1$	D (kpc)	z (pc)
NGC-40	WC8	11.0	+9	− 4.8	14.8	9	1400
He2-99	WC9	–	− 4	− 4.4	–	–	–
NGC-5315	WC6	14.8	− 4	− 4.4	18.2	44	3100
NGC-6751	WC6	13.3	− 5	− 4.4	16.7	22	1900
MI-67	WN8	10.2	+3	− 6.2	15.4	12	630
BD + 30° 3639	WC9	10.1	+5	− 4.4	13.5	5	440

TABLE Ia

Ring Nebulae

MR	Sp. Type	v	b	D (kpc)	z (pc)
6	WN5	6.9	− 10.1	1.6	284
7	WN5	11.7	− 0.1	6.9	12
21	WN5	11.2	− 1.0	3.6	63
34	WN8	7.8	− 4.8	4.0	336
49:	WN6–C	10.9	+ 0.2	4.8	17
60	WN6	11.4	− 1.5	4.0	105
97[a]	WN7	12.3	+ 1.7	1.0	1.0
100[a]	WN6	8.3	+ 1.6	1.9	53
102	WN6	7.7	+ 2.4	1.4	59

[a] Crampton, D.: 1971, *Monthly Notices Roy. Astron. Soc.* **153**, 303.

TABLE Ib

Planetary Nebulae

Name	Sp. Type	v	b	M_v	$v − M_v − 1$	D (kpc)	z (pc)
NGC-40	WC8	11.0	+9	− 4.8	14.8	9	1400
He2-99	WC9	–	− 4	− 4.4	–	–	–
NGC-5315	WC6	14.8	− 4	− 4.4	18.2	44	3100
NGC-6751	WC6	13.3	− 5	− 4.4	16.7	22	1900
MI-67	WN8	10.2	+3	− 6.2	15.4	12	630
BD + 30° 3639	WC9	10.1	+5	− 4.4	13.5	5	440

SECTION V

CHAIRMAN: H. JOHNSON

EVOLUTIONARY ASPECTS OF WOLF-RAYET STARS

B. PACZYŃSKI

Institute of Astronomy, Polish Academy of Sciences, Warsaw, Poland

1. Introduction

Problems related to Wolf-Rayet stars were recently reviewed by Underhill (1968) and were discussed at length during the symposium held at the Joint Institute for Laboratory Astrophysics in Boulder (Gebbie and Thomas, 1968). Therefore, I shall not review here the historical development of ideas, and I shall not give detailed references in those cases when all the information is available in the sources mentioned above. I shall discuss first those observational data that are most important for the understanding of the evolutionary status of these stars. Later I shall discuss the problem of the interior structure and the origin of WR stars. And finally theories of the driving force for the observed mass loss will be presented.

2. Observations

There seem to be three kinds of Wolf-Rayet stars known: nuclei of some planetary nebulae, lighter components of massive binary systems, and single Population I objects, frequently surrounded with ring nebulae. Nuclei of planetary nebulae are believed to have masses close to $1 M_\odot$, but there is no direct observational evidence to substantiate this belief. WR components of spectroscopic binaries have masses close to $10 M_\odot$ and their companions are O or B type stars of $\sim 30 M_\odot$. Single Population I WR stars are probably massive objects as they are young, but it is not possible to make a more specific statement.

Absolute visual magnitudes of the Population I WR stars are between -4 and -6 and nuclei of planetary nebulae are considerably fainter. All these stars are hot and radiate most of their energy in the ultraviolet. For this reason the most reliable (at least in principle) estimates of the effective temperature are available for the objects surrounded by nebulosities. According to Morton (1969, 1970) these temperatures are 50000 K, 33000 K, 23000 K and below 35000 K for WN5, WN6, WN8 and WC8 stars respectively. Van Blerkom (1971) estimates effective temperatures for two nuclei of planetary nebulae, BD $+ 30° 3639$ (WC9) and NGC 40 (WC8), to be equal to 32000 K and 34000 K respectively. There are 9 additional planetary nebulae with WR nuclei (Underhill, 1968) for which O'Dell (1963) and Seaton (1966) give effective temperatures. The mean logarithm of the temperature for these stars is 4.64 according to O'Dell, and 4.86 according to Seaton. Temperatures assigned by Capriotti and Kovach (1968) are even higher. Unfortunately, only one of these nine stars has a truly WR-type spectrum according to Smith and Aller (1969).

The effective temperature for γ_2 Velorum (WC8 + O7) as measured by Hanbury

M. K. V. Bappu and J. Sahade (eds.), Wolf-Rayet and High-Temperature Stars, 143–162. All Rights Reserved.
Copyright © 1973 by the IAU.

Brown et al (1970) is equal to 29 000 K. According to Lindsey Smith (private communication) and Baschek and Scholz (1971) the WC8 star is considerably fainter
than its O7 companion. If this is true, then the interferometric measurements refer
to the O7 star rather than to the WC8 component of this binary system.

Absolute visual magnitudes given by Smith (1968) and effective temperatures given
by Morton (1969; 1970) and Van Blerkom (1971) may be used to obtain absolute
bolometric magnitudes for the Population I Wolf-Rayet stars. It is surprising that
$M_{bol} = -8.7 \pm 0.5$ is obtained for all the subtypes. The corresponding luminosity
is $2.4 \times 10^5 L_{\odot}$. It would be surprising if the accuracy of this estimate were better
than 1 mag., and in fact the error may be considerably larger. The mean bolometric
magnitude of the nine nuclei of planetary nebulae with WR-like spectra is -4.3
according to O'Dell and -6.0 according to Seaton (1966).

When this luminosity of $2.4 \times 10^5 L_{\odot}$ is compared with the main sequence luminosity
of a $10 M_{\odot}$ star $(L = 0.5 - 1.0 \times 10^4 L_{\odot})$ it becomes clear that Population I Wolf-
Rayet stars are highly overluminous for their masses. This statement is true for WR
binaries, but it is commonly applied to single WR stars as well. We should be aware
that even in the case of binaries a certain amount of extrapolation is involved. We
do know masses for binaries, but the best estimates of bolometric corrections may be
obtained for the stars surrounded with the nebulosities. And these stars are single. WR
nuclei of planetary nebulae with estimated luminosities of the order of $10^4 L_{\odot}$ are
certainly highly overluminous for any sensible mass that may be assigned to them.

I think that everybody would agree that even if masses and luminosities of Wolf-
Rayet stars are not very well known, these parameters may be obtained with fair
accuracy for some objects at least. It is far less obvious that we can meaningfully
assign a particular radius to any given WR object. For example, Castor and Van
Blerkom (1970) estimate the photospheric radius of HD 192 163 (WN6) to be 13
$R_{\odot}$ at λ 5500, while the line emitting region is supposed to be five times larger. Given
the density of He II ions and the population of different energy levels, it is possible to find
that line emitting region is optically thick in the helium Balmer and Lyman continua.
In this case the photospheric radius at λ 228 and λ 912 may be as large as $70 R_{\odot}$.

Wolf-Rayet stars show evidence of mass outflow with velocities in the range of
500–2000 km s^{-1} as estimated from the violet displaced absorption components of
some spectral lines. The wide emission lines are most readily interpreted as originating within the radially expanding envelope (see, e.g., Castor, 1970; and Castor and
Van Blerkom, 1970). This envelope must be much larger than the star itself as no
occultation effect is observed. Acceleration of the outflowing gas must take place
within this extended envelope as a correlation is observed between line width and
ionization potential (see, e.g., Smith and Aller, 1971), and no lines were reported to
show an occultation effect. The rate of mass loss is believed to be in the range of
$10^{-6} - 10^{-4} M_{\odot}$/yr.

Planetary nebulae and the so-called 'ring nebulae' (Johnson and Hogg, 1965; Smith
and Batchelor, 1970) are visible around some single WR stars. When the expansion
velocity of a nebula is known it is of the order of 30 km s^{-1}. Masses of the nebulae

are very uncertain, and estimates may range from a fraction of a solar mass up to 100 $M_\odot$ and more. Nebulosities classified as planetary nebulae are believed to have been ejected from their central stars. It is obvious that the ejection mechanism must have been different from that which is responsible for the gas outflow observed now, as there is huge difference between the observed velocities: 30 km s^{-1} vs. 1000 km s^{-1}. Nebulosities observed around Population I WR stars are called ring nebulae and are believed to have been formed from interstellar matter swept by gas streaming from the central stars. Perhaps it is possible to explain these nebulae in terms of mass loss by the central stars only. Some objects traditionally classified as nuclei of planetary nebulae may in fact be Population I WR stars. The best known, and the most controversial object of this kind is BD + 30°3639. NGC 40 may be another candidate. I believe the nature of these stars will be discussed during this symposium. The difference between the two kinds of single Wolf-Rayet stars may not be so large as is commonly assumed.

Wolf-Rayet stars seem to be hydrogen deficient (see, e.g., Sljusariev, 1955; Castor and Van Blerkom, 1970), but probably not all the investigators would agree with this statement. According to Castor and Nussbaumer (1971) the WC8 component of γ_2 Velorum is overabundant in carbon by a factor of ten compared to normal stars. The existence of the two spectroscopic sequences may indicate that there is a dichotomy in the carbon to nitrogen ratio among the WR objects. This is a very controversial statement too. Let me hope that everybody may accept the point of view that there is no strong observational evidence against the hypothesis that Wolf-Rayet stars are hydrogen-poor, and have peculiar CNO abundances.

3. Evolutionary Status

Let us attempt to find the evolutionary phases that could be fitted to at least some of the known characteristics of the WR stars. Let us not consider here the very difficult problem of the origin of their spectra. It should be emphasized that nobody has succeeded so far in explaining why some stars develop such a peculiar spectrum. Let us consider such parameters as mass, luminosity and effective temperature. These are most easily compared with the models of stellar interiors. It should be realized that ordinary evolutionary computations may provide us with time variations of these parameters when the initial mass and chemical composition are specified, but these computations do not tell us how the stellar spectrum should look. Almost all the published evolutionary sequences refer to the models in hydrostatic equilibrium. At the same time the observed layers of the WR stars are very far from hydrostatic equilibrium, and to make things worse their radii are not well defined. Fortunately, it is reasonable to assume that almost all the energy being lost by these stars is lost in a form of electromagnetic radiation. Therefore, we may assume that the luminosity of a static theoretical model may be meaningfully compared with the observed luminosity of a Wolf-Rayet star. It is also reasonable to assume that the mean radius of an unstable star is not smaller than the equilibrium radius, and therefore, the mean effective temperature does not increase as a result of instability. These assumptions indicate that we should look for theoretical models with effective temperatures equal to or larger than those assigned to the observed WR stars.

The evolutionary behaviour of single and binary stars as referred to in this chapter is based on the published review articles (Iben, 1967; Paczyński, 1971b), and on some of my papers (Paczyński, 1970a; Paczyński, 1970b; Paczyński, 1971a).

Let us concentrate first on Population I Wolf-Rayet stars. Let us look for a model with a mass of 10 $M_\odot$, a luminosity of $2 \times 10^5 L_\odot$, and an effective temperature higher than 30000 K. It is clear that such a model cannot be in a pre-main sequence contraction (cf. Iben, 1965; Larson and Starrfield, 1971). It is also by far too luminous to be on a hydrogen main sequence.

About 80 or 90% of the entire lifetime of a massive star is spent on the main sequence, i.e. in a core hydrogen burning phase. This lifetime depends mainly on the stellar mass and ranges from 10^7 yrs for 15 $M_\odot$ to 10^6 yrs for $M > 100\ M_\odot$. It is also sensitive to the initial helium content. High helium content leads to high luminosity and short lifetime. The luminosity and lifetime are not sensitive to the abundance of heavy elements, at least for $Z < 0.05$. There is no reason why the Wolf-Rayet stars should be younger than 10^6 or 10^7 yrs, and the observationally estimated ages are consistent with the 'post main sequence' hypothesis (Mikulášek, 1969). It is important to note that the lifetime of the WR phenomenon may be orders of magnitude shorter than the age of the corresponding star as counted from the pre-main sequence contraction phase. It should be noted too that it is entirely impossible to distinguish observationally, by means of age estimates, in which part of the post main sequence evolution a particular group of stars may be. If there is an age difference between two subtypes of WR star it is most likely due to the variations in the initial mass and/or initial helium abundance. This is so because the duration of the whole post main sequence evolution is ten times shorter than the main sequence lifetime. This applies to binaries as well as to single stars.

It is well known that the luminosity to mass ratio changes very little during the evolution of a massive star, provided the stellar mass is constant and provided there is no large scale mixing that makes the star chemically homogeneous. This kind of evolution cannot explain high overluminosities of the Wolf-Rayet stars.

The luminosity of a homogeneous model increases strongly with the increasing mean molecular weight within the star. If a stellar model is being completely mixed it may become highly overluminous. In this case the main sequence lifetime is longer than if there were no mixing. The luminosity of a model that is not continuously homogenized depends mainly on its initial mass and initial helium content. If such a model is subject to a mass loss in a post main sequence evolution, then the luminosity does not change significantly as long as only the hydrogen-rich layers are removed. In this case the luminosity to mass ratio is increased too, and overluminous models are obtained. If considerable fraction of stellar mass is lost during the main sequence lifetime the picture is more complicated. The luminosity may decrease considerably, and the lifetime is increased.

Let us consider now a massive star that is a member of a close binary system. As soon as the star fills its Roche lobe, mass transfer to the companion will take place. If mass exchange takes place after hydrogen exhaustion in the core of the primary (the so-called case B) then this component will resemble a pure helium star after the termination of the mass transfer process. It has been suggested that this type of evolution leads to binaries with Wolf-Rayet components (Paczyński, 1967; Kippenhahn, 1969; Barburo et al., 1969). In this case the overluminosity of the WR component is due to the mass loss. This star is more advanced in evolution than its companion as it was originally the more massive of the two. Large scale mixing may be excluded as a source of overluminosity in this case because it would lead to a very slow evolution of the would-be WR star, which could not become overluminous within the main sequence lifetime of the OB companion.

A star of 10 $M_\odot$ on the helium main sequence has a luminosity of $1-2 \times 10^5 L_\odot$, and that is just what we need for a Wolf-Rayet component in a massive binary. If we believe in the universal Fermi interactions we can say that these stars are unlikely to be in the evolutionary phases following core helium burning. In this case the lifetime of the WR phenomenon would be less than 10^3 yrs as a result of severe energy losses due to neutrino emission (Paczyński, 1971a). Therefore, we may be almost certain that Wolf-Rayet components of massive binary systems are helium main sequence stars, perhaps with some hydrogen left in their envelopes. If there is no neutrino emission, then carbon burning cannot be excluded as an energy source for these stars. It is interesting that the relation between the helium models and WR stars was already suggested two decades ago by Salpeter (1952).

There is no direct observational evidence that single Population I Wolf-Rayet stars are overluminous as their masses are not known. If it were possible to demonstrate that these stars are hydrogen defficient than we could be sure that they are also overluminous. According to Castor and Van Blerkom (1970), HD 192163 is a single, hydrogen deficient WN6 star. Hydrogen deficiency might be due either to large scale mixing or to large mass loss. If it were possible to demonstrate that ring nebulae are ejected from their central stars, or that BD + 30°3639 is a Population I object, we would have evidence that massive hydrogen-rich envelopes have been lost by those stars. Notwithstanding such a poor knowledge of the most basic parameters for single WR stars, a large number of theoretical papers make the assumption that these objects are overluminous and helium-rich. It is mass loss rather than mixing that is currently used to produce the overluminosity. Vibrational instability (Simon and Stothers, 1970; and the papers referred to therein) and radiation pressure (Bisnovaty-Kogan and Nadyozhin, 1972) have been suggested as possible mechanisms responsible for that mass loss. Neither of these has yet been generally accepted as physically plausible. However, it should be emphasized that there seems to be a deficiency of luminous red supergiants (see, e.g., Schild, 1970). This deficiency was interpreted either as evidence of existence of the universal Fermi interactions and neutrino energy losses (Stothers, 1969), or as evidence of large mass loss from the luminous red supergiants (Barburo et al., 1969). Some mass loss is indeed observed (see, e.g., Gehrtz and Woolf, 1971), though it is not clear if it is sufficiently rapid to be important for stellar evolution. It is tempting to suggest that very massive stars lose their hydrogen-rich envelopes as soon as they become red supergiants. The helium core of such a star is left as a hot and overluminous object, presumably a single Wolf-Rayet star (Bisnovaty-Kogan and Nadyczhin, 1972). Such an evolutionary pattern is not possible for a component of a close binary, as in that case a red supergiant phase cannot be achieved because of mass exchange.

Let us consider now those WR stars that are found in planetary nebulae. It is not known whether all the central stars of planetary nebulae develop Wolf-Rayet type spectra at some evolutionary phase. But it is reasonable to suppose that all these objects have masses not very different from $1\ M_\odot$. Their high luminosities may be most easily explained by models with degenerate carbon-oxygen cores surrounded with one or two shell sources. Models of this kind were studied by Rose and Smith (1970), Paczyński (1970a, 1970b), and Uus (1970). This kind of stellar structure is obtained if one follows evolution of a model all the way from the main sequence through the hydrogen and helium exhaustion in the core, up to helium shell burning phase, in the red supergiant region of the HR diagram. At present, the theory cannot explain how the envelopes of such stars are ejected to form planetary nebulae, or why some nuclei develop Wolf-Rayet type spectra. However, the observed luminosities of the central stars are so high that it would be most surprising if young nuclei of planetary nebulae were not in a helium shell burning phase. The origin and evolution of these stars has been recently reviewed by Salpeter (1971).

4. Mass Loss From WR Stars

According to the most reasonable interpretation of the Wolf-Rayet spectra these stars are losing mass at the rate of $10^{-6} - 10^{-4}\ M_\odot/yr$. The electron temperature of the expanding envelope is estimated to be of the order of 5×10^4 K or 10^5 K. A large number of the observed spectral characteristics may be explained in terms of such a model (see, e.g., Castor, 1970b; Castor and Van Blerkom, 1970; Castor et al., 1970; and papers referred to therein). As mentioned in one of the previous chapters, the gas is accelerated outwards within the extended line emitting region. It is possible that the photospheric radius is much larger in the ultraviolet than in the visual. The kinetic energy flux associated with the mass outflow is of the order of $10^2 - 10^4 L_\odot$, that is about one per cent of the total stellar luminosity. There is no satisfactory theory that could explain the mechanism responsible for this outflow of gas, and this is perhaps the most challenging of the problems raised by WR stars. It should be emphasized that the cause of this mass loss may be completely different from the cause of the large mass loss that presumably took place previously, when the progenitors of Wolf-Rayet stars had massive hydrogen rich envelopes. Let us consider various suggestions for the driving force of the gas outflow observed now.

According to Limber (1964) mass outflow from Wolf-Rayet stars might be due to 'forced rotational instability'. This model cannot explain the origin of violet displaced absorption components visible at the edges of some emission lines, and therefore, it has not found support among other investigators. Since Castor (1970a) has shown that the so called transit-time effect is not important for WR binaries there seem to be no serious objection left to the model with a radially expanding envelope. Let us consider it more closely now.

It has been suggested that WR stars are vibrationally unstable helium stars (Paczyński, 1967; Simon and Stothers, 1969; Simon and Stothers, 1970; Ungar, 1971). This instability is supposed to be due to the very high temperature sensitivity of the triple alpha reaction, and at large amplitude could lead to the mass outflow. Similar instability appears in models of massive hydrogen rich stars, and in that case might be responsible for the P Cygni phenomenon (Schwarzschild and Harm, 1959; Appenzeller, 1970a). Recent non-linear hydrodynamic computations (Appenzeller, 1970a; Appenzeller, 1970b; Ziebarth, 1970; Talbot, 1971a; Talbot, 1971b; Ungar, 1971) indicate that mass loss driven by the pulsational instability is possible, though the results are not conclusive yet. Talbot and Papeloizou (Trimble, 1971) find that pulsational instability almost certainly does not blow the star apart, because pulsation energy is dissipated in shock waves and fed into stable harmonics. Rather, material seems to be lost gradually. It is very interesting that at the limiting amplitude the mean radius of a massive hydrogen rich star increases by a factor of 4 compared with the equilibrium value, and as a result the effective temperature decreases by a factor of two. No computations of this kind have been published for helium stars, but the picture should be similar. Pulsational instability induced by the triple alpha reaction may appear in chemically homogeneous helium stars with masses exceeding approximately 15 $M_\odot$ or so (Stothers and Simon, 1970), with the period of oscillations close to half an hour. This limiting mass was previously thought to be close to 8 $M_\odot$ because only the opacity due to electron scattering was used by Boury and Ledoux (1966). If there is a small hydrogen rich envelope present at the top of the helium core then the critical mass is considerably increased (Van den Borght, 1969; Stothers and Simon, 1970). It is difficult to reconcile these rather large masses with those observed in WR binaries. However, it is interesting that vibrational instability driven by the triple alpha reaction was also found in some helium shell burning models (Rose, 1967), which may be relevant for the WR nuclei of planetary nebulae.

Radiation pressure in the continuum was suggested by Rublev (1964b, 1965b) to be responsible for the mass from WR stars. Rublev (1964a; 1965a) estimated the effective temperatures to be close to 10^5 K. Because of very large bolometric corrections the estimated luminosities of WR stars were found to exceed the critical luminosity which is defined as

$$L_{cr} = \frac{4\pi Gc}{\kappa} M.$$

With electron scattering taken for the opacity we have

$$\frac{L_{cr}}{L_{\odot}} = \frac{65000}{(1+X)} \frac{M}{M_{\odot}},$$

where X is the hydrogen content by mass. For $L > L_{cr}$ stellar gravity cannot keep the atmosphere in hydrostatic equilibrium, and mass outflow is forced by the radiation pressure. Unfortunately, it is not possible to construct an interior model in hydrostatic equilibrium with such a high luminosity. I know only one evolutionary sequence in which the surface luminosity exceeded the critical value (Rose, 1969) but it was due to the improper surface boundary condition used. No selfconsistent model of mass outflow driven by radiation pressure in the continuum has been published so far, but there have been many attempts to do this (see, e.g., Bisnovaty-Kogan and Zeldovich, 1968; Bisnovaty-Kogan and Nadyozhin, 1972; Kutter *et al.*, 1969; Finzi and Wolf, 1970, 1971).

A few years ago mass outflow from early type supergiants was discovered by Morton (1967a, 1967b). Radiation pressure in resonance lines was found to be responsible for this phenomenon (Lucy and Solomon, 1970; Lucy, 1971). This mechanism may lead to a mass loss rate that does not exceed L/c^2 (Lucy and Solomon, 1970), corresponding to 10^{-8} $M_{\odot}$/year for a Population I WR star. This is a few orders of magnitude less than the rate observed in these stars.

From time to time one hears the suggestion that some kind of stellar activity may produce a hot corona around a WR star. Such a corona may generate a stellar wind much more powerful than the solar wind we know. Gas outflow with the velocity close to 1000 km s^{-1} indicates that the coronal temperature is in excess of 10^7 K. This temperature is two orders of magnitude higher than that indicated by the observed spectra of the expanding Wolf-Rayet envelopes. The same problem arises in every model in which the gas pressure is supposed to be a driving force. It is likely that this difficulty may appear, at closer inspection, in the vibrating star model mentioned earlier.

It is clear that there is no satisfactory explanation for the observed mass loss from Wolf-Rayet stars.

5. Summary and Concluding Remarks

The available observations indicate that Wolf-Rayet stars are objects with very high luminosity to mass ratio. This indicates that there is very little, if any, hydrogen left within these stars. Apparently a large fraction of mass has been lost in the past either to the nearby companion or to interstellar space as indicated by the presence of planetary and ring nebulae. These nebulae expand with a velocity of about 30 km s^{-1}, and it is probable that single WR stars lost their envelopes when their ancestors were red supergiants. This problem may be closely related to the origin of planetary nebulae.

Population I stars that appear as Wolf-Rayet objects are most likely burning helium in their cores, while disc population Wolf-Rayet stars are almost certainly in a helium shell burning phase of evolution. These stars are losing mass at the rate of $10^{-6} - 10^{-4}$ $M_{\odot}$/yr with the velocities of outflow of the order of 1000 km s^{-1}.

Expanding envelopes are responsible for spectral peculiarities, but the driving force is not known. It may be due to the high L/M ratio, peculiar abundances, vibrational instability of the star, or most likely radiation pressure may be responsible for this phenomenon.

Acknowledgements

It is a great pleasure to acknowledge that my attendance at the Symposium has been made possible by generous travel grants from the Argentinian National Research Council and from the International Astronomical Union. I am also greatly indebted to Dr. Virginia Trimble for a careful reading of the manuscript.

References

Appenzeller, I.: 1970a, *Astron. Astrophys.* **5**, 355.
Appenzeller, I.: 1970b, *Astron. Astrophys.* **9**, 216.
Barburo, G., Dellaporta, N., and Fabris, G.: 1969, *Astrophys. Space Sci.* **3**, 123.
Barburo, G., Giannone, P., Giannuzzi, M. A., and Summa, C.: 1969, in M. Hack (ed.), *Mass Loss from Stars*, D. Reidel Publ. Co., Dordrecht-Holland, p. 217.
Bashek, B. and Scholz, M.: 1971, *Astron. Astrophys.* **11**, 83.
Bisnovaty-Kogan, G. S. and Zeldovich, Ya. B.: 1968, *Astron. Zh.* **45**, 241.
Bisnovaty-Kogan, G. S. and Nadyozhin, D. K.: 1972, *Astrophys. Space Sci.* **15**, 353.
Boury, A. and Landoux, P.: 1965, *Ann. Astrophys.* **28**, 353.
Capriotti, E. R. and Kovach, W. S.: 1968, *Astrophys. J.* **151**, 991.
Castor, J. I.: 1970a, *Astrophys. J.* **160**, 1187.
Castor, J. I.: 1970b, *Monthly Notices Roy. Astron. Soc.* **149**, 111.
Castor, J. I. and Van Blerkom, D.: 1970, *Astrophys. J.* **161**, 485.
Castor, J. I., Smith, L. F., and Van Blerkom, D.: 1970, *Astrophys. J.* **159**, 1119.
Castor, J. I. and Nussbaumer, H.: 1971, *Bull. Am. Astron. Soc.* **3**, 378.
Finzi, A. and Wolf, R. A.: 1970, *Astrophys. Letters* **5**, 63.
Finzi, A. and Wolf, R. A.: 1971, *Astron. Astrophys.* **11**, 418.
Gebbie, K. G. and Thomas, T. N. (ed.): 1968, *Wolf-Rayet Stars*, U.S. Government Printing Office, Washington, D.C.
Gehrz, R. D. and Woolf, N. J.: 1971, *Astrophys. J.* **165**, 285.
Hambury Brown, R., Davis, J., Hebison-Evans, D., and Allen, L. R.: 1970, *Monthly Notices Roy. Astron. Soc.* **148**, 103.
Iben, I., Jr.: 1965, *Astrophys. J.* **141**, 993.
Iben, I., Jr.: 1967, *Ann. Rev. Astron. Astrophys.* **5**, 571.
Johnson, H. M. and Hogg, D. E.: 1965, *Astrophys. J.* **142**, 1033.
Kippenhahn, R.: 1969, *Astron. Astrophys.* **3**, 83.
Kutter, C. S., Savedoff, M. P., and Schuerman, D. W.: 1969, *Astrophys. Space Sci.* **3**, 183.
Larson, R. B. and Starfield, S.: 1971, *Astron. Astrophys.* **13**, 190.
Limber, D. N.: 1964, *Astrophys. J.* **139**, 1251.
Lucy, L. B.: 1971, *Astrophys. J.* **163**, 95.
Lucy, L. B. and Solomon, P. M.: 1970, *Astrophys. J.* **159**, 879.
Mikulášek, Z.: 1969, *Bull. Astron. Inst. Czech.* **20**, 215.
Morton, D. C.: 1967a, *Astrophys. J.* **147**, 1017.
Morton, D. C.: 1967b, *Astrophys. J.* **150**, 535.
Morton, D. C.: 1969, *Astrophys. J.* **158**, 629.
Morton, D. C.: 1970, *Astrophys. J.* **160**, 215.
O'Dell, C. R.: 1963, *Astrophys. J.* **138**, 67.
Paczyński, B.: 1969, *Acta Astron.* **17**, 355.
Paczyński, B.: 1970a, *Acta Astron.* **20**, 47.

Paczyński, B.: 1970b, *Acta Astron.* **20**, 287.
Paczyński, B.: 1971a, *Acta Astron.* **21**, 1.
Paczyński, B.: 1971b, *Ann. Rev. Astron. Astrophys.* **9**, 183.
Rose, W. K.: 1967, *Astrophys. J.* **150**, 193.
Rose, W. K.: 1969, *Astrophys. J.* **155**, 491.
Rose, W. K. and Smith, R. L.: 1969, *Astrophys. J.* **159**, 903.
Rublev, S. V.: 1964a, *Astron. Zh.* **41**, 224.
Rublev, S. V.: 1964b, *Astron. Zh.* **41**, 1063.
Rublev, S. V.: 1965a, *Astron. Zh.* **42**, 347.
Rublev, S. V.: 1965b, *Astron. Zh.* **42**, 718.
Salpeter, E. E.: 1952, *Astrophys. J.* **115**, 326.
Salpeter, E. E.: 1971, *Ann. Rev. Astron. Astrophys.* **9**, 127.
Schild, R. E.: 1970, *Astrophys. J.* **161**, 855.
Schwarzschild, M. and Harm, R.: 1959, *Astrophys. J.* **129**, 637.
Seaton, M. J.: 1966, *Monthly Notices Roy. Astron. Soc.* **132**, 113.
Simon, N. R. and Stothers, R.: 1969, *Astrophys. J.* **155**, 247.
Simon, N. R. and Stothers, R.: 1970, *Astron. Astrophys.* **6**, 183.
Sljusarev, S. G.: 1955, *Astron Zh.* **32**, 346.
Smith, L. F.: 1968, in K. G. Gebbie and R. N. Thomas (eds.), *Wolf-Rayet Stars*, U.S. Government Printing Office, Washington, D.C., p. 23.
Smith, L. F. and Aller, L. H.: 1969, *Astrophys. J.* **157**, 1245.
Smith, L.F. and Aller, L. H.: 1971, *Astrophys. J.* **164**, 275.
Smith, L. F. and Batchelor, R. A.: 1970, *Australian J. Phys.* **23**, 203.
Stothers, R.: 1969, *Astrophys. J.* **155**, 935.
Stothers, R. and Simon, N. R.: 1970, *Astrophys. J.* **160**, 1019.
Talbot, R. J. Jr.: 1971a, *Astrophys. J.* **163**, 17.
Talbot, R. J., Jr.: 1971b, *Astrophys. J.* **165**, 121.
Trimble, V.: 1971, *Nature* **232**, 607.
Underhill, A. B.: 1968, *Ann. Rev. Astron. Astrophys.* **6**, 39.
Ungar, S. G.: 1971, *Bull. Am. Astron. Soc.* **3**, 242.
Uus, U.: 1970, *Astr. Council, Acad. Sci. U.S.R.R., Scientif. Inf.* **17**, 35.
Van Blerkom, D.: 1971, *Astrophys. J.* **166**, 343.
Van der Borght, R.: 1969, Preprint: *Vibrational Stability of Pure Helium Stars Surrounded by Pure Hydrogen Envelopes.*
Ziebarth, K.: 1970, *Astrophys. J.* **162**, 947.

DISCUSSION

Westerlund: Paczyński talked about the red supergiants being the cause of all planetary nebulae and during the coffee break I objected to it because the normal red supergiants are stars that have just evolved from blue supergiants and this would mean that the planetary nebuale would belong to the extreme Population I and hence their distribution should be quite different from what it is in the Galaxy. I understand, however, that Paczynski does not refer to those stars only so I would like to ask him to clarify the point.

Paczyński: By supergiants I do not necessarily mean MK classification Ia. I referred to red supergiants on two occasions. First, when I was discussing solar mass stars which are believed to produce planetary nebulae at the end of their evolution; these stars, according to model computations, reach very high luminosities of the order of 10^4 $L_\odot$. Such high luminosities are indeed observed in Mira variables which are obviously disc population objects and must have masses around 1.5 $M_\odot$. I am not sure whether the observers will classify Miras as red supergiants, perhaps the best is to call them Mira supergiants. They are supergiants of low mass, bolometric absolute magnitude about -5, and, most likely, the progenitors of planetary nebulae.

The second reference to red supergiants was related to very massive, very luminous red supergiants, in the 30 $M_\odot$ class. These objects should appear in very young open clusters, but in open clusters or associations, even younger than h and χ Persei, we find almost none of such red supergiants while relatively large numbers of blue supergiants are observed. And this apparent lack of red supergiants has been referred to quite frequently in the literature and different hypotheses were suggested to explain it. One possible explanation is to assume that as soon as a very massive, very luminous star approaches the red supergiant stage of evolution the envelope is lost by the star. This picture would fit rather nicely those observations that indicate that single, Population I Wolf-Rayet stars are frequently associated with ring nebulae.

So, perhaps these stars are very massive, single objects that have lost their hydrogen-rich envelopes. Ring nebulae are conventionally assigned to extreme Population I and there are so few of them because there are so few massive Population I stars.

Johnson: If you would explain the deficiency of red giants in some specific place, for example h and χ Persei, would you then expect to find there some of the ring nebulae which you want to bring forth from the red giants. I don't think that h and χ Persei has any ring nebulae or any kind of nebulae. So is there a contradiction between the deficiency argument and the lack of ring nebulae.

Paczyński: I do not think so. As far as I remember, Lindsey Smith at the Boulder Conference a few years ago, claimed that we do not observe ring nebulae around those Wolf-Rayet stars that are close to other early type stars. So, we should not expect to see nebulae in any association as there are so many stars which ionize the surrounding. However, do we observe any Wolf-Rayet stars in h and χ Persei?

General answer: No!

Johnson: It is an interesting fact that though there are OB stars in h and χ Persei there are no H II regions. In other words, it seems to be deficient in interstellar gas.

Paczyński: I would like to concentrate on the last topic of my talk, and this is the problem of the mass loss from Wolf-Rayet stars that we do observe now. And please, distinguish this mass loss from the hypothetical mass loss that has taken place in the past. These are perhaps two entirely different phenomena. This is most clearly demonstrated in the case of nuclei of planetary nebulae. In the planetary nebulae we do observe large masses of hydrogen rich gas expanding at a low velocity of about 30 km s^{-1}. At the same time Wolf-Rayet nuclei of those nebulae are losing gas at a velocity which is close to 500 km s^{-1} and it is hard to believe that the same mechanism might be responsible for these two kinds of mass loss which differ so greatly in the outflow velocity.

Van Blerkom: Can we not have the outer material decelerated by interaction with the interstellar medium?

Paczyński: No, because the gas density in planetary nebulae is so much larger than the density of interstellar medium.

Thomas: What is roughly the radius of a planetary nebula?

Paczyński: The largest is 0.6 parsecs; a typical one is about 0.2 parsecs.

Thomas: Could it be one parsec? Why not?

Paczyński: Because it is not observed.

Thomas: If you take your computations you have roughly a continuous ejection of between 10^{-5} and 10^{-6} solar masses, and in a rough way your star slows down to the order of 30–40 km s^{-1} after the amount of the mass ejected equals the mass encountered in the interstellar medium and that occurs at about one parsec.

Johnson: In the case of the best study (NGC 6888) it was 2.6 parsecs.

Thomas: No, that is the outer region.

Johnson: The shell is very thin.

Thomas: Sure, but if you start out with 1500 km s^{-1} and you take the point where it drops below 100 km s^{-1} (it is roughly a parsec or a little bit less) and thereafter it moves out very slowly because you are putting on no mass at all to the back of the big thing.

Johnson: At the present time the stuff is supposed to move at a rather constant velocity from the star to the shell which is very thin.

Thomas: 1500 km s^{-1}.

Johnson: Yes. It gets there after 2.6 parsecs. The rings really have a larger radius by at least a factor of 3 than the largest planetary radii, and if you do your calculations...

Thomas: You will find that your velocity drops to the order of less than 100 km s^{-1} after something less than one parsec. So, you assume that the planetary is at a much earlier stage of evolution. I do not see this.

Paczyński: According to the observations a typical electron density in a planetary nebula is 10^4 electrons cm^{-3}. Are you prepared to say that the density of interstellar matter...

Thomas: This is the swept up density.

Paczyński: The planetary nebulae are not thin shells. This is almost a uniform density.

Westerlund: No!

Paczyński: I would say that in the typical young planetary nebulae, like those which have the Wolf-Rayet type nuclei, the part of the volume occupied by the gas is not much smaller than one. I mean fraction of the volume. If you go through all the planetary nebulae you find all kinds of filamentary structure.

Westerlund: No. I feel that the high densities for many planetary nebulae are not the average densities. I think (I do not remember any statistics now), that if you look at the Perek Catalogue or at my paper with Henize, you will see that a very great number of the planetary nebulae are very small spots of extremely large density. That is what is observed, there is no mean density obtained for a very large number of planetary nebulae, I am sure.

Paczyński: In the case of the few planetary nebulae associated with Wolf-Rayet nuclei, the estimates of electron density are available and those range from 10^4 upwards. I have done the computations many times, so I can tell off hand that you cannot put enough matter from the interstellar medium around the star to make the planetary nebula. Simply because there is not enough matter for that. So, you must assume that almost all the gas that you see is gas ejected from the star. Now, if you would like to eject this gas at a high velocity you need more mass from the interstellar gas to slow it down, and that would mean that almost all the nebular gas is from the interstellar medium.

So, I do not think there is any way to avoid the conclusion that in the case of the planetary nebulae with Wolf-Rayet nuclei we see two modes of mass outflow, one which took place some thousands of years ago, and another which is taking place now. Well, this is the most straightforward and simple, case, but you have to keep in mind the possibility that the same situation may be true for other Wolf-Rayet stars as well. In the case of binaries you can be quite sure that the major fraction of stellar mass was transferred to the companion and that this transfer had nothing to do with the mass loss that we do observe now. I am spending so much time on the subject because unfortunately it is quite frequent to find people confusing these two phenomena: the present mass loss which is observed now and more or less hypothetical previous mass loss. Now I am going to speak only about this mass loss which we do observe, now, the high velocity mass outflow. We do not know whether this phenomenon has an evolutionary significance, while we assume that the hypothetical previous mass loss was of a major evolutionary significance.

Thomas: Is the present information on mass loss coming from the widths of emission line profiles or do you include the violet absorption edges?

Paczyński: The observed rates of mass loss are in the range of 10^{-6} up to 10^{-4} solar masses per year.

Underhill: How do you get that? You need a density! Where do you set the density from?

Paczyński: I shall speak about this in a few minutes. At the moment we may just say that this is the value most frequently found in the literature. The velocity of outflow as observed in different objects is in the range of 500 to 2000 km s^{-1}. Let us take 1000 km s^{-1} as a typical value. This velocity is well established as the violet displaced absorption edges are observed. This velocity is independent of any model of the formation of the emission lines. In the model which I like most, and which will be presented in more detail tomorrow, the mass outflow takes place in a more or less spherically symmetric fashion, and the velocity of mass outflow may be deduced from the width of the emission lines. I know that this is a controversial model, and it is better to remember the violet displaced absorption lines, which as far as I know have never been interpreted in any other way. There is a correlation between the ionization potential of a given ion and the observed velocity of outflow. The low ionization corresponds to high velocity. This is conventionally interpreted in terms of excitation decreasing outwards.

Perhaps, the strongest argument of the outward increase of the velocity of outflow follows from the interpretation of the lines which show flat-topped profiles. As far as I remember those show the lowest excitation. And these show the largest velocities of outflow, so there can be little doubt that the gas is accelerated outwards.

Sahade: This last point is not settled yet. So, you are making an assumption.

Paczyński: No, it is not an assumption. I shall try to repeat the reasoning. We do observe the flat-topped profiles of some emission lines, and those do show the violet displaced absorption edges. The rectangular emission line profile indicates, that the line is optically thin, and is formed in the uniformly expanding region of low density. It is reasonable to expect that these lines are formed at a large distance from the star. These lines correspond to the highest expansion velocities as they are the widest of all, and they belong to the lowest ionization stages. The lines coming from the higher ionization stages are narrower implying lower velocities of expansion. I do not think it is possible to put those relatively narrow Gaussian shape profiles into a region outside of that in which the flat-topped lines are formed. I do not think there is any serious argument against the model of radially expanding envelope being accelerated outwards.

Sahade: In the case of binaries there is evidence that seems to suggest that the matter around the WR star decelerates outward; there is however, an acceleration further out which is indicated by the expanding evelope that shows in lines sensitive to dilution effect.

Underhill: We can discuss it more certainly after we have listened to the discussion of the theory of the spectra by Van Blerkom.

Paczyński: Let us consider now a radially expanding evelope that is accelerated outwards. I would like to stress a certain point that has not been emphasized in the past, but which I believe, is true. We observe that the profiles of unblended lines are symmetric. There is no indication of an occultation effect. It means that the line emitting region must be at least two or three times larger than the star. We have to assume that all the lines are formed in a region large compared to the star. At the same time we see the correlation between the ionization potential of a given element and the line width, i.e. the outflow velocity. It means that gas is being accelerated within that region in which the lines are emitted. Whatever is the force that pushes the gas outwards, it must operate within the line emitting region, and this is very important.

One more thing. In order to derive the mass loss rate we need the velocity of outflow, and this is fairly reliable; we also need the density of matter and the radius at the point where the density is estimated. The value of 10^{-4} solar masses per year is obtained of one takes the numbers given by Castor and Van Blerkom in their model of WN star envelopes. If someone does not like the high value of the rate of mass loss, this may be lowered a little. But I do not think that it may be lowered by more than a factor of ten in the particular case considered by Castor and Van Blerkom. If some other stars have less dense envelopes they may have smaller rates of mass loss. This lower density may be indicated by the lack of any bump in the Pickering decrement in some stars.

In any case it would be extremely difficult to have less than 10^{-6} solar masses per year, and that is much more than the rate observed in Of stars. This is most clearly demonstrated by the comparison of the WR and Of type sectra. These are similar in the ultraviolet. However, when you consider the visual spectra, you see nothing in the Of stars that would resemble the rich emission line spectra of WR stars.

The most obvious reason for this difference is the large difference in the gas density in the expanding envelopes. Wolf-Rayet envelopes are much denser, and as the outflow velocities are about the same, the rates of mass loss must be considerably higher in the Wolf-Rayet stars. What may be the mechanism for the observed mass loss from WR stars? Limber suggested the rational instability. This cannot explain the origin of the violet displaced absorption edges. In Limber's model the absorption should occur at the centre of the emission line, not at the edge. Another possibility is some kind of a hot corona, like the one we have in the Sun. By the corona I mean that part of the extended atmosphere where the gradient of the gas pressure is important and large enough to push the matter outwards, just as it is the case with the solar corona. The expansion velocity of the Wolf-Rayet envelope is of the order of 1000 km s⁻¹, and that implies a very high temperature, of the order of 10^7 K for the corona, if the gradient of the gas pressure is to be important. And this is independent of the gas density. We know from the observations that the acceleration takes place within the line emitting region, where the electron temperature is of the order of 10^5 K as indicated by the observed ionization equilibrium.

Therefore, we have here the discrepancy of two orders of magnitude in the temperature. Unless you are prepared to invent an exceedingly complicated model with dense low temperature clouds in which the lines are formed, and very hot intercloud medium which accelerates those clouds, you cannot explain the observed mass outflow by any mechanism involving the gas pressure.

Thomas: Without arguing for or against 10^7, how do you know it cannot be 10^6? Do you have enough observations to really exclude coronal ions like in the Sun?

Paczyński: If you have 10^6, how would you want to preserve the relatively low ionization?

Thomas: I do not have a uniform atmosphere, of course.

Paczyński: So you are in the position I have mentioned. You have to introduce a hot continuous medium with cool dense clouds.

Thomas: All I have to do is to have a radial distribution.

Paczyński: No.

Underhill: What is wrong with that? The observations push you there.

Paczyński: Only if you are willing to accept such a complicated picture.

Thomas: Why do I need it?

Paczyński: Because otherwise you will not be able to produce lines in the same region in which the matter is being accelerated.

Thomas: I do not think we can claim consistency until we have solved the fluid mechanical problems coupled to the excitation better than we have.

Paczyński: I do not understand. You mean, I cannot have in the same region 10^7 degrees?

Thomas: No. You are saying that you cannot have the velocity excitation correlation that you observe and at the same time have anywhere in the atmosphere, regions at a million degrees. To me it is far more critical to ask observationally whether I can exclude a million degrees until such time as you have solved the fluid mechanical problem, to really show the correlation between velocity and excitation under some kind of a generalized stellar wind solution.

Paczyński: I am not sure I understand your remark.

Kuhi: I do not think there is any observational evidence for any coronal lines of any kind in a Wolf-Rayet star.

Thomas: Do you have rocket spectra?

Kuhi: No, I am talking about the optical lines identified in some symbiotic stars.

Thomas: Sure. In the solar optical spectra, whatever you see, you would not have seen it from a long way away.

Underhill: You see the solar coronal lines in one or two symbiotic stars but I believe you need not only very high electron temperatures but also low electron density, 10^9, or so.

Thomas: That is a quite reasonable thing.

Underhill: That is what we are talking about. When we, stellar observers, talk about coronal lines in the visible...

Thomas: Is there any evidence yet? Do I have enough rocket observations to exclude anything ionized higher than O VI?

Underhill: Not in the UV.

Thomas: Then my mind is open. That is all.

Underhill: So we have to refer our evidence to the forbidden lines.

Thomas: That is right.

Paczyński: You may look at the equation of motion for gas dynamics, and you will find that you have terms that correspond to the gas pressure, and kinetic energy of the gas. At a large distance from the star the gas pressure is negligible and you may estimate the kinetic energy from the observed velocity of outflow. Because of energy conservation, this kinetic energy has to come from somewhere. If you want to use the gas pressure to accelerate the matter you have to store this huge kinetic energy as the internal energy of gas in the region close to the star. No matter how complicated a picture you are going to introduce, you cannot avoid this problem.

Johnson: If you have 10^7 degrees in an atmosphere as large as the Wolf-Rayet star, and as dense, you should see it as an X-ray source, and I think so far these stars have not been found as X-ray sources although several people have written papers making tentative identifications but never a completely convincing one so far. There are papers by Wallerstein and by Stecher. It was a good idea.

Underhill: The amount of X-rays in those stars will never get to us from what we think are reasonable distances for these stars.

Kuhi: With regard to the paper by Wallerstein about X-ray sources, I do not believe at all in Wolf-Rayet stars as X-ray sources.

Paczyński: The kinetic energy flow associated with the observed mass loss from WR stars is approximately one per cent of the stellar luminosity. It is considerable, but it is not a dominant fraction of the total energy flux.

Thomas: It seems to me consistent to have mechanical energy a couple of per cent of the radiative.

Paczyński: Yes, but this indicates the existence of an efficient mechanism for transformation of the thermal energy into kinetic energy.

Apparently, not everybody is convinced that we may rule out the gas pressure as the agent causing the mass outflow. Let us consider how the radiation pressure may act on matter and cause its outflow: these are the radiation pressure in the lines and in the continuum. According to Lucy and Solomon, the radiation pressure in resonance lines is responsible for the mass outflow from the Of stars. They put a very strict upper limit to the rate of mass loss that may be achieved in their model. It is given as $-\mathrm{d}M/\mathrm{d}t \leqslant L/c^2$. This is independent of the velocity of outflow. The detailed computations made by Lucy and Solomon are consistent with this upper limit. It is unfortunate, that while using the resonance lines we may take out from the radiation field momentum rather than energy.

Morton: But many absorption lines participate in the momentum transfer.

Paczyński: If you have a few resonance lines you may increase the rate by a factor of few.

Morton: There is one point in that connection. Their theory predicts that usually only one absorption line dominates at a given temperature class, but the observations suggest that as many as 5 or 10 lines could be important so that there is a difference there between their very simple theory and the observations. A factor 10 correction may be in order.

Paczyński: This means that perhaps we may push the rate of mass outflow from 10^{-8} up to 10^{-7} solar masses per year in a star as luminous as Wolf-Rayet stars are believed to be, i.e. about 10^5 or 2×10^5 solar luminosities. Let us consider now the radiation pressure in the continuum. As far as I remember it was originally suggested by Rublev. He thought that Wolf-Rayet stars are so luminous that the radiation pressure on free electrons is responsible for the observed mass loss. The radiation pressure on free electrons can balance the gravity of a 10 solar mass star if the luminosity exceeds about 4×10^5 or 6×10^5 solar luminosities, depending on the chemical composition. It is not much more than the observed luminosity of a typical Wolf-Rayet star. From the observational point of view, considering the uncertainty in the estimates of masses and luminosities of those stars, we cannot be entirely sure that the luminosity is not higher than the critical value given above. However, if you look into the problem more closely from the theoretical point of view, you will find that it is practically impossible to achieve such a situation. First of all, none of the theoretical models that had been computed with the proper boundary conditions at the stellar surface ever achieved the surface luminosity higher than the critical. There is, in fact, a very good safety valve within the star: every time the luminosity approaches the critical value the star expands and the luminosity is lowered and the critical value is not reached.

Smith: I do not quite understand the safety mechanism.

Paczyński: Let us suppose that we managed to exceed the critical luminosity somewhere within the star. It may happen, for example, during the helium flash. And now we want this high heat flux to diffuse through the star to the surface. The flux heats up the matter on its way, the gas expands, the gas pressure is lowered, and the energy output is lowered too. We may look on to this problem from a different direction too. The mass outflow from the Wolf-Rayet stars continues for at least 10^2 yrs. and, therefore, it is more or less stationary. We may integrate the equations of stellar structure from the surface inwards, assuming there is a stationary outflow of mass. It turns out that if the radiation pressure in the continuum is responsible for mass outflow, the so called critical point of the flow is below the photosphere, at a large optical depth. And then it turns out that it is practically impossible to fit such an envelope to the stellar interior that is assumed to be in a hydrostatic equilibrium. At least nobody has succeeded so far in doing this in a consistent manner and it is almost certainly impossible, if the luminosity exceeds the critical value.

Morton: Do you have such a problem whether or not you were above the critical luminosity?

Paczyński: As far as I know no case was found so far. Perhaps the matching may be possible for the luminosity below the critical value.

Morton: That argument tells us that some of the mass outflow may not be included in our current theory.

Paczyński: The mass outflow is driven by the radiation pressure in the continuum. In principle, if a star had a luminosity above the critical value the mass would flow out. However, in such a case the whole star would be blown out on a dynamical time scale. It is so, because it is not possible to match a solution for the outflowing envelope with the stellar interior that is in hydrostatic equilibrium. I believe that simple minded models with the radiation pressure in the continuum as a driving force are out of question. I feel now that we have almost proven that Wolf-Rayet stars cannot exist, and I believe that it is time for me to stop!

Conti: I have heard several allusions to a Wolf-Rayet star observed for the last 2000 yrs. Could someone tell me which star this is?

Thomas: Gamma Velorum is in the Almagest.

Conti: But the O star dominates the spectrum; the Wolf-Rayet star is much fainter.

Thomas: You mean that one of my standard illusions is being shattered? You mean you can vary Gamma Velorum by a factor of 2 over the 2000 yrs history?

Conti: You can vary it by a factor of 2 and you would not change the visual magnitude.

Van Blerkom: Can you say what was the dynamical time scale?

Paczyński: It is essentially the same as the pulsation time scale, or the rotation time scale (i.e. a few hours or days). I should like to make clear that the theory is not yet in a position to prove convincingly what kind of abundance anomalies there should be present in Wolf-Rayet stars. We may expect hydrogen deficiency in those stars that have undergone mass exchange. It is difficult to make a quantitative statement because very little is known about semi-convection. The deficiency may be by a factor of 2, which would not be detectable, or by a factor of 10, which perhaps might be detectable, or even more than a factor of 10, if a subsequent mass loss, the one we do observe now, has removed all the hydrogen-rich envelope. Well, as this point is not clear I would like to concentrate on the carbon-nitrogen-oxygen abundances. We know that cosmic abundances are such that oxygen is more abundant than carbon, and nitrogen is less abundant than carbon. When the matter is processed through the CNO cycle these elements are redistributed among themselves. First, carbon is almost entirely transformed into nitrogen. The equilibrium abundance ratio is of the order of 1 to 100. Then on a much slower time scale oxygen is transformed into nitrogen. The first step takes place when less than one per cent of hydrogen is burned into helium. It means that at the very beginning of hydrogen burning almost all carbon is transformed into nitrogen. The second process is much slower. Almost all hydrogen has to be burnt into helium before oxygen is transformed into nitrogen. The zero-age main sequence star is believed to be chemically homogeneous, and to have cosmic abundances. Therefore, carbon, nitrogen, and oxygen abundances are in the ratio of 3:1:9 approximately. Combined they contribute about one half of the total mass of the elements heavier than helium. At the end of the main sequence life time, almost all carbon is transformed into nitrogen within the inner half of the stellar mass, i.e. within the region considerably larger than the convective core. The outer half of the stellar mass has the original cosmic abundances. Within the convective core all hydrogen is burnt into helium, and not only carbon, but also oxygen are transformed into nitrogen. Therefore, within the helium core we have a lot of nitrogen.

Now we enter the core helium burning phase. Within the helium burning region all nitrogen that has been left by the CNO cycle is transformed into oxygen. Also, new carbon and oxygen is produced from helium. Therefore, in the region where helium is burning we have no nitrogen, and plenty of oxygen and carbon, and we are closer to the original abundance ratios.

Let me clarify the problem of the time variations of the chemical composition by making a schematic table. At the top we have the initial abundances (by mass), at the end we have the products of helium burning (see Table I).

The abundance variations shown in the Table may be interpreted either as the variations from the stellar surface (top) to the helium exhausted stellar core (bottom), or as the time variations within the stellar core. In order to show the abundance anomalies at the stellar surface we have to remove a lot of mass from the envelope, or we have to mix the stellar matter thoroughly. It is important that as the

TABLE I

Variation of chemical composition with stellar evolution

Hydrogen	Helium	Carbon	Nitrogen	Oxygen	Heavy Elements
0.69	0.27	0.006	0.002	0.018	0.014
0.68	0.28	0.000	0.008	0.018	0.014
0.00	0.960	0.001	0.024	0.001	0.014
0.0	0.959	0.002	0.000	0.025	0.014
0.0	0.5	0.40	0.0	0.08	0.02
0.0	0.0	0.48	0.0	0.48	0.04

nucleosynthesis progresses we have either mostly carbon plus oxygen or we have mostly nitrogen. The carbon to nitrogen ratio changes by many orders of magnitude, and the change is always rapid. It is like the appearance of the Wolf-Rayet type spectra, where either carbon and oxygen lines or nitrogen lines are prominent.

Underhill: But you still have to get these from the center out to the atmosphere.

Paczyński: You have either to mix it or to remove the surface layers.

Underhill: And how to do this nobody has ever explained.

Paczyński: If we have a binary then mass exchange removes the surface layers. Once you remove more than one half of the stellar mass, and this is easy to do in all the models with mass exchange, then you are likely to see the products of the carbon-nitrogen-oxygen cycle, and you are likely to find some anomalies in the abundance ratio of these three elements. If hydrogen is very strongly under abundant then you may expect to see more nitrogen than either carbon or oxygen. If you have no hydrogen and you see more carbon and oxygen than nitrogen, it means that this carbon and oxygen are the products of helium burning. No matter how difficult it may seem to get the products of helium burning to the surface, that is what must have happened. May I mention that R Coronae Borealis and the so-called helium stars are observed realities. In these stars hydrogen is underabundant by a factor of 10^5, and carbon is considerably overabundant. In these stars we see the products of helium burning mixed up to the surface. We do not know how the star is able to do this, but it obviously does so. The problem of mass removal or mixing is not solved for the Wolf-Rayet stars. Nevertheless we should be aware that the C:N:O abundance ratios are likely to be anomalous if either mass loss or mixing took place within a star.

Wood: I wonder about the universal Fermi interactions. What is the nature of their effect on your calculations?

Paczyński: There is a hypothesis that all the particles, no matter what their nature is, interact weakly with each other, just as they interact gravitationally. It is assumed that the constant which determines the strength of these interactions is the same for all the particles. Neutrino emission is a result of these interactions. It is very strong when the temperature and density are high. This is a different neutrino emission process than the one that appears in normal nuclear reactions. The latter will hopefully be detected from the Sun directly. It comes from the nuclear reactions which are reasonably well studied and their existence is beyond any doubt. However, if you consider the neutrino emission due to the universal Fermi interactions you find that it is not yet confirmed by any direct experiments.

Wood: What is the effect on normal reaction rates if these neutrino fluxes exist?

Paczyński: Neutrinos escape freely from the stellar interior. They carry out energy and, therefore, they act as a cooling agent. Strangely enough, if we have this cooling agent the star becomes hotter, just because it has to shrink more to compensate for the additional energy losses. In the carbon burning phase the amount of energy emitted in the form of neutrinos is a factor of 100 larger than the photon luminosity of the star. As a result the evolutionary life time gets shorter by a factor of 100 or so, and that is why the neutrino emission affects the late stages of stellar evolution so much.

Wood: These universal Fermi interactions are effective throughout all temperatures regimes?

Paczyński: Yes, but the phenomenon becomes really important once you are in the helium exhausted region of a star.

Thomas: What I want to do is to put together what it seems to me you have come up with, and to contrast that with what we had some time ago. I want to do this on the basis of the atmospheric framework that I proposed the first day: a classical photosphere, a non-classical photosphere, a chromosphere, a corona, and an exosphere. In the latter we have a mass transfer; in the corona, a momentum transfer; in the chromosphere, an energy transfer; in the non-classical photosphere, we have only population effects; and lowest there is the classical model with neither transfer nor population effects. It seems to me that you are calling for a mass loss ranging from 10^{-4} to 10^{-8}, considering the various estimates that have been made. But these various estimates are strongly dependent upon the momentum transfer; and here I worry both about the density and the density gradient, and I have a choice as to the kind of momentum transfer or support possible. We can have wholly a kinetic temperature support or some kind of a random macroscopic velocity; or an expansion. If we have an expansion, then the velocity of expansion has to be like ϱ/R^2. So we have these three possibilities. And in the energy situation, I must have somehow an energy supply sufficient to support the spectrum, and it seems to me that this must produce ionization exceeding O VI. I think that is the highest excitation we have seen in the visible spectrum. That is all one really knows. When we come down to the classical photosphere, it appears that we had better have the effective temperature in excess of 30000 degrees, if indeed we have a helium star. Is that correct?

Paczyński: That is provisional.

Thomas: All right. But it seems to me that we have trouble with the helium star, because various people continue to talk about something as low as 20 or 25 thousand degrees for the photosphere temperature.

Paczyński: The trouble with the helium stars is that they are too hot.

Thomas: But this kind of a minimum effective temperature is some kind of a concensus that people are talking about.

Paczyński: Yes, this is purely observational, not theoretical.

Thomas: All right, if you have a pure helium star, what do you like as a minimum? I thought you said one day a minimum effective temperature was 30000 degrees.

Paczyński: Yes, but from an observational point of view .Theoretically you cannot get an effective temperature of a massive helium star lower than one hundred thousand degrees if the star is in hydrostatic equilibrium and has no hydrogen envelope.

Thomas: Let me continue the empirical evolution in thinking. A long time ago I argued vigorously against an outward increase in expansion velocity, arguing rather for a decrease. And I argued for increase in ionization outward, rather than an outward decrease, with those two things observationally coupled. I argued for a high kinetic temperature in the outer atmosphere and probably an atmosphere supported either by a compound of material both moving up and moving down, or some kind of a turbulence. I never liked this word so I never liked the suggestion. The alternative if you wanted only a high kinetic temperature, was something near 10^7 degrees and I rejected that. I did not believe it. I have no basis for rejecting it, but it did seem to me an awfully big temperature. The argument for the suggestion that the degree of excitation and ionization increase outwards rather than decreasing was based on analogy with my thinking in the solar atmosphere. At that time I suggested the spicules were heating the solar atmosphere; and then observationally one had the interesting situation that the spicular velocity decreased outwards, from an original ejection, possibly a deceleration under gravity. At that time it worked reasonably well so far as explaining the solar atmosphere, because I would have an initial rise in ionization and excitation coming from the introduction of a mechanical heating, and then an eventual flow coming from the decrease in the spicular velocity. At that time there were two theories of heating mechanisms: one the spicular kind of thing; the other, the convective acoustic waves sort of thing. The accoustic waves seemed more realistic than the spicules and were generally adopted for the overall atmosphere. The relation between spicules and non-uniform heating of the chromosphere is, however, still a strong item of question. The simple acoustic wave mechanism of some years ago has now given the way to a discussion of how you avoid heating the lowest parts of the atmosphere; and it now appears that the production of aerodynamic oscillations occur much deeper in the atmosphere than the regional "surface bubbling" picture, and are stored most of them not propagating in the region of the temperature minimum. Then we get a certain amount of mechanical energy propagating upwards from the region of the temperature minimum outwards. Now I emphasize this picture for the following reason: Instead of talking about the production of the heating mechanism and propagation of the pressure waves at $\tau \sim 0.8$, what one does is to put in a kind of instability deep down in the solar atmosphere, and the mechanisms of the instability now being looked at include exactly also the same ones as cause the Cepheid variation, the opacity variation, the gradient in the ionization, etc. So the situation as regards origin of aerodynamic energy becomes much more similar between solar type and other types of stars. Now, what is the other evolution? It is simply that instead of having the corona beginning at a very great height (you remember people were talking of 50000 km, 20 years ago, then 5–10000 km). it now appears that at about 2000 km we have a rise to coronal conditions.

Chromospheric conditions begin already at something like 1000 km. So the heating and an abrupt rise to excitation conditions corresponding to 10^6 K in the atmosphere occur very very soon, within a couple of thousand kilometers from the surface. They occur within a few hundred km from where the propagation of aerodynamic energy sets in, at the temperature minimum. So with this model then what do we have for the picture? Over the majority of the non-classical atmospheres we have all corona and exosphere; the chromosphere is a very small region.

The second thing that happens in this same region, the region of the very abrupt rise, is that we no longer have any kind of a uniform atmosphere. In these inhomogeneities we find differences in electron temperature, by factors of two and three, ranging between something less than a million degrees to something near four and five million degrees. This reflects the energy balance in the atmosphere. Then we must admit the existence of an outflow of mass from the Sun in the form of the solar wind, with the driving mechanism being just this hot corona bounded by the interstellar medium.

Return now to the problem of distribution of ionization in the atmosphere of the WR star, how rapidly it falls off and how rapidly it increases. It seems to me that Kuhi has presented a very strong case for the picture that over a great extent of the WR atmosphere it looks as though one has a decrease of ionization outward, and an increase in velocity field outward, but the absolute level of kinetic temperature is very uncertain.

We have talked about 10^5 K in the region of line formation. We have talked about 30 000 K for the electron temperature in the lower atmospheric region where it seems to me that we are more and more coming to the same general picture for the WR stars as for the Sun. We see the photosphere in the continuum. We have a very small region where T_e rises abruptly. Then we have an extended corona-exosphere. The difference between stars lies in how much mass is in these various regions. Thus, how easy it is and what you need to observe the several regions, varies from stellar type to stellar type. The same holds in the exosphere and mass outflow. For all very hot stars apparently we measure outward velocities of the order of a thousand kilometers per second. Not a couple of hundred but of the order of a thousand kilometers per second; some non-zero mass loss.

It is not an extended atmosphere in the sense of that height where you get the heating, divided by the radius of the star, is very large. On the contrary, it is very small for the Sun, and not obviously large for the WR stars. It is an extended atmosphere phenomenon in that the extent of the whole atmosphere out to the interstellar medium, is large.

So when I ask details of helium photosphere, I am clamped down here in a very small region. When I ask for details of mass loss and momentum transfer I have an enormous region, even over the solar atmosphere. Now what we get from Paczyński's talk yesterday is one strong thing. It is not clear what is the mechanism for getting these high velocities, nor mass loss for the hot stars. Equally obviously we have trouble with the mechanisms of heating; that has not even been discussed, because we are not involved with it. But I do have trouble with the high velocities.

Paczyński: With the rate of mass loss?

Thomas: I have trouble with the mass loss rate but I also have trouble with getting these high velocities. You discarded the pressure mechanism. If I am going to have any kind of a stellar wind argument, any kind of a pressure mechanism, the upper limit in velocity is going to be set by $(2/\gamma - 1)^{1/2}$ times the sound velocity.

If I want to ignore interchange between internal energy and macroscopic energy, the gamma is about five thirds, and the velocity is about $\sqrt{3}$ times the sound velocity, a very small number. If I want to tap the internal reservoir and say I would start out, for example, with all ionized helium and then expand the gas and let the helium recombine, gamma goes towards one and you can pick up about a hundred kilometers per second. If you want to go all to oxygen so the whole atmosphere is oxygen and take like a thousand volts for the ionization of oxygen, you can increase your velocities somewhat more. But you have to go to very heavy elements in your atmosphere before you can get something like a thousand kilometers expansion velocity.

The other alternative is to go to some catastrophic thing like the supernovae or the novae kind of explosion. That is something deep inside. It seems very hard to see how you can have a continuous catastrophe to provide this kind of ejection. So the real problem here, in addition to specifying what excitation level I really have, which I do not think you are going to know till you have rocket UV observations, is really the corona and the exosphere in terms of mass loss and momentum transfer.

I have always believed in Kopal and Mrs. Shapley's arguments as to the kind of density gradients you have in the outer atmosphere.

Now Su-Shu Huang has a paper which shakes the foundations for this. But it is an observational point as to what density gradient do we need here, just as we have to find out what excitation level we need here, just as the argument related to the mass loss is very fundamental. It is indeed true that all these mechanical transfer effects represent a very small fraction of the radiation energy of the star. But what you should compare it to is the luminosity of the star that can be tapped. I cannot just dam up the radiation field and use all of that to expel material from the star. So, saying it is a small value does not help. I must have a good model, energetically and thermodynamically, to know how much energy I can tap. I suggest that that is the real phenomenon the Wolf-Rayet star is: find out what we mean by the corona, the region where we have a non-thermal supported density gradient, and the exosphere, where we have mass ejection, although we do not know where it comes from.

Underhill: You mean the exosphere motions are just details of the kinetic energy distribution.

Thomas: In some way it is very vague. All I say is look at the evolution of our thinking here. We are pushing back close to the surface of the star all the increasing excitation phenomena.

Is O vi really the highest excitation in the WR atmosphere? I just do not believe it, but then you come down to the real problem you suggested yesterday, if the emission lines occupy such a big volume relative to the photospheric radius, how do I get that volume maintained involving mass loss.

Paczyński: I would like to clear up some misunderstanding about the 10^{-2} times the luminosity. I think that this number is small in the sense that the total luminosity of the star is predominantly in the form of radiation. This additional flux of kinetic energy does not affect the bolometric magnitude of the star. However, this number is very big if you want to explain it. We cannot find so far an adequate mechanism that could convert one percent of luminosity into kinetic energy flux. So I do not want to deny the importance of this.

Thomas: I was not arguing with you, I was just stressing again may be what you have stressed in another way. That it looks as though you have a big reservoir but that you do not. And I agree with you that it does not look as though it can bother much the bolometric magnitude, atleast that is the way it looks now. But all the recent work on IR bolometric corrections do bother me.

Underhill: The real problem, it seems to me, is to get a close estimate of the actual plasmas. What you are saying is that theoretically it is very difficult, if not impossible, with present models, to find a mechanism that will transfer enough of the available energy to give the kinetic energy that is carried off if the mass loss is significantly greater than 10^{-8} solar masses per year.

Thomas: Not in this. We are talking about velocity per particle.

Underhill: I am not bothering about velocity per particle. We observe it. We lose so much energy. We have decided on really very uncertain grounds that the mass loss is considerably larger upto 10^{-4} rather than 10^{-8} solar masses per year. Now it is quite conceivable that we are interpreting our observations of line strengths quite incorrectly. We see material leaving the star but we have really no estimate of the density of that material. We make somewhat plausible arguments about the density and size of the sphere possible but if those are out by a couple of factors of ten, then we are in trouble. If we cannot make a sufficiently efficient mechanism by any conceivable physical process, I would go back to our estimate of density and size of sphere because this is the region where we are indeed uncertain.

Thomas: This is the point. I do not care what the actual value of the mass loss is. To get this kind of an energy per particle is awfully hard.

Paczyński: When the rate of mass loss is as low as 10^{-8} solar masses per year, the radiation pressure in resonance lines probably can accelerate matter to these velocities. There is no fundamental difficulty as long as you keep the rate of mass loss at such a low value.

Morton: I think we have at least one feasible mechanism for hot stars, that will give these velocities. We have enough problems without worrying about how to get the high velocities. In the observational determination of high mass loss rates, we must remember that we measure the velocity in a few particular ions and then make some crude estimates about the relative populations of the other ion states and other elements which are not observed.

SECTION VI

CHAIRMAN: A. B. UNDERHILL

THEORY OF WOLF-RAYET SPECTRA*

DAVID VAN BLERKOM

Dept. of Physics and Astronomy, University of Massachusetts, Amherst, Mass., U.S.A.

1. Introduction

From the time of their discovery in 1867, Wolf-Rayet stars have been objects of great interest for theoretical astrophysicists. Their spectra, dominated by extremely wide emission lines, set them apart from the run-of-the-mill stars. Moreover, the classic assumptions that have enabled accurate model atmospheres to be constructed for hot main sequence stars: hydrostatic equilibrium, radiative equilibrium, local thermodynamic equilibrium, and negligible curvature effects, are probably all violated in Wolf-Rayet stars. As early as 1894, Scheiner proposed that

an enormous gaseous envelope (of unknown composition) surrounds the absorbing atmosphere, and produces bright lines in the spectrum by supplying to the slit of the spectroscope a greater quantity of light than the star's photosphere, in spite of the higher temperature of the photosphere.

In 1929 Beals presented evidence to support the hypothesis that the great width of the emission lines is due to the Doppler effect in a rapidly expanding envelope. Based on earlier work by Milne (1926), he proposed that radiation pressure in the spectral lines provided the propulsive force to drive the outflow. This work stimulated several papers on line and continuum formation in extended atmospheres. However, as it became realized that the theories were too rudimentary to apply to real stars, remarkably little theoretical work was done over the next thirty years. Only recently, with the availability of electronic computers, have quantitative studies of the expanding envelope model been made. These form the bulk of the present review.

Three alternatives to this model have been proposed. Thomas (1949) envisioned a type of super chromosphere supported by nonisotropic macroscopic motions in which the electron temperature T_e exceeds the radiation temperature T_r. Code and Bless (1964) advocated prominence-like activity in which streams of material are ejected into a thermalized shell. Finally, Limber (1964) discussed the possibility that forced rotational ejection of matter forms the circumstellar envelope. Whatever the strengths or weaknesses of these three hypotheses, no theoretical work exists to review beyond the original suggestions. We therefore limit discussion in the following sections to the expanding envelope hypothesis. As the results in Part IV show, this model has some success in describing the emission observed in Wolf-Rayet stars.

In attempting to present a coherent view of the theoretical interpretation of Wolf-Rayet lines and continua, there is a problem in that the number of investigations is quite small. This is especially true of theories of continuum formation. I have therefore endeavored to supplement the material available with calculations of an illustrative nature made for this review. It is hoped that the reader will not be too impatient with

* Contribution No. 121 of the Five College Observatories.

M. K. V. Bappu and J. Sahade (eds.), Wolf-Rayet and High-Temperature Stars, 165–202. All Rights Reserved.
Copyright © 1973 by the IAU.

this material, in particular the comparison of the gray and non-gray model atmospheres in Part II. Finally, those empirical arguments that have been raised over the years for various aspects of Wolf-Rayet spectra are not discussed. If it cannot be expressed as an equation, it is not here.

2. The Continuous Energy Distribution

Although it is the spectacular emission line spectrum which is the outstanding characteristic of a Wolf-Rayet star, the continuous energy distribution is anomalous in a more modest fashion. Compared to hot main sequence stars, there is

(1) a slight ultraviolet excess for WN stars

(2) a large infrared excess, especially pronounced for the WN stars.

The result of this behaviour is that the color temperature T_c depends on wavelength: the longer the wavelength, the cooler the star appears (Kuhi, 1966; Kuhi, 1968). This accounts for the extremely low color temperatures for Wolf-Rayet stars quoted throughout the literature, when all other indications pointed to very high temperatures.

The construction of a model atmosphere to account for the continuous distribution in Wolf-Rayet stars is fraught with difficulty. The usual assumptions of radiative and hydrostatic equilibrium, local thermodynamic equilibrium (LTE) and plane-parallel geometry, which have enabled very satisfactory model atmospheres to be computed for hot main sequence stars (e.g. Mihalas, 1964), may all be invalid for Wolf-Rayet stars. It is not unexpected, therefore , that true model atmospheres for Wolf-Rayet stars do not now exist. There are, however, studies which relax one or another of the classical assumptions and whose results may be relevant to the Wolf-Rayet phenomenon. In this category there are but two papers, separated in time by nearly forty years. The major achievement of both is that plane-parallel geometry is not assumed, rather the atmosphere is taken to be spherically symmetric. Both investigations assume radiative equilibrium and LTE. First in time is the paper by N. A. Kosirev (1934) on the radiative equilibrium of extended photospheres. Since Kosirev meant his theory to apply to Wolf-Rayet and P Cygni stars, he assumed steady outflow of matter with constant velocity. He also assumed a gray opacity law of the Kramer's type; $\kappa \sim \varrho T^{-4}$. With these assumptions, an opacity variation $\kappa \varrho \sim r^{-1.5}$ resulted. Independently, and exactly at the same time, Chandrasekhar (1934) published his investigations of spherical atmospheres governed by the opacity law $\kappa \varrho \sim r^{-n}$, although without application to specific stars.

Since the assumptions of radiative equilibrium, LTE and gray opacity seem hardly applicable to Wolf-Rayet atmospheres, there would appear to be no virtue in following this approach further. Nevertheless, the Kosirev and Chandrasekhar method does lead to some interesting, if expected, results in a rather straightforward manner, and sets the stage for the modern computer models. Moreover, the techniques they introduced are finding use in recent investigations of radiative transfer in extended atmospheres. We will therefore give some space to the solution of the transfer problems in a gray spherical atmosphere.

To start, an opacity law of the form

$$\kappa\varrho = c_n/r^n \tag{1}$$

is assumed, so that the optical depth is

$$\tau = \int_r^\infty \kappa\varrho dr = \left(\frac{1}{n-1}\right)\frac{c_n}{r^{n-1}}; \tag{2}$$

Rather than use the Eddington approximation to solve for the mean intensity of radiation, an expression derived by Larson (1969),

$$J = \frac{3H_0}{r^2}\left(\frac{n-1}{n+1}\right)\left[\tau + \frac{1}{3}\left(\frac{n+1}{n-1}\right)\right], \tag{3}$$

is employed. In a spherical atmosphere $H(r)/r^2$ is a constant (denoted H_0), where $H(r)$ is the first moment of the intensity, i.e

$$H(r) = \tfrac{1}{2}\int_{-1}^{1} I(r,\mu)\,\mu d\mu$$

and μ is the cosine of the angle between the radius vector and the direction of propagation. Hummer and Rybicki (1971) have given accurate numerical solutions for atmospheres satisfying Equation (1), and find Equation (3) to be a rather good approximation. From the assumption of radiative equilibrium it follows that

$$J = B(T) = \sigma T^4/\pi. \tag{4}$$

If T_1 is the temperature at $\tau = 1$, it is easily shown from Equations (2), (3), and (4) that

$$T(\tau) = T_1 \tau^{1/2(n-1)}\left[\frac{\tau + \frac{1}{3}\left(\frac{n-1}{n+1}\right)}{1 + \frac{1}{3}\left(\frac{n-1}{n+1}\right)}\right]^{1/4} \tag{5}$$

The intensity of radiation is found most simply using a Cartesian coordinate system we will call the (p, z) representation. Figure 1 shows a ray which passes a distance p from the center of the sphere. The emergent intensity is $I_\nu(p, \infty)$ and the energy emitted at frequency v in all directions per unit time is

$$F_v = 4\pi\int_0^\infty I_v(p, \infty)\,2\pi p dp. \tag{6}$$

It is convenient to introduce as a variable the angle ϕ between the ray at any point

　　　　　　　　　　　　DAVID VAN BLERKOM

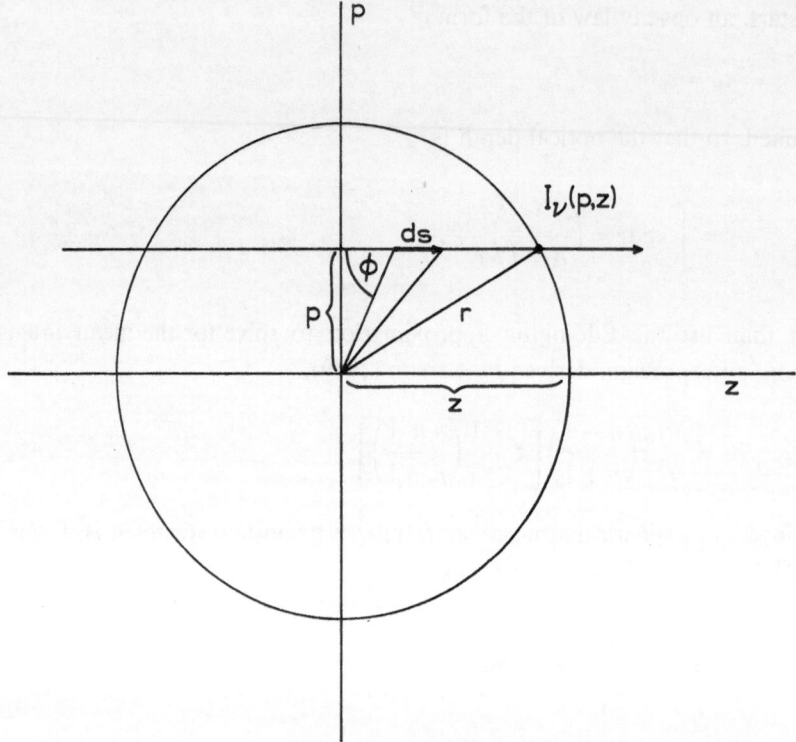

Fig. 1.　The transfer equation is solved for radiation propagating along the ray $p = $ const in the
(p, z) representation.

along the line of flight and the radius vector to that point. Then Kosirev's analysis
results in the following expression for F_v:

$$F_v = 4\pi^2 R_1^2 \int_0^\infty B_v(T)\, \tau^{-2/(n-1)} \Phi_n(\tau)\, d\tau \tag{7}$$

where R_1 is the radius at which $\tau = 1$ and the function $\Phi_n(\tau)$ is defined by

$$\Phi_n(\tau) = 2 \int_0^\pi \exp\{-(n-1)\csc^{n-1}\phi\, \psi_n(\phi)\,\tau\} \sin\phi\, d\phi \tag{8}$$

and

$$\psi_n(\phi) = \int_0^\phi \sin^{n-2}\phi'\, d\phi' \tag{9}$$

Equations (5), (7), (8) and (9) give the solution of the spherical gray atmosphere in
radiative equilibrium and in LTE. We apply these results to a group of stars which
are distinguished by different values of n in the opacity. All the atmospheres extend
to infinity; however if n is high the optical thickness of most of the atmosphere is
negligible. Thus different values of n determine the extension of the atmosphere, with

lower values of n corresponding to more extended atmospheres. In the limit $n \to \infty$, the plane-parallel gray atmosphere should be obtained.

By demanding LTE in an extended atmosphere, a large fraction of the atmosphere is forced to a temperature lower than the boundary temperature in a plane parallel-model. Compared to a plane parallel atmosphere, fluxes from extended atmospheres with the same values of T_1, should exhibit increased emission in the infrared and decreased emission in the violet. This does not imply that observations of an extended atmosphere would show an infrared excess and ultraviolet deficiency. In usual prac-tise, the observed flux is compared to that of a black body (or model atmosphere if available) which best matches its behavior in the visible, at say 5000 Å. It is relative to this that excesses or deficiencies are said to exist.

We therefore compute the gradient

$$\phi_c = 3\lambda - \frac{d}{d(1/\lambda)}(\ln F_v)$$

from Equation (7) at $\lambda = 0.5\ \mu$, and compute the temperature T_c which a black body would need in order to give the same gradient. This is found from the expression

$$\phi_c = (c_2/T_c)/[1 - \exp(-c_2/\lambda T_c)]$$

where $c_2 = 1.43879$ cm deg (Allen, 1960). Table I gives ϕ_c and T_c for spherical models in which $T_1 = 50,000$ K and the extension parameter n is varied.

TABLE I

Color temperatures and gradients for spherical model
atmospheres with different
extensions

n	ϕ_c	$T_c(10^4$ K$)$
2	1.32	1.20
3	0.90	2.18
4	0.77	3.05
5	0.73	3.54
7	0.70	3.99
10	0.69	4.26
∞	0.66	5.00

It should be noted that the effect of increasing the extent of the atmosphere (decreasing n) is to make the continuous spectrum appear cooler in the visible. It is possible, therefore, that a hot extended model and a more compact cool model would look alike. For example, $\phi_c = 0.90$ for $T_1 = 30000$ K and $n = 5$ matches the gradient for $T_1 = 50000$ K and $n = 3$. A comparison of the fluxes from these models shows that they are nearly identical over all frequencies of interest.

Figure 2 shows the behavior of F_v/F_{v_1}, where v_1 corresponds to a wavelength of 5000 Å, for a star characterized by $T_1 = 5 \times 10^4$ K and $n = 2$ (great extension). This is compared to $B_v(T(\tfrac{2}{3}))/B_{v_1}(T(\tfrac{2}{3}))$, which approximates the relative flux from a gray

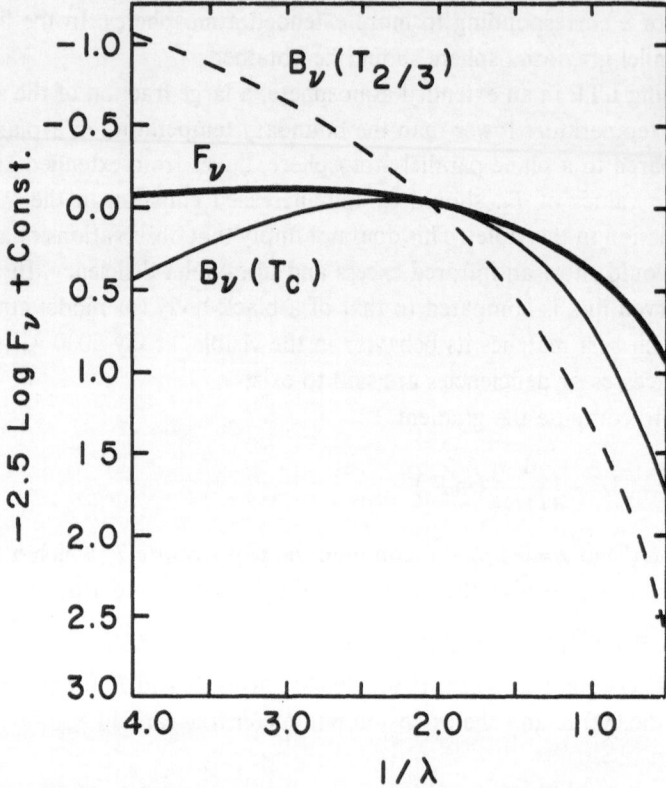

Fig. 2.   The emergent flux from a gray extended atmosphere with $T_1 = 5 \times 10^4$ K and $n = 2$ compared with black body curves at $T(\frac{2}{3})$ (dashed line) and T_c. The black body curve at T_c matches the slope of the computed flux distribution at 5000 Å. In the infrared and ultraviolet the computed flux exceeds the black body value.

plane parallel atmosphere. The infrared excess and ultraviolet deficiency are striking. When compared to a black body distribution $B_\nu(T_c)/B_{\nu_1}(T_c)$, where $T_c = 1.2 \times 10^4$ K, there is now an ultraviolet, as well as an infrared, excess, reminiscent of the WN stars.

Let us now consider the very recent work of Cassinelli (1971a; 1971b). Cassinelli considered the Kosirev problem but without the gray opacity assumption. However, since he was concerned with extended, but stable, atmospheres, he assumed hydrostatic equilibrium. Thus, for Wolf-Rayet atmospheres, this represents one step forward and one step back. The solution of the non-gray problem is not amenable to analytic methods and involves numerical techniques for a computer. We therefore present only the results of this investigation. Cassinelli finds that (1) models which have the same temperature, $T(\frac{2}{3})$, but different geometrical extensions, can produce very different flux distributions and (2) a hot star with a very extended atmosphere has an optical continuum similar to that of a star with a cooler less extended atmosphere. Cassinelli measures the geometrical extension by the ratio $R(\tau = 0.001)/R(\tau = \frac{2}{3})$, where τ is a mean optical depth scale. His results are shown in Figure 3 and illustrate the above conclusions.

Since Cassinelli has assumed hydrostatic equilibrium, his atmosphere models are

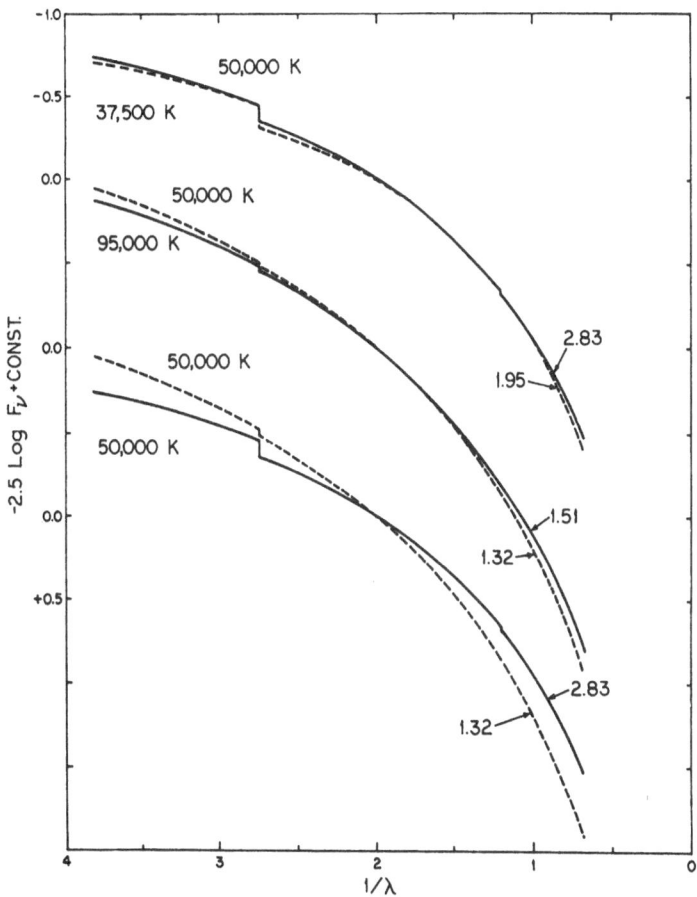

Fig. 3.   Emergent fluxes in non-gray extended atmospheres after Cassinelli (*Astrophys. Letters* **8**, 108). The parameter labeling each curve is a measure of the extension, $R(0.001)/R(\tfrac{2}{3})$. The resemblance of hot extended to cooler, more compact models is illustrated.

compact compared to the gray calculations illustrated. Nevertheless, both the gray and non-gray calculations lead to the same qualitative results. As can be seen from Figure 4, the agreement is also quantitative, for apart from an emission jump at the Balmer limit, the gray and non-gray results are nearly indistinguishable. The non-gray calculation illustrated here has $T(\tfrac{2}{3})=48\,865$ K, $T_1 = 54\,450$ K and $R(0.001)/R(\tfrac{2}{3})=1.89$ Both gray models have the same value of T_1 as the non-gray model.

Cassinelli compared the results of this non-gray model with the observed continuum flux (Kuhi, 1968) from the WN 6 star, HD 191765 and found excellent agreement. In view of the similarity of the gray and non-gray models, the gray atmosphere also fits the observed points. However, Smith and Kuhi (1971) have made substantial corrections to the original observational data, and consequently the theoretical curves are not nearly as convincing. Because of the rapidity of the gray atmosphere calculations, it is possible to generate a large number of models. In Figure 5, the uncorrected observations (open circles) are seen to be well described by the curve corresponding to

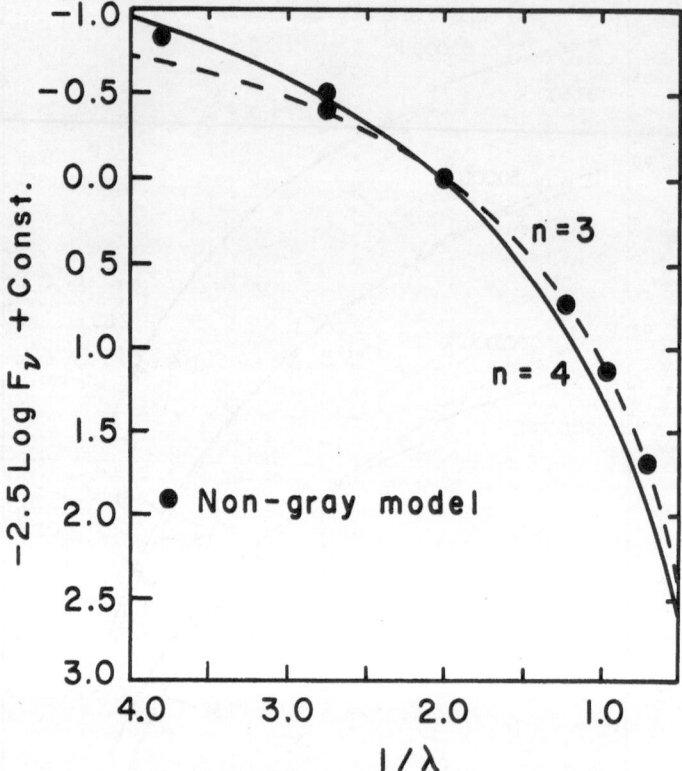

Fig. 4. Comparison of two gray models at $T_1 = 54450$ K with the non-gray model at the same value of T_1 and $R(0.001)/R(\frac{2}{3}) = 1.89$ (dots). The agreement is quite remarkable.

$T_1 = 5 \times 10^4$ K and $n = 5$. The corrected data (filled circles) follow the curve for $n = 2.5$ and for the same value of T_1. It is likely, therfore, that a non-gray model, of greater extension than the one used initially, can be constructed which will adequately describe the observations.

Although the relative flux distributions of WR stars can be simulated by the model atmospheres discussed, there is real doubt as to whether the physical parameters of the models are representative of those actually found in WR stars. What is greatly suspect here is the assumption of LTE, since in an extended atmosphere matter density becomes low and photon mean free paths large. Both effects inhibit an approach to LTE. Thus, a non-LTE model for the continuous energy distribution in stars with spherically symmetric atmospheres is certainly required before a satisfactory explanation of WR spectra will be obtained. Nevertheless, the calculations presented here show sufficient agreement with the observations to be at least suggestive of the processes operative in a real atmosphere. The infrared continuum *may* be formed deep enough in the envelope that the geometrical extension of the atmosphere is not great, and radiative equilibrium and LTE may not be as egregious approximations as might at first be surmised. The line forming region is believed to lie outside that in which the continuum is produced. Since the temperature in that region must be about 5×10^4 K

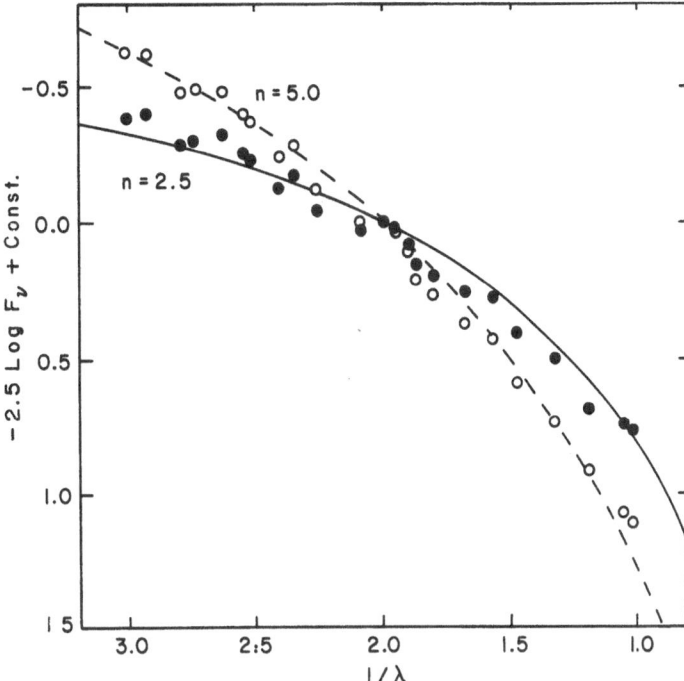

Fig. 5. Emergent flux from two gray models with $T_1 = 5 \times 10^4$ K and $n = 5$ and 2.5. These are compared to observations of HD 191765 both before (open circles) and after (closed circles) additional reddening corrections made by Smith and Kuhi.

(in a WN 6 star) to produce the degree of excitation observed, the possibliity exists that a temperature inversion occurs within the envelope. The continuum would be formed mainly in the inner region of declining temperatures, the lines further out, with perhaps additional ultraviolet continuum also arising from these outer regions. This is, of course, entirely speculative, but with so little in the way of theoretical work to fall back on, one can do little more than speculate.

3. Line Formation in an Expanding Atmosphere

3.1. INTRODUCTION

Although by no means extensive, the literature on line formation in moving atmospheres is far more substantial than that on the continuous energy distribution. The early papers by Gerasimovic (1934), Chandrasekhar (1934), Wilson (1934) and Bappu and Menzel (1954) and Chapter XX in Rosseland's textbook (1936) incorporate many of the techniques used at present. However, the assumption implicit in these investigations, of complete transparency of the atmosphere, makes them unsuitable for Wolf-Rayet models. The theory developed by Sobolev in his monograph *The Moving Envelopes of Stars* (1960) and in Chapter 28 of Ambartsumian's text *Theoretical Astrophysics* (1958) has been employed in several studies of Wolf-Rayet and other stars believed to be losing mass. One should especially note the papers by Rublev

(1961; 1963) on Wolf-Rayet atmospheres and another in the same spirit by Van Lyong (1967). Sobolev's method has recently been extended by Castor (1970) and subsequently applied to the excitation of helium lines in Wolf-Rayet stars by Castor and Van Blerkom (1970). It seems worthwhile therefore to review this theory in a somewhat leisurely manner.

The star is assumed to eject mass from a well defined surface at a radius r_c. That part of the star with $r \leqslant r_c$ we will call the core. Since the actual physical process which causes the mass ejection is not known with certainty, the velocity distribution in the atmosphere must be regarded as somewhat *ad hoc*. The comments of Chandrasekhar (1934) are still relevant today:

If one postulates that the parent star is continually ejecting atoms then from a dynamical point of view there are not many possibilities of the ways in which this could happen. The ejection process could, in fact, take place in one of two ways:

(A) At the boundary of the star the atoms (presumably only those with a relatively small but finite outward velocities) are 'repelled' by some kind of force which is, say, f times the gravitational attraction. Unless f is very nearly unity we could reasonably assume that f is a constant, i.e. the repulsive force – whatever its nature – falls off like gravity inversely as the square of the distance.

... this hypothesis includes, as a special case, the emission of particles arising from unbalanced radiation pressure,...

(B) The atom at the boundary of a star might receive a large initial outward velocity (either in a single process or in stages), and in escaping from the star be continually de-accelerated in the gravitational field of the star, the atom either escaping from the star with a finite outward velocity, or after ascending a certain distance begin to fall back towards the parent star. We could have an atmosphere of high-speed particles set up in this way.

The dependence of line width on excitation potential originally discovered by Beals (1929) and recently reinforced by Smith and Aller (1971) indicates that process A occurs in Wolf-Rayet atmospheres. Then, if g is the acceleration of gravity at r_c

$$\frac{d^2 r}{dt^2} = (f - 1)\, g r_c^2 / r^2 .$$

It follows easily that

$$v(r) = [2(f - 1)\, r_c g\, (1 - r_c/r)]^{1/2}$$

if $v(r_c) = 0$. Since the velocity at infinity is

$$v_\infty = [2(f - 1)\, r_c g]^{1/2}$$

the velocity distribution takes the form

$$v(r) = v_\infty (1 - r_c/r)^{1/2} . \tag{10}$$

This law was used e.g. by Castor (1970) in his study of line formation and, in a slightly modified form, by Lucy (1971). We will therefore use Equation (10) for $v(r)$ in all subsequent calculations.

Consider a line emitted by an expanding atmosphere which has a central frequency, measured in the laboratory, of v_0. Since different parts of the atmosphere approach and recede from a stationary observer, the radiation received will be spread in frequency due to the Doppler effect. If $v(v)$ is the photon frequency as seen by an observer moving with the material at a velocity v and $v(0)$ is the frequency seen by a stationary observer,

$$v(0) = v(v) + \frac{v_0}{c} \mathbf{n} \cdot \mathbf{v} \tag{11}$$

where $\mathbf{n}$ is the direction of propagation of the photon. The maximum frequency displacements occur for radiation emitted by material with the limiting velocity v_∞. Because $v(v) \approx v_0$, i.e. emission occurs locally close to line center, the spectral line has a total half width of

$$\Delta_s = v_{max}(0) - v_0 = \frac{v_0}{c} v_\infty . \tag{12}$$

It is convenient to measure frequency displacement in units of Δ_s; so a dimensionless frequency parameter is defined by

$$x = [v(0) - v_0]/\Delta_s . \tag{13}$$

The observed line extends from $x = -1$ (red) to $x = +1$ (violet).

If thermal broadening is assumed, then from Equation (11), the normalized absorption coefficient of an element of gas moving with velocity v, for a photon having frequency $v(0)$ in the stationary observer's frame is

$$\phi[v(0), v] = \frac{1}{\sqrt{\pi \Delta}} \exp \left\{ - \left[v(0) - \frac{v_0}{c} \mathbf{n} \cdot \mathbf{v} - v_0 \right]^2 / \Delta^2 \right\} \tag{14}$$

where Δ is the local value of the Doppler width,

$$\Delta = \frac{v_0}{c} v_{th}, \quad v_{th} = (2kT/m_A)^{1/2} . \tag{15}$$

Let

$$\begin{aligned} \delta &= \Delta/\Delta_S \\ u &= v/v_\infty \\ \mu &= \mathbf{v} \cdot \mathbf{n}/|v| . \end{aligned} \tag{16}$$

Then the normalized absorption coefficient can be written in terms of the dimensionless variable x as

$$\phi(x - u\mu) = \frac{1}{\sqrt{\pi \delta}} \exp[-(x - u\mu)^2/\delta^2] . \tag{17}$$

The notation here is that of Hummer and Rybicki (1968).

An important difference exists between lines formed in atmospheres which are stationary and those which are rapidly expanding. In the former, a strong line is formed very near the geometrical boundary of the star. In an expanding atmosphere, however, contributions can come from matter distributed across the entire envelope. One can see this in the following simple example. Consider a ray passing a distance

p from the center of the star, as shown in Figure 1, where the radiation is seen by a stationary observer to be displaced to the violet of line center by an amount x. Let T be constant, so Δ does not vary through the envelope. The optical depth along the ray $p = $ constant is

$$\tau(x, p, z) = \int_z^\infty k(r)\, \phi(x - u\mu)\, dz \tag{18}$$

where $\mu = z/r$ and $r = (p^2 + z^2)^{1/2}$. The absorption coefficient $k(r)$ contains only atomic constants and the population density of absorbing atoms $N_A(r)$. The continuity equation states that the total number density of atoms $N_{tot}(r) \sim r^{-2}v^{-1}$. For simplicity, we assume that $N_A(r)/N_{tot}(r)$ is constant throughout, so $k(r) \sim r^{-2}v^{-1}$. Finally

$$\tau(x, p, z) \sim \int_z^\infty \frac{\phi(x - u\mu)}{vr^2}\, dz \tag{19}$$

where the constant of proportionality is determined by the normalization $\tau(x, p, -\infty) = 1$.

Equation (19) is integrated numerically for a model in which $v_\infty = 2000$ km s^{-1}, $T = 5 \times 10^4$ K, $x = 0.5$ and $p = 2r_c$. The composition is taken to be pure helium, so that with these parameters, $v_{th}/v_\infty = 7.2 \times 10^{-3}$. The results are shown in Figure 6. The optical depth makes a rather abrupt jump from a value near zero to its final value $\tau(x, p, -\infty) = 1$, the transition occurring within a geometrical distance of only 0.1 r_c. Outside of the transition zone, the atmosphere is transparent to radiation of the given frequency. The sharpness of transition is due to the narrowness of the line profile, which falls off on a scale of atomic Doppler widths Δ. As seen from Equation (17), the profile function peaks when

$$x = u_z$$

where $u_z = u\mu$ is the z component of the velocity distribution. Expressing μ in terms of p and z, this becomes

$$x - u[(p^2 + z^2)^{1/2}]\, z/(p^2 + z^2)^{1/2} = 0. \tag{20}$$

For a given value of p, that value of z which satisfies Equation (20) is denoted z_0. In the specific example considered, $z_0 = 1.655\, r_c$, and as shown in Figure 6, z_0 lies in the middle of the transition zone. When Equation (20) is solved, for different values of p but a given frequency x, the roots $z_0(p)$ define a surface in the expanding envelope, such that the atmosphere is transparent except within a small region about this surface. The surface may be termed a surface of constant (line of sight) velocity, since u_z is constant there. In the remainder of this paper, the subscript zero refers to a variable evaluated on a surface of constant velocity: thus, $r_0 = (z_0^2 + p^2)^{1/2}$. Different surfaces correspond to different frequency displacements x, and a number of such surfaces is shown in cross section in Figure 7 for the velocity distribution (10). These surfaces, distributed over all r, contribute to the observed spectral line.

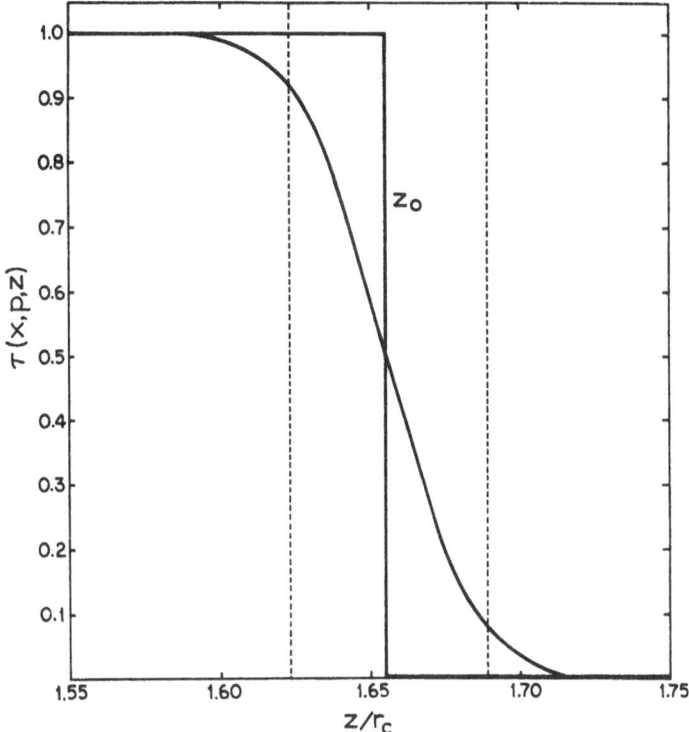

Fig. 6. Optical depth along a ray $p = $ const at a fixed frequency x and normalized such that the total optical thickness across the ray is unity. The step function approximation, with a jump at z_0 on the constant velocity surface is also shown. The dashed lines correspond to the locations of the constant velocity surfaces at $x \pm \delta$.

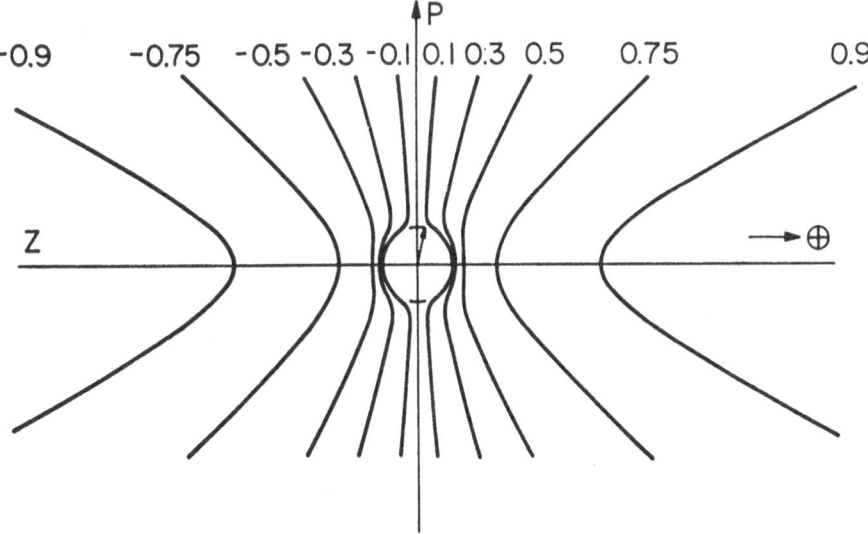

Fig. 7. Cross section of the constant velocity surfaces in a star with a velocity distribution as given by Equation (10). From Castor (*Monthly Notices Roy. Astron. Soc.* **149**, 112).

3.2. ESCAPE PROBABILITY METHOD

In a rapidly expanding atmosphere only a geometrically narrow zone is non-transparent to radiation of a given frequency, so that a transfer problem exists only in this region. The abruptness of the jump in opacity along a ray at a constant velocity surface may be exploited to yield an approximate solution of the transfer equation. In the limit of vanishing profile width, $\delta \to 0$, the optical depth approaches a step function

$$\tau(x, p, z) = \tau(x, p) \, y(z) \tag{21}$$

where $\tau(x, p)$ is an abbreviation for $\tau(x, p, -\infty)$, the total optical depth along the ray $p = $ constant, and

$$y(z) = \begin{cases} 1 & z < z_0 \\ 0 & z > z_0 \end{cases} \tag{22}$$

We now assume this limiting case applies for profiles of finite width, and evaluate $\tau(x, p)$. Because the approximation involves infinitesimally narrow lines the actual profile does not enter. Sobolev (1958) assumes $\phi(x)$ to be rectangular, while Castor (1970) does not specify a line shape; both find easily that

$$\tau(x, p) = k(r_0) \left[\left(\frac{\partial}{\partial z} \right)_{x, p} u_z \right]^{-1} z_0$$

$$= k(r_0) \left[1 + \frac{z_0^2}{r_0^2} \left(\frac{d \ln u}{d \ln r} - 1 \right) \right]^{-1} r_0 / u(r_0). \tag{23}$$

The absorption coefficient for a line transition between levels $n = 1$ and $n = 2$ is

$$k = \frac{\pi e^2}{mc} (gf)_{12} \left(\frac{N_1}{g_1} - \frac{N_2}{g_2} \right) \left(\frac{c}{v_0 v_\infty} \right) \tag{24}$$

where f_{12} is the oscillator strength, g_1 and g_2 are the statistical weights and N_1 and N_2 are the population densities of levels $n = 1$ and $n = 2$ respectively. The last term in parentheses converts k to the dimensionless frequency scale x. It should be remarked that $\tau(x, p)$ can be very large in spite of the geometrical narrowness of the line forming region.

The equation of transfer along a ray $p = $ constant takes the familiar form

$$-\frac{\partial I(x, p, z)}{\partial z} = k(r) \, \phi(x - \mu u) \left[I(x, p, z) - S(r) \right] \tag{25}$$

where the source function is assumed to be independent of frequency and isotropic in all frames. The calculation of the intensity must take into account the two possible lines of sight in the atmosphere: (i) rays with $p < r_c$ which encounter the core and (ii) rays with $p > r_c$ in which case the core is bypassed. We shall assume that the core radiates only a continuous spectrum I_c, which for the sake of simplicity, is assumed to be independent of v and p, i.e. there is no limb darkening.

Expressions for $I(x, p, z)$ are given by Castor (1970) and as would be expected reflect the step function behavior of the opacity. In particular, the emergent intensity $I(x, p, \infty)$ is

$$I(x, p, \infty) = \begin{cases} S(r_0)(1 - \exp[-\tau(x, p)]) & (p > r_c) \\ S(r_0)(1 - \exp[-\tau(x, p)\,y(z_c)]) + \\ \quad + I_c \exp[-\tau(x, p)\,y(z_c)] & (p < r_c) \end{cases} \tag{26}$$

where

$$z_c = (r_c^2 - p^2)^{1/2}. \tag{27}$$

The terms involving the step function $y(z_c)$ are simply interpreted. Radiation of frequency $x>0$ (shortward of line center) originates in the hemisphere nearest the observer, so that $z_0 > z_c$ and $y(z_c) = 1$. Radiation longward of line center, $x<0$, originates in the far hemisphere where $z_0 < z_c$ and $y(z_c) = 0$. Thus, for $p < r_c$

$$I(x > 0, p, \infty) = S(r_0)(1 - \exp[-\tau(x, p)]) + I_c \exp[-\tau(x, p)] \tag{28}$$
$$I(x < 0, p, \infty) = I_c.$$

The intensity shortward of line center is composed of a part emitted by the envelope proportional to $S(r)$, and a part emitted by the core, proportional to I_c. The continuous radiation is absorbed in the region of the atmosphere directly between the observer and the core. This accounts for the absorption component which often appears in the violet wing of Wolf-Rayet emission lines. In the expression for the intensity of radiation longward of line center, the envelope emission term does not appear. This is due to the occultation by the core of that region of the far hemisphere which lies directly behind it. Equations (26) therefore describe the basic physical processes which effect the emission line profiles in expanding atmospheres: envelope emission, absorption by material in front of the core, and occultation of material behind it. The power emitted by the star, per unit frequency interval, is then

$$F_x = 4\pi \int_0^\infty 2\pi p \; dp I(x, p, \infty). \tag{29}$$

Let us consider a particular atomic transition which takes place between levels 1 (lower) and 2 (upper). The source function is (e.g. Avrett and Hummer, 1964)

$$S = \frac{N_2 A_{21}}{N_1 B_{12} - N_2 B_{21}} \tag{30}$$

where N_1 and N_2 are population densities and the other terms are the Einstein rate coefficients. The source function is assumed to be isotropic and independent of frequency. Magnon (1968) has indicated that these assumptions which appear to give accurate results in stationary atmospheres, may be less satisfactory in moving atmospheres. However, there is little evidence to corroborate this assertion, and we will continue to use Equation (30) for the line source function.

The rate of downward transitions $2 \rightarrow 1$ is given by

$$R_{21} = N_2 A_{21}$$

while the rate of upward transitions (diminished by stimulated emission) is

$$R_{12} = (N_1 B_{12} - N_2 B_{21}) \, \bar{J}.$$

The term $\bar{J}$ represents the intensity of radiation integrated over the line profile and averaged over angle:

$$\bar{J} = \tfrac{1}{2} \int_{-1}^{1} \mathrm{d}\mu \int_{-\infty}^{\infty} \phi(x - u\mu) \, I(x, p, z) \, \mathrm{d}x. \tag{31}$$

Because $\phi(x - u\mu)$ is zero except for a narrow range of frequencies about $x = u\mu$, a photon of frequency x can be absorbed only when the equality is satisfied. This, however, defines the constant velocity surface for frequency x, from which all the envelope emission at that frequency arises. Therefore, (apart from the core radiation) a photon which is absorbed at a point (p, z) must have been emitted at that point. Expressed differently, a photon emitted in the envelope is either reabsorbed at the same spot or escapes from the region entirely. Let β be the escape probability, i.e. the fraction of photons emitted in the transition $2 \rightarrow 1$ which escape. A fraction $(1 - \beta)$ are therefore reabsorbed, so

$$R_{12} = (1 - \beta) \, R_{21}$$

or

$$(N_1 B_{12} - N_2 B_{21}) \, \bar{J} = (1 - \beta) \, N_2 A_{21}$$

With Equation (30) for the source function, this becomes

$$\bar{J} = (1 - \beta) \, S. \tag{32}$$

The mean intensity $\bar{J}$ may be formally evaluated by substituting the intensities found previously with the narrow line approximation into Equation (31). This has been done by both Sobolev (1958) and Castor (1970) with the result that

$$\beta = \int_{0}^{1} \mathrm{d}\mu \, [1 - \exp(-\tau(x, p))]/\tau(x, p). \tag{33}$$

We have ignored, however, the upward transitions caused by absorption of the continuous core radiation. When this is included, Equation (32) is replaced by

$$\bar{J} = (1 - \beta) \, S + \beta_c I_c \tag{34}$$

where, as Castor (1970) shows, if $\mu_c = (1 - r_c^2/r^2)^{1/2}$,

$$\beta_c = \tfrac{1}{2} \int_{\mu_c}^{1} \mathrm{d}\mu \, [1 - \exp(-\tau(x, p))]/\tau(x, p). \tag{35}$$

To a fair approximation,

$$\beta_c = W\beta \tag{36}$$

where

$$W = \tfrac{1}{2}\left(1 - (1 - r_c^2/r^2)\right)^{1/2} \tag{37}$$

is the usual dilution factor.

3.3. THE TWO LEVEL ATOM

In order to calculate the emergent intensity of radiation in a spectral line from Equation (26), it is necessary to know the run of the line source function. Because of the difficulty in determining $S(r)$, workers attempting to describe real atmospheres relied on dubious assumptions. Rublev (1963) for example took $S(r) = \text{const}$, while Van Lyong (1967) assumed $S(r) = S_0(r_c/r)^t$. There is one case, however, in which the source function can be obtained in a more reliable manner; that is for an atmosphere composed of atoms with only two bound levels. It is worth studying the two level atom for this reason, although as is often the case, application to Wolf-Rayet stars is not immediate.

The source function for the two level atom, in the case of complete frequency redistribution, is given by Equation (30). This is combined with the equation of statistical equilibrium,

$$(B_{12}\bar{J} + C_{12})\, N_1 = (A_{21} + B_{21}\bar{J} + C_{21})\, N_2 \tag{38}$$

to yield

$$S = (1 - \varepsilon)\,\bar{J} + \varepsilon B(T) \tag{39}$$

where

$$\varepsilon = C_{21}/[C_{21} + A_{21}(1 - \exp(-h\nu_0/kT))^{-1}] \tag{40}$$

is the probability per scattering that a photon is lost from the line by a collisional de-excitation of the excited state, and $B(T)$ is the Planck function at line center frequency ν_0. Equations (34) and (39) yield

$$S = \frac{(1 - \varepsilon)\,\beta_c I_c + \varepsilon B(T)}{(1 - \varepsilon)\,\beta + \varepsilon} \tag{41}$$

which gives the source function in terms of known variables.

With a view towards applying this result to actual model atmospheres there is a disconcertingly large number of variables which must be supplied. Runs of temperature and population densities are required to fix ε, $B(T)$ and the absorption coefficient $k(r)$. The model which calls for fewest external parameters is one in which pure line scattering ($\varepsilon = 0$) occurs. Nevertheless, this simple case is one of considerable current interest in the interpretation of emission lines from quasi-stellar objects.

Scargle, Caroff, and Noerdlinger (1970) suggest that the profile of the C IV resonance line λ 1548.2 in PHL 5200 ($z = 1.98$), which is observed to have an abrubt and deep absorption trough shortward of line center, is formed by scattering in a very rapidly expanding envelope ($v_\infty = 10000$ km s^{-1}) surrounding a continuum emitting core. The model is thus precisely the same as ours for Wolf-Rayet stars. Let us therefore employ the escape probability method to this case, and use the physical model given by Lucy (1971).

The velocity distribution (10) is assumed along with the equation of continuity:

$$-\frac{dM}{dt} = 4\pi r^2 \varrho v \qquad (42)$$

where $-dM/dt = 2M_\odot$ yr^{-1} is the rate of mass loss. Carbon is postulated to have the cosmic abundance and 10% of the atoms are C IV throughout the flow. The core radius $r_c = 5$ pc and $v_\infty = 10000$ km s^{-1}. In terms of atomic constants

$$k(r) = \frac{\pi e^2}{mc} f N$$

where $f = 0.2$ is the oscillator strength (Allen, 1955) and N is the number density of C IV ions. From Equations (23) and (24) the total optical depth along a ray $p = $ constant is

$$\tau(v, p) = \tau_0(r_0) / \left[1 + \frac{z_0^2}{r_0^2}\left(\frac{d\ln v}{d\ln r} - 1\right)\right]_{r_0}. \qquad (43)$$

where

$$\tau_0(r) = \frac{\pi e^2}{mc} f N c r / v_0 v(r). \qquad (44)$$

The quantity $\tau_0(r)$ is a convenient measure of the distribution of scatterers.

With the above assumptions, it is found that $\tau_0(1.1r_c) = 8.0$ and decreases monotonically, so that, for example $\tau_0(5r_c) = 0.2$. Figure 8 shows the flux F_v/F_c computed by the escape probability method. The computed profile does not resemble the observed profile in that (i) the central intensity is only 1.5 the continuum value while it is observed to be nearly 4 times the continuum intensity and (ii) the absorption is far too gradual compared to the observed absorption which is nearly complete just shortward of line center. In this, we agree exacly with Lucy (1971). Other choices of physical parameters do not change this result significantly, and apparently the CIV resonance line is not formed by scattering in a spherically symmetric, expanding envelope.*

Perhaps the most informative calculation that can be made for Wolf-Rayet stars is one that demonstrates the wide range of profiles possible from a moving atmosphere. As Rublev (1963) and Castor (1970) have shown, these encompass all the line shapes observed on Wolf-Rayet spectra. In Castor's computations, ε and $B(T)$ are assumed to be constant throughout the envelope and different opacity distributions are chosen.

---

* Recently, Caroff, Noerdlinger, and Scargle (1971, preprint) argue that the observed central intensity can be obtained with a very different model than that considered above. In their model, the parameter $Q_0 = \tau_0(r)/d\ln v/d\ln r)$ goes suddenly to zero as the core is approached, while in the above model Q_0 remains finite.

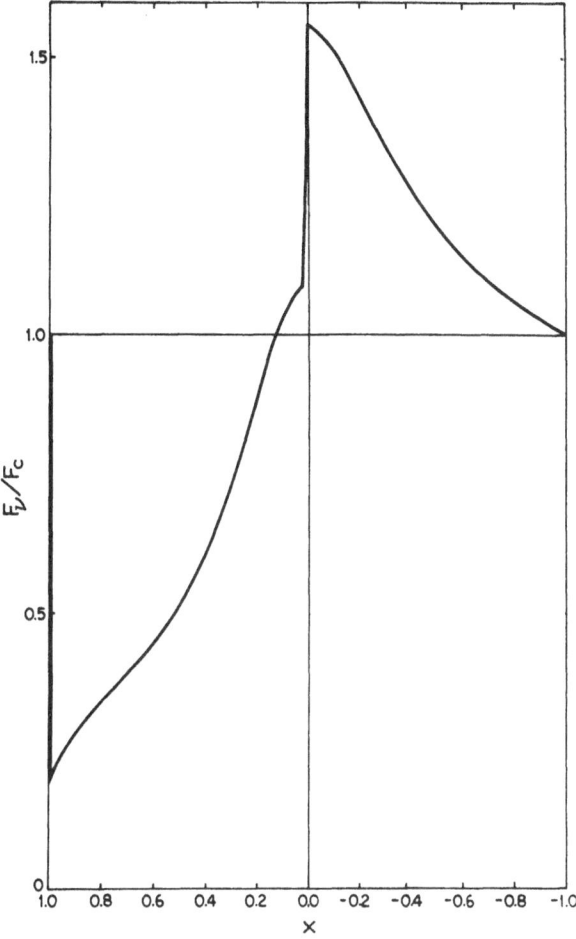

Fig. 8. Spectral line produced by a pure scattering expanding envelope with the parameters appropriate to the C IV resonance line $\lambda 1548.2$ in PHL 5200. The theoretically generated profile does not resemble the observed profile and casts doubt on this interpretation of the line.

Figure 9 given one example of a resulting line profile. It demonstrates that the rounded appearance of most Wolf-Rayet emission lines does not perforce rule out their formation in an expanding atmosphere. In particular, we do not require a turbulent region with random velocities in excess of 1000 km s^{-1} as suggested by Underhill (1968).

Nevertheless, one possible flaw in the theory should be pointed out. The optical depth parameter $\tau_0(r)$ used in Figure 9 has been chosen to be small very near the core, when $v(r) \leqslant 0.2v_\infty$. If, however, $\tau_0(r)$ is a monotonically increasing function of r as $r \to r_c$, e.g. in the quasi-stellar model treated above, a profile quite asymmetric at line center results. This is evident in Figure 8 for the quasi-stellar line. No line in a Wolf-Rayet star has this appearance, and one must conclude either that (i) the two level atom and escape probability method fail to present a physically accurate picture of line formation near the core or (ii) the density of absorbing atoms or ions vanishes near the core. If the temperature increases monotonically as $r \to r_c$ the atoms in the highest

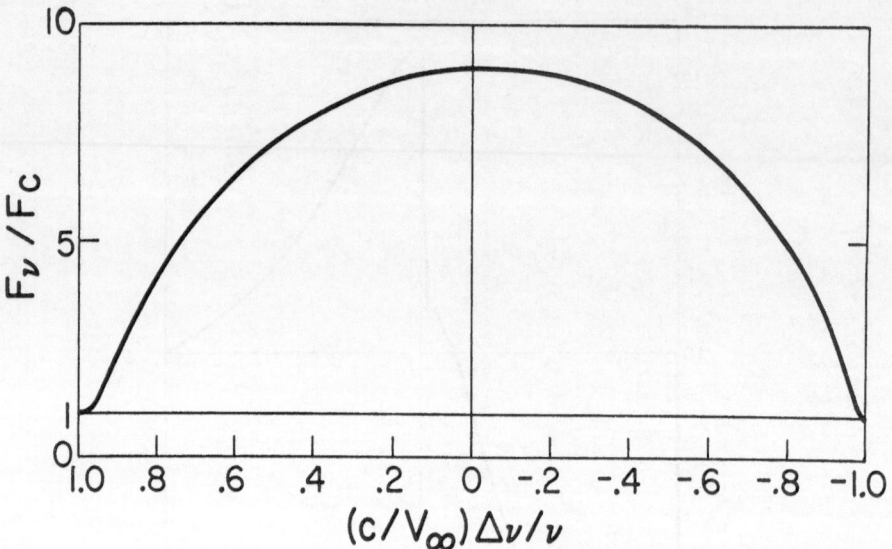

Fig. 9. A representative line shape of the rounded type. The parameters are $\varepsilon = 0.021$ and $B(T)/I_c = 0.105$. From Castor (*Monthly Notices Roy. Astron. Soc.* **149**, 124).

stages of ionization should exist in the vicinity of the core. The lack of strongly asymmetric lines from these ions suggests, if the theory is not itself at fault, that the temperature rise inward is not monotonic. One may recall the speculation at the end of Section I that a temperature inversion occurs between the continuum and line forming regions. It is then conceivable that the ionization equilibrium in a narrow zone about the core shifts to low stages of ionization. The asymmetries in the lines produced in this part of the envelope are masked by the emission from the more extensive sections of similar ionization further out.

4. Application to Wolf-Rayet Stars

4.1. Laser action in a Ciii line

The theory of line formation in a moving atmosphere developed in Part III has so far been merely descriptive. We have shown that the commonly observed Wolf-Rayet line shapes can be reproduced by an appropriate choice of parameters. A more useful diagnostic approach is the inverse, i.e., to deduce the physical parameters from the observed spectrum. Rublev (1963) and Van Lyong (1966) have attempted to do this, and derive a velocity distribution and a limb darkening law for the continuous radiation. The analysis is too detailed to be reproduced here, however the results should be regarded with some caution due to the unsatisfactory choice of source function mentioned previously. Nevertheless, their investigations show the type of information which may be obtained from a detailed comparison of line profiles with theoretical predictions.

A somewhat less ambitious project is to attempt to account for the total energy in a line rather than its shape. West (1968) has reported some preliminary calculations of this nature for the C_{III} intercombination line $\lambda 1909$. The line is observed to be rather strong in γ Vel, i.e. if

$$R = \int_{-1}^{1} (F_v/F_c)\,dx$$

is the ratio of the total line to continuum flux, $R \approx 4$. Because the oscillator strength is very low, $gf = 3.1 \times 10^{-7}$ (Garstang and Shamey, 1968), there are intimations of laser action. West's results appear to require population inversions with amplification $(\tau_0 < 0)$ if the star has a normal carbon abundance and if the total density at the base of the envelope is not to exceed 10^{12} cm^{-3}. However, for reasons not clear to this reviewer, West limited the line emission to only that part of the envelope directly between the observer and the core, thus removing the greater part of the emitting volume. Only shortward displaced radiation is produced in this region and the total flux in the line cannot therefore be found. West could compare only the line center intensity ratio F_{v_0}/F_c with R, which is of questionable value.

Let us therefore consider the problem in the context of the theory developed here. The core radius is taken to be $r_c = 5R_\odot$, and the C_{III} line emitting region in which carbon is all C^{+2}, is assumed to lie within $5r_c$. The density decreases outward from the core as r^{-2}, and at $r = 5r_c$, $v(r) = 1500$ km s^{-1}. The continuous radiation field is that appropriate to a 40000 K main sequence model, or $I_c = 4.78 \times 10^{-3}$ ergs cm^2/s/Hz. This is West's model, except the velocity distribution (10) is used in place of his linear law.

The population ratio N_2/N_1 of the upper to lower levels of the line is postulated, as is the total density $N_1 + N_2$ at the base of the envelope. The latter is chosen such that the corresponding hydrogen density at r_c would be in the neighbourhood of 10^{12} cm^{-3} if the cosmic abundances of the elements were present. With N_1 and N_2 specified, the source function follows from Equation (30) and the optical depth, including stimulated emission is

$$\tau_0(r) = \frac{\pi e^2}{mc}(gf)_{12}\left(\frac{N_1}{g_1} - \frac{N_2}{g_2}\right)\frac{r\lambda_0}{v(r)}. \tag{45}$$

Note that S is independent of r since the ratio N_2/N_1 is assumed constant within the emitting region. Equations (26) yield the emergent intensity $I(v, p, \infty)/I_c$ and quadratures over p and v give the desired ratio R.

Figure 10 shows R as a function of N_2/N_1 for several values of $N_1 + N_2$. The existence of laser action is neither confirmed nor denied by these results. Apparently, all that can be stated is a lower limit on the density, since even with amplification $R = 4$ is not attainable for $N_1 + N_2 = 1 \times 10^8$ cm^{-3}.

The problem is certainly that too much information is being wrested from one line. The C_{III} spectrum of γ Vel is very fully developed, however, and a more reliable analysis would incorporate many lines. This has only just now been done by Castor and Nussbaumer (1971). We discuss this work in more detail later, and simply note

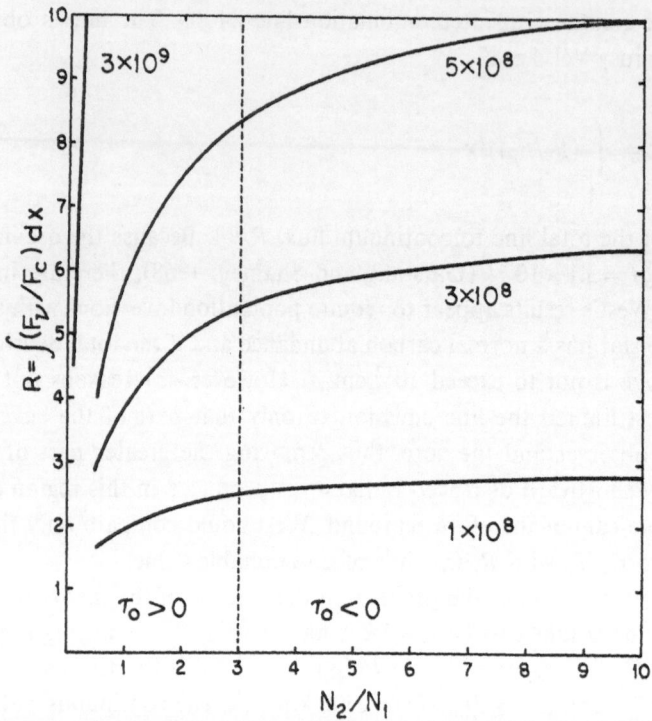

Fig. 10. Ratio of total line to continuous flux in the C III intercombination line λ1909. The curves are generated using the escape probability method and West's model of γ Vel. Laser action is suggested only in the low density models.

here that the lowest five levels of C^{+2} are found to be nearly in LTE, at an electron temperature $T_e = 2.2 \times 10^4$ K. Thus, the population ratio, $N_2/N_1 \approx 0.1$, is thirty times less than that required for laser action. If this ratio is used in West's model, a total density $N_1 + N_2 \approx 3 \times 10^9$ cm^{-3} is necessary in order that $R \approx 4$. The population density of C^{+2} ions is approximately the sum of the ground state densities of the singlets and triplets. Since the term statistical weight g_2^t of the state $2p^3P^0$ is 9, while the statistical weight of the upper level of $\lambda 1909$ is 3, we have $N(C^{+2}) \approx N_1 + 3N_2$. The density of C^{+2} ions at the base of the envelope is therefore around 3.5×10^9 cm^{-3}. Castor and Nussbaumer find $N_e = 4 \times 10^{11}$ cm^{-3} at $r = 3.6r_c$. This corresponds to a density $N_e \approx 5.2 \times 10^{12}$ cm^{-3} at the base of the envelope when an r^{-2} dependence is assumed. If the electron density is due entirely to hydrogen ionization, the ratio $N(C)/N(H)$ at $r = r_c$ is roughly 6×10^{-4}, or twice the cosmic abundance ratio.

4.2. He II LINES, A COARSE ANALYSIS

Because of the prominence of He II lines in the spectra of Wolf-Rayet stars, an analysis of these lines should be a strong test for any theory of line formation. In attempting to extract physical information from the line spectrum one encounters the usual 'inverse problem' in radiative transfer, viz. to determine the source function from the emergent flux. The following treatment is in the spirit of a coarse analysis.

We consider a given line transition for which the frequency independent parts of the emissivity and absorptivity are $j(r)$ and $k(r)$ respectively. Since the emission is assumed to be isotropic, and a fraction $\beta(r)$ of the energy released locally escapes we find

$$E = 4\pi \int j(r)\,\beta(r)\,dV \tag{46}$$

for the power emitted by the whole atmosphere in the line. If the angle dependent terms in Equation (33) for β are ignored, then

$$\beta(r) = (1 - \exp[-\tau_0(r)])/\tau_0(r). \tag{47}$$

The optical depth is related to $k(r)$ through the equation

$$\tau_0(r) = \frac{cr}{v_0 v(r)}\,k(r) \tag{48}$$

and by definition, $j(r) = k(r)S(r)$.
Equation (46) thus becomes

$$E = (8\pi^2 v_0/c) \int_0^\infty S(r)\,v(r)\,(1 - \exp[-\tau_0(r)])\,dr^2. \tag{49}$$

As a rough approximation it is assumed that the integrand is constant within a line emitting region of radius r_E, and zero outside of it. Then

$$E \approx (8\pi^2 v_0/c)\,r_E^2 v(r_E)\,S(r_E)\,(1 - \exp[-\tau_0(r_E)]). \tag{50}$$

It should be noted that Equation (50) ignores any occultation or absorption of continuum radiation. A quantity more likely to be tabulated by an observer is the equivalent width W_λ. This may be found by dividing Equation (50) by the continuous flux in wavelength units at the position of the line

$$F_c = 4\pi^2 r_c^2 I_c c/\lambda_0^2$$

where, as used previously, I_c is in frequency units and c/λ_0^2 effects the conversion. Then

$$W_\lambda = 2\,(r_E/r_c)^2\,(\lambda_0 v(r_E)/c)\,S(r_E)\,(1 - \exp[-\tau_0(r_E)])/I_c. \tag{51}$$

We may examine the type of errors associated with the use of Equation (50) or (51) by calculating E or W_λ exactly for a model atmosphere in which all processes are included. West's model for the C_{III} emission in γ Vel, which was discussed in the preceding section, will serve this purpose. Accordingly we take $r_E = 5r_c$ and $v(r_E) = $ $= 1500$ km s^{-1}. The source function depends on the ratio N_2/N_1. This is assumed to have its LTE value at 22000 K, so $N_2/N_1 = 0.1$ and $S = 0.41$. All the remaining parameters are determined once $N_1 + N_2$ is specified. Table II gives $\tau_0(r_E)$, W_λ (exact), and the ratio of the approximate to exact equivalent widths.

TABLE II

Comparison of approximate and exact equivalent widths for a
simple W–R model

$N_1 + N_2$	τ_0	W_λ(exact)	W_λ/W_λ(exact)
1×10^9	0.064	14.8	0.826
3×10^9	0.192	38.0	0.905
5×10^9	0.320	55.2	0.977
7×10^9	0.448	68.1	1.043
1×10^{10}	0.640	82.1	1.133
5×10^{10}	3.200	116.8	1.615

The agreement between the exact and approximate calculations is very good, and in fact better than one might have anticipated.

Following Castor and Van Blerkom (1970) this coarse analysis is applied to the He II lines in the spectrum of the WN 6 star HD 192163. Relative line intensities $E_{n, n'}/E_{4, 3}$ are taken from the observations of Smith and Kuhi (1970). Of particular importance are the $\lambda 4686$ $(4 \rightarrow 3)$, $\lambda 3203$ $(5 \rightarrow 3)$ and $\lambda 10124$ $(5 \rightarrow 4)$ lines since they couple the levels $n = 3, 4$ and 5. It is assumed, and later shown to be consistent, that these three lines are optically thick, so that $(1 - e^{-\tau_0}) \sim 1$ and

$$E_{n, n'}/E_{4, 3} = v_{n, n'} S_{n, n'}/v_{4, 3} S_{4, 3}. \tag{52}$$

The source function is given by Equation (30), which may be written as

$$S_{n, n'} = \frac{2hv_{n, n'}^3}{c^2} \left[\frac{N_{n'}/g_{n'}}{N_n/g_n} - 1 \right]^{-1}. \tag{53}$$

Equations (52) and (53) determine the ratios $N_3 g_4/N_4 g_3 = 4.65$ and $N_4 g_5/N_5 g_4 = 4.04$.

For levels $n > 5$, the lines in the Pickering series are used. Then if $y_n = N_n g_4/N_4 g_n$

$$E_{n, 4}/E_{4, 3} = \left(\frac{v_{n, 4}}{v_{4, 3}} \right)^4 \frac{y_3 - 1}{1/y_n - 1} [1 - \exp(-\tau_{4, n})]. \tag{54}$$

The optical depths are not known a priori, and must be assumed. A scale factor A is defined such that

$$\tau_{4, n} = A \frac{N_4/g_4 - N_n/g_n}{N_4/g_4} (gf\lambda)_{4, n} = A (1 - y_n) (gf\lambda)_{4, n} \tag{55}$$

where

$$A = \frac{\pi e^2}{mc} \frac{N_4/g_4}{v(r_E)} r_E. \tag{56}$$

Equations (54) and (55) are solved for y_n for several values of A. The resulting population ratios are shown in Figure 11. One must now decide which of the curves in the figure are physically meaningful.

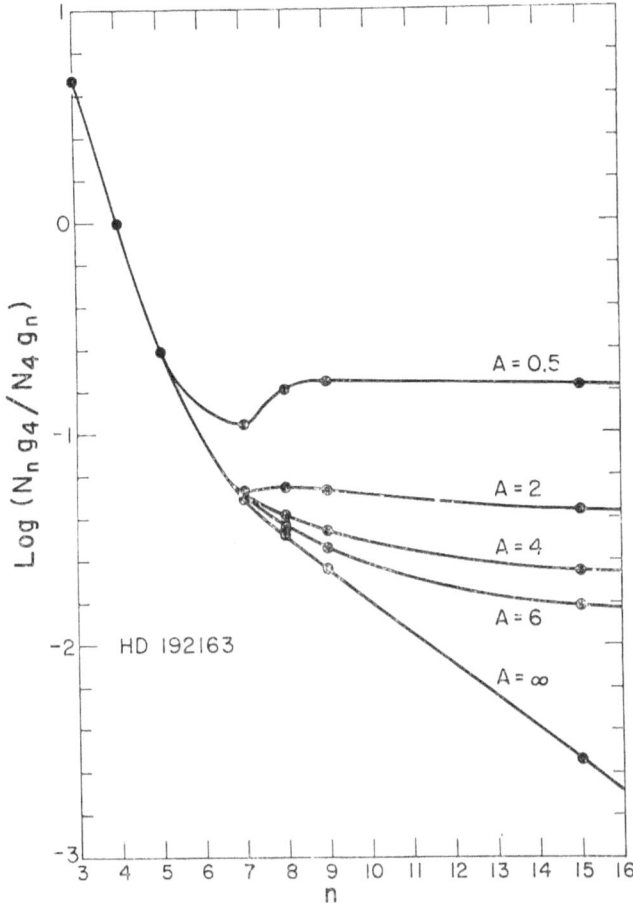

Fig. 11. Relative atomic energy level populations of He II as derived by a coarse analysis of the WN 6 star HD 192163. Physically meaningful behavior is shown by curves with $3 \leqslant A \leqslant 6$.

Electron-ion collisions act to bring the level populations to their LTE values N_n^* at the local electron temperature, where

$$N_n^*/g_n = \frac{1}{2}\left(\frac{h^2}{2\pi m k T_e}\right)^{3/2} \exp\left(h v_n/k T_e\right) N_e N_i \qquad (57)$$

and $h v_n$ is the ionization potential for an atom in level n. Griem (1963) has derived a useful formula to determine those levels which are nearly in LTE by comparing the total collisional upward rate out of level with the sum of the radiative downward rates. If the collisional rate exceeds the radiative decay by a factor of 10, the population should be within 10% of its LTE value. According to Griem, this condition is met if $n \geqslant n_0$, where

$$n_0^{17/2} = \frac{7.4 \times 10^{18} z^6}{N_e}\left(\frac{k T_e}{E_H}\right)^{1/2}. \qquad (58)$$

E_H is the ionization potential of hydrogen, and $z=2$ for He II. Taking representative values of $T_e = 1 \times 10^5$ K and $N_e = 10^{12}$ cm^{-3} for a W–R envelope, we find $n_0 \approx 10$. Thus, collisional processes strongly dominate over radiative processes for $n \geqslant 10$. For such levels $\exp(h\nu_n/kT_e) \approx 1$ for any reasonable envelope temperature. Thus, N_n/g_n will be essentially constant for $n > 10$. In Figure 11 the curves which satisfy this condition, along with the requirement that no population inversions occur, lie in the range $3 \lesssim A \lesssim 6\,\mu^{-1}$.

If we take $A = 4$, for example, the Table III gives the resulting values of y_n, $\tau_{4,n}$ and T_{ex}, where

$$y_n = \exp\left(- h\nu_{n,4}/kT_{ex}\right) \tag{59}$$

defines the excitation temperature of each line.

TABLE III

Results of a coarse analysis of the WN 6 star HD 192163

n	$N_n g_4/N_4 g_n$	$\tau_{4,\,n}$	T_{ex}	$\log_{10} b_n$
5	0.247	1.01×10^2	1.02×10^4	0.908
6	0.136	1.30×10^1	1.10	0.711
7	0.048	4.32×10^0	0.88	0.306
8	0.041	1.93×10^0	0.92	0.256
9	0.035	1.05×10^0	0.94	0.205
15	0.0195	1.31×10^{-1}	0.93	0.008

The excitation temperatures are about 5 to 10 times lower than the electron temperature T_e one would expect on the basis of the observed ionization of the atmosphere. If b_n is in the usual departure coefficient, $N_n = b_n N_n^*$, then one may show that

$$\log_{10} \frac{b_4}{b_n} = \frac{0.625}{\lambda_{4,\,n}} \left(T_{ex}^{-1} - T_e^{-1}\right). \tag{60}$$

Since we have argued that $b_{15} \approx 1$, Equation (60) may be used to find b_4, and subsequently the other values of b_n. With $T_e = 5 \times 10^4$ K as a reasonable guess of T_e, we obtain $\log_{10} b_4 = 1.39$. The final column of the table lists $\log_{10} b_n$ for $n > 4$.

We may obtain information of the extent of the emitting region from knowledge of the line source functions. From Equations (26) and (29), the flux in a line relative to that in the underlying continuum may be obtained. If the angle factor in $\tau(\nu, p)$ is ignored and the product $S(r)\,(1 - \exp[-\tau_0(r)])$ is assumed constant for $r < r_E$ and zero for $r > r_E$, as in the derivation of Equation (50), then

$$(F_\nu - F_c)/F_c = \left(\frac{r_E}{r_c}\right)^2 \frac{S(r_E)}{I_c} \left(1 - \exp[-\tau_0(r_E)]\right). \tag{61}$$

Read off the spectrogram, $(F_\nu - F_c)/F_c \approx 8$ for $\lambda 4686$, and since the line is optically thick, the exponential term vanishes. For the continuous radiation field, $I_c = B_\nu$

(40000 K) is probably adequate, and not, at any rate, very critical to the argument. With $N_3 g_4 / N_4 g_3 = 4.65$ as computed previously, $S(\lambda 4686)/I_c(\lambda 4686) = 0.32$, and $r_E = 5 r_c$. This is of sufficient size to account for the lack of occultation or absorption effects in the He II spectrum. It should also be pointed out that the large tabulated value of $\tau_{4,5}$ justifies the original assertion that $\lambda 10124$ is optically thick.

The coarse analysis of the He II lines from HD 192163 is thus seen to lead to a self-consistent model of the Wolf-Rayet envelope. Other tests, e.g., prediction of the strengths of lines other than those in the Pickering series, do not contradict the results obtained. Of course, a detailed description of the run of physical variables with depth is not produced. Only an idea of the state of the atmosphere at a 'typical' point in the emitting region derives from this approach. Nevertheless, the simplicity of the coarse analysis is its strength since almost any attempt to go beyond it involves one in far more demanding tasks.

4.2. SOLUTION OF THE MICROSCOPIC RATE EQUATION

The atomic level populations in a stellar atmosphere are determined by a myriad of competing microscopic processes. If we postulate that statistical equilibrium obtains, then the level populations have reached a steady state such that

$$dN_n/dt = R_n + C_n = 0. \tag{62}$$

The terms R_n and C_n are the net rates at which level n is populated by radiative and collisional processes, respectively. The solution of Equation (62) for N_n is a task which is tedious at best and hopeless at worst. Atomic parameters, such as the collisional excitation and ionization cross sections, are poorly known even for the hydrogen atom. Moreover, the transfer equation for every transition must be solved iteratively with the statistical equilibrium equations in order to evaluate the radiative rates.

Fortunately the existence of a radial outflow of matter at high velocity makes the microscopic rate approach more tractable than would be the case for a stationary atmosphere. This is because the radiation field in every line transition can be found by the escape probability method in terms of only local values of the physical parameters. The rub is that radiation in the bound-free continuum is only slightly affected by the presence of a large velocity field. Nevertheless, if one is not too scrupulous in handling the transfer problems in the continua, a start at least can be made on the microscopic determination of the level populations.

The first such calculation was done by Castor and Van Blerkom (1970) for He II in the WN 6 star HD 192163, for which a coarse analysis already provided information on the state of the atmosphere. The rate equations are solved at one 'representative point' in the emitting region, chosen to lie midway between r_c and r_E at $r \approx 40 \, R_\odot$. The first 30 levels of the He II ion are considered and every collisional and radiative transition coupling these levels and the continuum are included. A crude approximation for the radiation fields in the continua is formulated so that they are treated in a manner analogous to the line escape probability method. The level populations are guessed initially, and then iterated until convergence to a desired accuracy is obtained.

The calculation is done for a grid of assumed values of helium density $N(\text{He})$ and electron temperature T_e at the representative point.

The suitability of the resulting models can be assayed most easily be computing the parameter A from Equation (56). Only a few choices of $N(\text{He})$ and T_e give values of A in the observed range. Although one unique set of parameters cannot be found, the model with $T_e = 1 \times 10^5$ K and $N(\text{He}) = 2.5 \times 10^{11}$ cm^{-3} agrees rather well with the results of the coarse analysis. Since T_e exceeds the core temperature, a non-radiative energy source seems to be required to maintain the excitation in the emitting region.

A similar calculation has been attempted for C^{+2} in γ Vel by Castor and Nussbaumer (1971). Again, the statistical equilibrium equations are solved at one representative point, $r = 3.6 r_c = 55 R_\odot$. The equations are solved for the 14 lowest terms of C^{+2}, with the escape probability method used for the line transitions. Because of a deficiency of atomic data only bound-bound radiative and collisional transitions could be included. Thus, the coupling of the levels to the continuum by collisions and radiative processes (including in this case dielectronic recombination) is ignored. Several model parameters are determined by fitting the theoretically derived intensities of the UV lines to observations. The range over which these parameters may vary and still yield good results is found to be fairly narrow. The best fit to the UV line equivalent widths give $T_e = 22000$ K, $N_e = 4 \times 10^{11}$ cm^{-3}, and $N(C^{+2}) = 1 \times 10^9$ cm^{-3}. If the electrons come mainly from hydrogen ionization, the abundance of carbon is about 8 times the cosmic value. Also, interestingly, T_e is significantly lower than the temperature of the core, which is assumed to radiate a continuous spectrum $I_c = B_\nu(30000 \text{ K})$. This suggests that the line emitting region may be in radiative equilibrium.

5. Electron Scattering

The optical depth due to electron scattering in a Wolf-Rayet atmosphere is estimated by taking the product $\sigma_e N_e L$, where σ_e is the Thompson cross section and N_e and L are typical electron densities and lengths. An optical depth $\tau_e \sim 0.5$ is obtained in this way. It is important to recognize that although Thompson scattering is coherent in the frame of the electron, it is noncoherent to an observer who sees the electron in thermal motion. Thus, electron scattering causes frequency redistribution which the escape probability method ignores. The inclusion of this effect is a refinement which is only now being incorporated in the theory. We first resort to a very approximate treatment of noncoherent electron scattering originally developed by Münch (1950).

In Münch's procedure, a plane-parallel layer of free electrons of thickness τ_e and temperature T_e is irradiated by line photons falling on its inner boundary. Since both atomic absorption and electron scattering occur together, the results obtained must be considered as only qualitative. The redistribution function depends on the approximation used to obtain it, with somewhat different functions given by Münch (1950), Hummer and Mihalas (1967) and Weymann (1970). We use the expression given by

Hummer and Mihalas:

$$R(x', x) = \binom{1}{w} \text{ierfc}\left(\frac{|x' - x|}{2w}\right) \tag{63}$$

where ierfc(z) is the integral of the complementary error function and w is the ratio of the electron Doppler width Δv_e to the width of the incident line Δv_L. The parameter x measures the frequency displacement from line center in units of Δv_L. For $\tau_e < 1$, the profile of the line after it emerges from the electron layer may be described by the approximate expression

$$\psi(x) = (1 - \tau_e) \phi(x) + \tau_e \int_{-\infty}^{\infty} \phi(x') R(x', x) \, dx' \tag{64}$$

where $\phi(x)$ is the profile of the line before scattering.

A study of a spectral line showing the possible influence of electron scattering has been carried out by Castor, Smith and Van Blerkom (1970) for $\lambda 3483$ of N IV in the WN 6 star HD 192163. This strong emission line shows a violet displaced absorption component, shortward of which is an extensive emission wing. Such behavior cannot be reproduced by a radial expansion model. When the incident profile $\phi(x)$ is of the normal P Cygni type, the emergent $\psi(x)$ matches the observed line very closely. In this case, the best fit is obtained for $\tau_e = 0.5$ and $w = 4$. This value of w, taken literally, would imply temperatures in excess of 5×10^5 K. However, the deficiencies of the model make any quantitative statement suspect.

Underhill (1968) has presented detailed tracings of the line spectrum of HD 191765. She shows, in particular, that the Pickering lines $\lambda 4200$ $(11 \rightarrow 4)$ and $\lambda 5411$ $(7 \rightarrow 4)$ are both nearly Gaussian in shape. If the results of the coarse analysis of HD 192163 can be taken to apply to this WN 6 star as well, then $\tau_{4,11} < 1$ and $\tau_{4,7} \sim 4$. One would expect different profiles for the two lines, with the thin line showing a flat-topped behavior. For the velocity distribution given by equation (10), most of the envelope is moving at nearly constant velocity. If it is assumed that $v(r) = $ constant throughout, the emergent profile can be found analytically. Sobolev (1958) shows the line to be flat-topped if the envelope is transparent and parabolic if it is opaque. These calculations ignore occultation and absorption, i.e., the core is reduced to a point. In order to see whether a parabolic line can be made to appear Gaussian by electron scattering, $\phi(x)$ is taken to be a parabola in Equation (64) and $\psi(x)$ computed for various values of τ_e and w. It is found that the core of the electron scattered line still is parabolic and changes discontinuously into a flat wing emission. No emergent line profile appears in the least Gaussian.

6. Work Following Symposium

Dr Lawrence Auer of Yale University and the author have investigated the effect of electron scattering in expanding atmospheres, in which electrons are distributed throughout the envelope and share in its expansion. Although still preliminary, it seems appropriate to comment here upon this work because of its relevance to WR stars. The model used is the usual one of a core and an expanding envelope, but the transfer of photons is treated by a Monte-Carlo technique. In this method, one photon at a time is followed from its point of emission to its final destination. Very many photons must processed in order that the result be statistically meaningful.

Let us first suppose that $w=0$ throughout the atmosphere, i.e. no *thermal* frequency redistribution occurs at a scattering event. There is, of course, a frequency redistribution due to the macroscopic motion of the electrons. If a photon is emitted in the atmosphere at a frequency displacement x relative to a stationary external observer, it may be scattered as it traverses the envelope and emerge with a frequency displacement x'. For any model in which $v(r)$ is constant or monotonically increasing, $x'<x$, i.e. on the average the photon is shifted to the red. This is due to the fact that all points of the envelope recede from each other and the net effect of scattering is to decrease the photon energy. Thus, an extensive red wing (extending beyond $x=-1$) is formed, with no violet wing at all.

If thermal redistribution is operative ($w>0$) energy is transferred from line center to both wings. Again, however, the red wing is far more extensive than the violet. The following simple example demonstrates these effects.

Let $v(r)=$ const for a fully ionized envelope in which the line in question is formed by recombination, i.e. the emissivity $j(r) \alpha N_e^2$. Since the product $r^2 N_e v$ is constant for steady outflow of matter, $j(r) \alpha r^{-4}$. If the envelope is *transparent* to the line radiation, the emergent profile, in the absence of electron scattering, is found by integrating the emission along surfaces of constant velocity (see, e.g., Rosseland, 1936). The profile is found to be

$$\phi(x) = \begin{cases} 0 & |x| > 1 \\ 1 & 0 < x < 1 \\ (1-x^2)^{1/2} & -1 < x < 0 \end{cases} \tag{65}$$

in arbitrary units. To the violet of line center, the line is flat-topped, while to the red it is diminished by occultation.

We now include the effect of electron scattering by stipulating the radial optical depth in electrons from the core to infinity, τ_e, and the degree of thermal redistribution w. The Monte-Carlo technique is employed in this case. For $\tau_e=0$, the profile given by Equation (65) should be recovered. Figure 12a shows that with 100000 photons processed, the two calculations agree almost exactly.

Figure 12b shows the profile for an electron optical depth $\tau_e=1$ and *no* thermal redistribution. As discussed above, there is a pronounced red wing on the line, but no wing on the violet side. Thermal redistribution is 'turned on' in Figure 12c. The parameter w is the ratio of the electron Doppler width to the half width of the un-scattered line profile, i.e.

$$w = \left(\frac{2kT}{m_e}\right)^{1/2} / v_\infty.$$

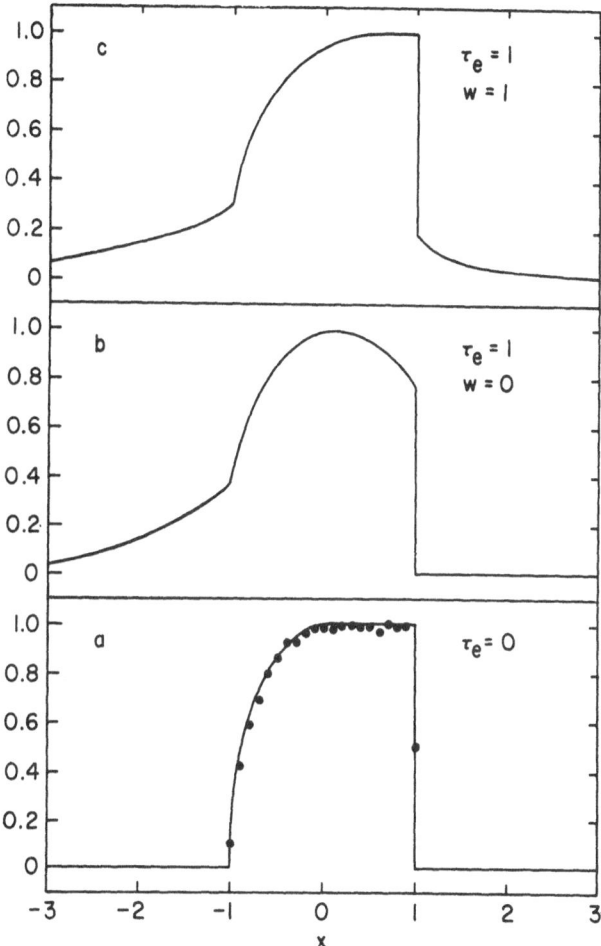

Fig. 12. Monte Carlo calculations of a line formed in an atmosphere expanding at constant velocity and having no line opacity. The exact profile is compared to the Monte Carlo result (filled circles) in Figure 12a. Electron scattering with and without thermal redistribution ($w=0$ and $w=1$) cause the profiles shown in Figures 12b and 12c respectively.

For $v_\infty \approx 1000$ km s^{-1} and $T \approx 5 \times 10^4$ K, $w \approx 1$. This case is shown in the figure. A violet wing is now apparent, and the red wing is somewhat enhanced over the $w=0$ case.

These results depend on the assumptions of constant outflow velocity, transparency of the envelope in the line and the specific line formation mechanism. A WR star would not be expected to satisfy any of these assumptions, so the line profiles shown in Figure 12 look nothing like observed WR lines. Nevertheless, the appearance of an extended red wing is a feature of electron scattering in an expanding atmosphere and is expected to show up when a more realistic model of line formation is employed. Because of the blending problem associated with lines of such widths, it is difficult to find an uncontaminated line in order to study its wing structure. A look at Dr. Underhill's line profiles (Underhill, 1968), however, does not show definite asymmetries which could be interpreted as due to electron scattering. Exactly what this signifies for WR stars is not yet certain. Perhaps estimates of electron densities in the envelopes are too high by nearly a factor of ten. Further study of this problem is under way.

References

Allen, C. W.: 1964, *Astrophysical Quantities*, The Athlone Press, London.
Avrett, E. H. and Hummer, D. G.: 1965, *Monthly Notices Roy. Astron. Soc.* **130**, 295.
Bappu, M. K. V. and Menzel, D. H.; 1954, *Astrophys. J.* **119**, 508.
Beals, C. S.: 1929, *Monthly Notices Roy. Astron. Soc.* **90**, 202.
Cassinelli, J. P.: 1971a, *Astrophys. J.* **165**, 265.
Cassinelli, J. P.: 1971b, *Astrophys. Letters* **8**, 105.
Castor, J. I.: 1970, *Monthly Notices Astron. Soc.* **149**, 111.
Castor, J. I. and Van Blerkom, D.: 1970, *Astrophys. J.* **161**, 485.
Castor, J. I., Smith, L. F., and Van Blerkom, D.: 1970, *Astrophys. J.* **159**, 1119.
Castor, J. I. and Nussbaumer, H.: 1971, preprint.
Chandrasekhar, S.: 1934, *Monthly Notices Roy. Astron. Soc.* **94**, 444.
Chandrasekhar, S.: 1934, *Monthly Notices Roy. Astron. Soc.* **94**, 522.
Code, A. D. and Bless, R. C.: 1954, *Astrophys. J.* **139**, 787.
Garstang, R. H. and Shamey, L. J.: 1967, *Astrophys. J.* **148**, 665.
Gerasimovic, B. P.: 1934, *Z. Astrophys.* **7**, 335.
Griem, H. R.: 1963, *Phys. Rev.* **131**, 1170.
Hummer, D. G. and Mihalas, D.: 1967, *Astrophys. J.* **150**, L57.
Hummer, D. G. and Rybicki, G. B.: 1968, *Astrophys. J.* **153**, L107.
Hummer, D. G. and Rybicki, G. B.: 1971, *Monthly Notices Roy. Astron. Soc.* **152**, 1.
Kosirev, N. A.: 1934, *Monthly Notices Roy. Astron. Soc.* **94**, 430.
Kuhi, L. V.: 1966, *Astrophys. J.* **145**, 715.
Kuhi, L. V.: 1968, in K. Gebbie and R. N. Thomas (eds.), *Wolf-Rayet Stars*, National Bureau of Standards Special Publication 307, p. 103.
Larson, R. B.: 1969, *Monthly Notices Roy. Astron. Soc.* **145**, 297.
Limber, D. N.: 1964, *Astrophys. J.* **139**, 1251.
Lucy, L. B.: 1971, *Astrophys. J.* **163**, 95.
Magnon, C.: 1968, *Astrophys. Letters* **2**, 213.
Mihalas, D.: 1964, *Astrophys. J. Suppl.* **9**, 321.
Milne, E. A.: 1926, *Monthly Notices Roy. Astron. Soc.* **86**, 459.
Much, G.: 1950, *Astrophys. J.* **112**, 266.
Rosseland, S.: 1936, *Theoretical Astrophysics*, Clarendon Press, Oxford, Chapter XX.
Rublev, S. V.: 1961, *Soviet Astron.* **4**, 780.
Rublev, S. V.: 1963, *Soviet Astron.* **6**, 686.
Scargle, J. D., Caroff, L. J., and Noerdlinger, P. D.: 1970, *Astrophys. J.* **161**, L115.
Schener, J.: 1894, *Astronomical Spectroscopy*, Ginn and Co., Boston, p. 277.
Smith, L. F. and Kuhi, L. V.: 1970, *Astrophys. J.* **162**, 535.
Smith, L. F. and Aller, L. H.: 1971, *Astrophys. J.* **164**, 275.
Sobolev, V. V.: 1958, in V. Ambartsumian (ed.), *Theoretical Astrophysics*, Pergamon Press, London, Chapter 28.
Sobolev, V. V.: 1960, *The Moving Envelopes of Stars*, Harvard Univ. Press, Cambridge, Mass.
Thomas, R. N.: 1949, *Astrophys. J.* **109**, 500.
Underhill, A. B.: 1967, *Bull. Astron. Inst. Neth.* **19**, 173.
Underhill, A. B.: 1968, in K. Gebbie and R. N. Thomas (eds.), *Wolf-Rayet Stars*, National Bureau of Standards Special Publication 307, p. 183.
Van Lyong, L.: 1967, *Soviet Astron.* **11**, 224.
West, D. K.: 1968, in K. Gebbie and R. N. Thomas (eds.), *Wolf-Rayet Stars*, National Bureau of Standards Special Publication 307, p. 221.
Weymann, R. J.: 1970, *Astrophys. J.* **160**, 31.

DISCUSSION

Thomas: Personally I think it futile to compare these continuum models to observations in any sense until the assumption of LTE is checked. If collisions should turn out to dominate photo-ionization, then the LTE assumption is valid. If photoionization dominates then instead of a rapid drop in temperature it begins to rise up again.

Van Blerkom: In an extended atmosphere, conditions are favourable for photoionization to dominate, since densities fall to very low values.

Thomas: Be sure to check it. Just before you put the infrared and ultraviolet excesses, ask where in terms of $\tau = 1$ in the visual, which you are talking about here, where in the atmosphere do three things happen:

(a) $\tau = 1$ in the UV,
(b) $\tau = 1$ in the infrared.
(c) the photoionization rates equal the collisional rate. Then right away you can answer the question posed.

Van Blerkom: Cassinelli originally computed his models for central stars of planetary nebulae, but realized that they described Wolf-Rayet continua rather well. It may be a fortuitous accident.

Morton: How do these models with the curved atmospheres compare with the plane-parallel non-grey models for O type stars?

Van Blerkom: Cassinelli's model would probably describe OB supergiants, since these have continua similar to Wolf-Rayet stars.

Morton: What happens to the Lyman continuum in Cassinelli's models?

Conti: It looks to me that Cassinelli's spherically extended model atmospheres might well explain the very large discrepancy between the interferometrically derived temperature and what we think is the real temperature in the case of ζ Puppis.

Morton: I wish I could be more optimistic that the limb darkening can have really that much effect, but the possibility is worth investigating.

Van Blerkom: I am very glad that Cassinelli's paper came out in 1971 because otherwise there would only have been Kosirev's. Cassinelli had to do this calculation in order to make Kosirev's results worth presenting.

Kuhi: In your integrations over the atmosphere, how do you avoid the discontinuity that results naturally in the density distribution whenever you get $r = r_c$. What do you do there?

Van Blerkom: If you let absorbing atoms exist all the way to the core of the star where $r = r_c$ you get rather severe discontinuities in the emergent line profile, because of strong occultation effects and absorption near the core. These are not observed. In order to get the line profiles that Castor gave in this paper, he had to assume that there were no absorbing or emitting atoms near the core. Either the escape probability method and the two level atom is incorrect near the core or else excitation conditions near the core are such that those atoms do not exist there. This might fit in with the view that there is a temperature inversion and it is getting too cold for atoms in the particular stage of ionization to be around.

Kuhi: These calculations always end up giving you parabolic profiles for the optically thick case. Can you suggest what you would have to do to get a Gaussian profile that seems to be the case mostly observed?

Van Blerkom: I imagine you have to play around with the velocity distribution to get a Gaussian.

Kuhi: My next question is for the P Cygni problem but applies to the WR stars as well. Most of the absorption edges that you get with this kind of calculation will give you intensities in the absorption component down to about 0.5, and it seems very hard to get the intensity down to zero when you have the intensity of the emission component say 5 or 10 times that of the continuum, which is typically what you see in Hα or Hβ. Can you suggest what you would have to do there?

Van Blerkom: That is the problem that Lucy ran across when he tried to explain the quasi-stellar case where he used the expanding atmosphere model we have used. You could not get both a deep absorption and a high emission at line center. The only explanation that one could think of then was that the part of the envelope that fills in the absorption has to be missing. In other words, you cannot have a spherically symmetric atmosphere. You must have some kind of a jet coming preferentially at you. I do not believe that.

Underhill: Why do you not believe in jets?

Van Blerkom: To have a jet preferentially directed towards you all the time, seems to me too fortuitous.

Underhill: I can see another way. The absorption can be strengthened by using scattering. In other words, you have a non-LTE like calculation. You can always fix it that lines get a strong absorption and not so strong emission. You bring your required emission intensity up by increasing the size of your shell.

Van Blerkom: These are non-LTE calculations.

Underhill: You have to work out more than for two levels.

De Groot: How certain are you that the emission line profiles of P Cygni type are really Gaussian? If the difference with the parabola is that we have a little more of a wing, may be you could do something with a little bit of electron scattering.

Van Blerkom: That was my hope when I attempted the calculation. You take a parabola, put it through an electron scattering layer and out it would come a Gaussian. It did not quite work out that way, but that might be the fault of the very crude way it was treated. It might be possible that the usual rounded parabolic type profiles with some electron scattering would look Gaussian.

De Groot: In P Cygni itself I had a lot of profiles and I do not think they are Gaussian. They had a wing on the red edge. It is also quite steep, may be not as steep as your model would indicate, but at least in between the two.

Kuhi: I did not mean to say that the profiles in P Cygni were Gaussian. I had two separate questions: Gaussian profiles in WR stars and the problem of the depth of absorption in the P Cygni stars.

Johnson: I just wanted to clarify whether your continuum comes from something you might call a photosphere, whereas when you talked about line emissions you went into an envelope. And I would like to ask what, if anything, the envelope may contribute to the continuum.

Van Blerkom: We considered that for the two WR stars that John Castor and I analyzed and we found that in the visible continuum the optical depth of the envelope is so small that there should be no observed continuum emission from the envelope.

Underhill: These calculations are for a two-level atom. In order to obtain the ionization balance between ions, you need at least a three-energy-level atom, a continuum plus two line levels. Have you any feeling for how possible it would be to extend the theory in that direction?

Van Blerkom: We did a thirty level atom for helium in which the continuum is included. You can do a multi-level atom because the line transfer is very simple, with the escape probability method. It is much simpler than if the atmosphere was standing still. The rub is that the continuum is not affected by the precence of the velocity field. Hence you have to solve the continuum problem in every single transiton. And that is a tremendous problem. We did a rough escape probability type method for the continuum which is very, very crude. We tried to include every single collisional and radiative transition.

Thomas: Why do you have to be so sophisticated on this? Certainly one or two lower levels for the ionization is sufficient, and treating the transfer problem for one continuum is quite straightforward. Anne Underhill's question is only about ionization equilibrium.

Van Blerkom: If you are just interested in ionization equilibrium, then you can get away with it.

Underhill: It seems to me you are being too elegant also. I would go along with Thomas. Every atom we have got has a continuum that we are interested in, a ground state that we do not observe lines from unless we have rocket UV-spectra, and some levels most of them clustered rather close to the continuum. The ones we observe are much closer to the continuum than they are to ground. We observe several stages of ionization, at least three. It seems to me a very plausible programme to set up the spherical moving atmosphere ionization balance, because you are taking in the major continuum of at least the three ions. Then you plot the fraction of each ion that is around. Then, depending on your density, which may be high enough, you can argue very convincingly that the levels that you want to observe from those populations are linked to the ion population of the next one by LTE with your assigned local temperature. And then you could perhaps pick this thing up. That may be the problem partly with P Cygni.

Van Blerkom: I might point out it is a non-trivial problem to treat one continuum in a spherical atmosphere even if you do not take into account the motion.

Thomas: It depends on the degree of sophistication you want. Are you talking about accuracies of a factor one per cent or a factor two?

Van Blerkom: A factor of two.

Underhill: Factor of two in the ionization balances would be quite acceptable. Then to calculate particular lines you might go back and be very fussy. If you want to match profiles you do have to be fussy.

Van Blerkom: The continuum optical depths for the very lowest levels are so high that the radiative rates are practically in detailed balance and so are the collisional rates that set things. The continuum really came into the higher levels when we wanted to compute their populations.

Thomas: It depends. Certainly in the regions where the strong resonance lines are formed, the continuum cannot possibly be in detailed balance.

Let me go now to your analysis of the He II lines. In the integration you have assumed that S times the bracket is constant. But S is going to decrease towards lower optical depth and so is the bracket. So that product is not a product of two factors, one of which increases and the other decreases. Both decrease, giving a systematic behaviour. It seems to me that the number of assumptions you have put in here are large enough that one worries about trusting the result.

Van Blerkom: This is a coarse analysis, but I think it is pretty good. Of course, I cannot prove it. What it does is to give you a feel of the average behaviour in the emitting region.

Thomas: The only reason I am worried comes from having put a lot of time on interpreting gradients, both in terms of emission decrements and of height gradients. I just find that you can really confuse things when you compare the observations and a theory in which you average things out. I remind you that we have changed the optical depth in the solar atmosphere by a factor of 50 just by being more presice. I am all for your procedure. It is just that I would be uncertain about how much I would trust my intuition on a coarse analysis as to what is most important, although I agree that a coarse analysis comes first.

Underhill: I think you are depending very strongly on the statement you made that you assumed all lines to be formed in the same region. If you consider that this is an expanding atmosphere and that the regions are defined by the constant velocity surfaces and by the on-the-spot hypothesis, then that assumption is not as difficult to accept as it is in the stationary atmosphere. I think that is what saves you. It is difficult for me to accept that all lines are formed in the same region. Now He II 4686 has a very much stronger f-value than one of the higher Pickering lines.

For a stationary atmosphere these lines are certainly not formed in the same region. But when you take into account the atmosphere is moving, then the only parts that count are those parts moving with just the right velocity to get you on the line center.

Van Blerkom: A line is formed in an expanding atmosphere over a large fraction of the envelope, that is, wherever the appropriate ions exist. That is why I say the lines are formed in the same region even though their f-values and optical depths are very different. Now, I would like to comment on the electron temperatures deduced from the theoretical models. The analysis of the WN6 star HD 192163 by Castor and myself indicated that the electron temperature in the envelope exceeded that of the core. Thus, a source of energy other than radiative is suggested. Castor and Nussbaumer have studied γ_2 Velorum. This is a very complicated system since the Wolf-Rayet is the fainter component. They found a best fit of their model to observations for a core temperature less than the electron temperature of the envelope. This might be an indication that radiative equilibirum holds.

Thomas: There is a very strong difficulty of supporting this atmosphere, and Castor says the only way he sees how to do it, is a random turbulence of some hundreds or thousand kilometers a second. So this is not exactly what you would call a self-consistent situation.

Underhill: That is right.

Morton: In regard to the analysis of the ultraviolet spectrum of γ_2 Velorum, if we accept the present point of view that the O star is brighter, one can assume that all the emission lines, except the $\lambda 1909$ line, in the UV spectrum are due to the O star. Now, what C III lines were actually analyzed?

Van Blerkom: Ten lines from the fourteen lowest terms of C III; they were mainly in the ultraviolet.

Thomas: What is the highest stage of ionization observed in this star?

Smith: Probably C IV.

Van Blerkom: Let us return to the question of support of the atmosphere. Castor and Nussbaumer used a spherically expanding envelope model, so the question of support does not arise.

Thomas: I asked Castor, what he needs in the way of supporting the atmosphere? He said he needs a random turbulence. It is conceivable I misunderstood, so I may be wrong.

Morton: Now, how does his support problem differ from your support problem?

Van Blerkom: You do not have a support problem if you are not supporting anything.

Morton: You have an outflowing atmosphere.

Van Blerkom: Right, but we do not know what is causing the outflow.

Morton: I do not see that that argues against this model in particular.

Thomas: No, the problem is this. If I am going to invoke either random motion or differential expansion motion on the order of several hundreds or thousand kilometers a second, it is very difficult for me to see how you get this, without some kind of mechanical input of energy to maintain this. So when the suggestion is made that may be we can do these models in radiative equilibrium, I am afraid I just sit back and smile.

Van Blerkom: The question was raised earlier about what lines were used in the model of γ_2 Velorum The lines $\lambda\lambda 2296$ and 1176 allow the core temperature and envelope electron temperature to be related. The intercombination lines $\lambda\lambda 1909$ and 2846 allow the density to be determined.

Conti: Lindsey Smith and I had been writing a paper on γ_2 Velorum, and she has just sent me a print of its ultraviolet spectrum; it looks just like a B0 supergiant.

Underhill: That is what it suggests to me, too.

Conti: Which is to say, it is the O9 supergiant we observe and all the lines that you see there belong to that star. None of them can be definitely WR lines, especially none of these.

Thomas: Let us just be sure I understand Conti's remark here. Is it what you are saying that the only line here you can be sure belongs to the WR component of γ_2 Velorum, is 1909 Å? Everything else is incidental?

Conti: Everything else is the O supergiant.

Thomas: So, my question about what was the highest level of ionization you observed, is irrelevant!

Conti: I do not think you can do it from γ_2 Velorum. If you want to look at the UV you must get a rocket spectrum of a single Wolf-Rayet star.

Morton: The Princeton OAO, with its high-resolution spectrometer should be able to sort out the system by determining which UV lines shift the same way as the visible absorption lines. O VI also occurs in the UV spectra of the OB supergiants, in absorption and possibly also in emission at 1032 and 1038 Å.

Thomas: Is the spectrum good enough to see that?

Underhill: As far as I know, the available spectra of γ_2 Velorum do not have adequate resolution in that region to show you anything.

Smith: Let me strike an intermediate position. It is quite clear that between about 1000 and 2000 Å which is the region that Conti and I compared with ζ Puppis, it is very difficult to tell the difference between the spectrum of γ_2 Velorum and of ζ Puppis (See also Stecher, in the last WR Symposium). Most of the lines in this region are strong resonance lines, and apparently |the O companion is chiefly responsible for these lines. But in the region from 2000 to 4000 Å, there are a large number of emission lines, and most of them are probably due to the Wolf-Rayet star. The emission lines can be conspicuous despite the disparity in the magnitudes, exactly the same way as they are in the visual region from 3000 to 6000 Å. Which lines Castor used I do not remember, but he had many C III lines to choose from.

The data used was from OAO observations, provided by Lilly. The OAO observations agree moderately well the observations of Stecher, but Stecher's have slightly better resolution. Below 2000 Å, the OAO observations are better than Stecher's because they are free of atmospheric absorption. In that region, the spectrum of γ_2 Velorum looks very much like that of ζ Puppis, so probably the very strong resonance lines are due mainly to the O9 supergiant. Below 1000 Å, we know nothing at all, but between 2000 and 4000 Å, there are many emission lines (whose equivalent width is greatly reduced because of the contribution from the O9 supergiant to the continuum) that come from the Wolf-Rayet star. I wish to say, Castor is aware of the final numbers on the relative luminosity, so that he is taken that into account.

Van Blerkom: Half his manuscript is concerned with correcting for the presence of the O star.

Underhill: That is a very difficult point. He actually uses, in the analysis two lines, between 2000 and 3000 Å. This problem really boils down to which lines are radiation dominated, and which are collision dominated. There are two different kinds of problems being solved that are of a different character. I think it is interesting, but I do not think it solves anything.

Smith: Does it need to be added here that C III $\lambda 1909$, is certainly not seen in the supergiant spectra? So, that line in particular, is coming from the Wolf-Rayet star. But in the region from 2000 to 4000 Å there are many C III lines, as was already mentioned by Van Blerkom.

Underhill: That line is seen in Wolf-Rayet stars but I do not think that Lindsey Smith is sure it is not coming from the O star, or from gas in the system.

Smith: All I said was that we do not see it from other supergiants, so I assume it is not coming from the O9 supergiant in this particular case.

Sahade: Well, it should be coming from the 'outermost' envelope, is it not so?

Underhill: Yes, it is coming from the general envelope of the system.

Van Blerkom: It is a calculated risk to try to analyze a binary. I would like to point out that Bappu's results showed unblended C IV lines in some single WC stars. It might be a more profitable thing to deal with them, because there are no ambiguities at all.

Paczyński: Coming back to WN stars, as far as I remember the paper of Castor and Van Blerkom, they have calculated about six different models with different electron densities and temperatures. If you plot the results on the electron temperatures – electron density plane, you find a line on this plane along which the models give the observed line ratios. For a high electron temperature you need a high density and vice versa. My impression is that it is difficult to decide whether the electron temperature should be high or low.

Van Blerkom: Yes, it is difficult to pick a unique model. This was just one that seemed to fit, but there might be others. I would like to try putting in lower values of the electron temperature and the electron density to see if one might get reasonable agreement with observation.

Underhill: This is a WN star, a WN6. I think, using the He II spectrum by itself, you cannot get a unique answer. You may very well be able to balance it out with a lower temperature, but He II does not really care what temperature, once it is above $50000°$. However, it is WN and you have N V very strong, as well as other N ions. If you consider the N V, you find 10^5 degrees is rather a nice temperature. You will not find anything seriously lower than that, adequate to give you the N V in emission.

Van Blerkom: What is the ionization potential necessary to get N V? It is substantially higher than for He II. So, it might be formed in a very different part of the envelope.

Underhill: I have some information which I intend to demonstrate on this point tomorrow, to illustrate it.

Van Blerkom: I am not saying that the entire envelope is in radiative equilibrium, and I do not think that Castor is saying that either. The results suggest that the region in which C III lines are formed, might be in radiative equilibrium. That does not apply to the entire envelope, by any means.

Underhill: And all the C III lines that you observe are not necessarily formed in the same region. In fact, you are almost certain $\lambda 1909$ is formed around the system. Some of it might be formed where $\lambda 2296$ is or $\lambda 1176$ or $\lambda 5696$. The real problem with these stars is, approximating the atmosphere, as you are forced to for an illustrative theory, with one point.

Paczyński: I would like to talk about one thing that was not mentioned in Van Blerkom's paper, and which I believe is important. If you take your and Castor's model for the envelope of a WN star, you can calculate the critical depth in different He II continua. You did this for the visual continua and you found them to be optically thin. Therefore, we should not expect any jumps in the visual part of the spectrum. It is possible to do similar computations for the three ultraviolet continua. If you take your numbers you find that Lyman and Balmer continua are very thick. Their optical depth is above 100. If you go to the third continuum, which has an edge at 2050 Å, you find that the optical depth is about 0.3. There are two important things that follow from that. The spectral region around 2050 Å is observable, and we may see an edge either in emission or in absorption at this wavelength. It would be very interesting to observe those WN stars which show a nice Pickering series and to see if there is any edge. This could help in deriving the electron densities and temperatures in the envelopes. And there is a second aspect. The temperature of the central star is assumed to be about 40000 K. The maximum of the Planck curve is just shortward of 800 Å. This means that a significant part of energy is emitted in the wavelengths in which the envelopes are optically thick. Therefore, the photospheric radius varies by a factor of 5 or 6 between visual and far ultraviolet. It will be very difficult to build a model of such an extremely non-grey and non-plane parallel atmosphere. This kind of model has never been studied from the point of view of the radiation pressure in the continuum acting as an agent for the mass outflow.

Van Blerkom: I thought Rublev studied radiation pressure in the continua beyond the principal series of He II and found that it was negligible compared to the electron scattering.

Paczyński: It depends on your model. In this case the electron scattering gives you an optical depth of about one, whereas the optical depth in the far UV is one hundred. This leaves a possibility of having radiation pressure in the continuum as the driving force for the mass outflow. There is no such model available in the literature now. Perhaps from the theoretical point of view Wolf-Rayet stars can exist, inspite of my lecture last afternoon!

Thomas: That theory depends on whether it increases or decreases the emission in the continuum. Just the fact that I have a high optical depth does not mean anything; it is whether the intensity increases or decreases relative to the black body, that is important.

Underhill: The black body is a very simple model.

Thomas: Be clear on the effect. If the intensity goes up, then I may have something; if it goes down, it just reinforces the conclusion of no effect.

Paczyński: Well, I just do not know, because I do not think that anybody has built a model atmosphere with a photospheric radius varying by a factor of 5 from one wavelength to another. I am not in a position to answer that.

Underhill: You can do that automatically in all the plane parallel atmospheres. It is the only geometry in which solutions have been obtained. To put opacity differences into a spherical atmosphere has not yet been possible, because of numerical difficulties. Although we are sounding very critical of the type of theory just presented indeed, there is a 40-year gap in which no progress was made. The new thing is a considerable step forward. I do not think at any time that the people who have offered these papers have really implied that they were more than a numerical experiment giving something that looks vaguely like a Wolf-Rayet star.

Paczyński: I really do not know. I just wanted to point out that in the particular model which fits very nicely the Pickering series observed in two WN stars you may calculate the optical depths in the three continua. And they come out to be large. One of these jumps is observable.

Thomas: It is always large if you go down deep enough. What you mean is large enough above a certain point.

Paczyński: It is large enough above the visual photosphere.

Underhill: It is larger than the model outer-atmosphere?

Paczyński: Yes.

Underhill: With the amount of gas at half the length you have there, you get vary large opacity in the ultra-violet continua. In some ways, that is why I would like to have some hydrogen there because it is not so opaque.

These are difficult problems and you have a life-time of work in front of you.

SECTION VII

CHAIRMAN: J. SAHADE

WOLF-RAYET BINARIES AND ATMOSPHERIC STRATIFICATION

LEONARD V. KUHI

Astronomy Dept., Univ. of California, Berkeley., Calif., U.S.A.

1. Introduction

Considerable effort has been expanded on both of these topics ever since the binary nature of most bright WR stars was revealed by radial velocity and eclipse observations and the basic correlation of line-width with ionization potential was discovered. I would like to describe briefly the methods of detection of binaries, to review the percentage of known binaries and to summarize the mass determinations that result from the study of WR binaries.

CV Serpentis, a former eclipsing binary will also be discussed prior to reviewing the history of the atmospheric stratification problem. I would then like to present what I think is the solution of that problem based on the behaviour of CIII 5696 with increasing temperature (i.e. earlier spectral sub-class) in the WC stars interpreted in terms of a radially expanding envelope.

2. Binaries

2.1. ARE ALL WR STARS BINARIES?

This question has been asked at every conference on WR stars and no doubt will continue to be asked at all future meetings because there can be no final answer. The question is of such importance in understanding the evolutionary history of WR stars that it seems worthwhile to summarize the methods of detection of WR binaries before discussing the results.

At present these methods involve the following:
(a) visual detection, i.e. presence of two stars,
(b) variations in radial velocity,
(c) eclipses,
(d) presence of another spectrum,
(e) small ratio of emission-line to continuum intensities.

Because of the great distances of most WR stars the first method is virtually useless. The second will detect only those systems for which the variations in radial velocity are large enough to be readily measured and hence will leave undiscovered those systems for which the companion has a considerably lower mass or the orbital inclination is close to zero. The third requires orbital inclinations of 80° to 90° and subsequently is a very selective method which would miss a large number of possible binaries. The fourth requires a favourable luminosity ratio, i.e. the companion should be no more than ∼1.5–2.0 mag. fainter than the WR star or it simply will not contrib-

M. K. V. Bappu and J. Sahade (eds.), Wolf-Rayet and High-Temperature Stars, 205–227. All Rights Reserved.
Copyright © 1973 by the IAU.

ute anything to the composite spectrum. Thus again low luminosity (which usually equals low mass) companions will be missed. The fifth method is really the same as the fourth but makes use of the continuum contribution of the companion rather than its absorption spectrum. This is especially useful when the companion is of very early spectral type and does not have any strong absorption features. However, it does contribute to the continuum of the composite spectrum and hence the emission lines will appear to have smaller intensities relative to the continuum than a single WR star.

This is a powerful method which has worked extremely well in detecting binaries but relies on two assumptions: (a) the ratio of line to continuum intensities does not vary with time nor from star to star of the same spectral type and (b) that a truly single star exists. Scanner evidence suggests that the ratios do vary in close binaries (which are readily detected by one of the other methods anyway) but probably not in the so-called single stars. In actual practice therefore any ratio that is more than $\sim 10\%$ (typical fluctuation in a close binary) lower than a single star can be considered as indicating a binary. Nothing can be done about the second assumption other than to say that the star with the highest ratio is taken to be single (it will usually have defied all other methods as well) and the other stars are compared to it. Again the method will fail to detect low luminosity low mass companions. If the minimum change detectable is taken as $\sim 10\%$ implying a luminosity ratio of $\sim 10:1$ the undetected mass can still be quite large. For example let us assume the standard mass luminosity law, take the typical mass of a detected O-type companion as 20 $M_\odot$ and its luminosity as 4 times that of the WR star. Then our luminosity limit is 1/40th that of the O star and corresponds to a mass of ~ 5 $M_\odot$. Since the mass of the WR star is ~ 5 $M_\odot$ such a large undetected companion would have a considerable effect on the radial velocity curve if the orbital inclination and separation were favourable. However, if it were not then once again no companion would be detected.

Bearing these difficulties in mind we can then attempt to answer the question posed by the title of this section. Approximately 105 WR stars are known in our galaxy and Smith's 1968 paper lists 40 of these as binaries. However all of these stars have not been studied to the same extent because of the difficulty of getting good radial velocity measurements of faint stars. We also note that a number of stars not listed as binaries by Smith were suggested as such by Kuhi (1968b) on the basis of low line-continuum ratios, e.g. HD 9974, 65865, 197406 and MR 119. Consequently her list represents a very non-uniform sample and it would be more realistic to take all stars brighter than some limiting magnitude, say $v = 10.0$ on Smith's system. There are then 33 stars of which she lists 19 as binaries. If we also allow for the fact that southern hemisphere stars have not been studied as fully as those in the north by eliminating all those with declinations south of $-25°$ then there remain 15 stars of which 11 are listed as binaries. That is to say when we have a uniformly studied sample the percentage of binaries is $\sim 73\%$. This is a very high percentage compared to field stars and if one allows for the non-detection of low-luminosity low-mass objects for the reasons cited above it may very well be that all WR stars are binaries. Arguing against this would be the fact that no star with a ring nebula appears to be a binary. Unfortunately this is as close as one can get to a definitive answer from the observations. Nothing further need be said.

2.2. MASSES

Binaries provide us with the only estimates of the masses of WR stars and consequently are of great importance in understanding the evolutionary history of WR stars. However, we stress again that unlike normal binaries the WR binaries are not well behaved in that different emission lines give velocity curves of different amplitude with different K-values. In fact the emission lines consistently give more positive K-values than the absorption lines of the companion: this is the so-called 'red-shift' and has had no satisfactory explanation. Clearly the presence of absorption on the violet side of the line could account for some of the shift as could the contribution from gas streams in the system. However, a less *ad hoc* explanation might be the following: the eclipse observations of V 444 Cygni (Munch, 1950) as well as the work of Castor *et al.* (1970) suggest that the optical depth in electron scattering is ~ 0.5. The data presented below indicate that the velocity increases with radius; hence an electron in the outer parts of the envelope scattering a photon from the inner part always has a larger velocity outwards than the emitting atom. Consequently the scattered photon will be shifted to the red in wavelength and the resultant emission profile should also be shifted to the red. The amount of the shift would depend on the exact location of the line-emitting region and the optical depth in electron scattering which probably varies from line to line. In close binaries the distribution of emitting atoms is not spherically symmetric; material exists in gas streams as well so that a detailed calculation becomes very difficult.

Such radial velocity difficulties should be borne in mind when discussing the masses of WR stars; *they are not at all well determined.* Table I summarizes the present data available on the masses of WR stars. It lists not only the most recent estimates of different authors but also the different values derived by the same author from different emission lines. References to earlier work can be found in Smith's article in the Boulder Symposium volume. HD 228766 was originally left out of the table given by Smith (1968b) because she had classified it as an Of star. However, the WN 7 (Hiltner, 1951) classification is definitely confirmed by the Kuhi and Smith Atlas. In addition Bracher (1967) suggests that the inclination may be quite large because of the large mass function. Kuhi and Smith also note a weakening of absorption lines at time of conjunction which may indicate the presence of a weak eclipse.

The masses found range from $\geqslant 4$ to $\geqslant 20 \, M_\odot$ with the best determined value (for V 444 Cygni) being $\sim 10 \, M_\odot$. The mass ratios M_{WR}/M_{OB} range from ~ 0.2 to 0.4 suggesting a much smaller luminosity ratio L_{WR}/L_{OB} then observed. The resulting conclusion is that the WR stars are overluminous for their mass. Lindsey Smith has compiled the orbital separations of WR binaries using the value of sin i determined from eclipsing systems or estimating it by assuming a mass for the WR star of 11 $M_\odot$. Table II lists the relevant data. She also draws attention to the fact that the WC binaries seem to have a significantly larger separation than the WN's (if one excludes

TABLE I
Masses of Wolf-Rayet stars

HD	Name	Sp. Type	$\frac{M_{WR}}{M_{OB}}$	$M_{WR}\sin^3 i$	$\frac{M_{WR}}{M}$	Reference
68273	γ_2 Vel	WC8 + O9I[b]	0.22	19.7	$\geqslant 19.7$	(1)
			0.26	14.5	$\geqslant 14.5$	
			0.28	13.0	$\geqslant 13.0$	
152270		WC7 + O5–8	0.24	1.6		(2)
168206	CV Ser	WC8 + O	0.24	8.0	$\geqslant 8.0$	(2)
			0.31	5.2	$\geqslant 5.2$	
			0.23	8.1	$\geqslant 8.1$	(3)
186943		WN4 + B	0.28	6.0		(2)
			0.42	3.4		
190918		WN4 + O9I[a]	0.16	0.5		(2)
228766		WN7 + O6[a]	0.19	5.4	$\geqslant 5.4$	(2)
			0.22	4.2	$\geqslant 4.2$	
			0.21	4.6	$\geqslant 4.6$	(4)
193576	V444 Cyg	WN5 + O6	0.40	9.3	10.0	
			0.43	8.4	9.0	(5)
			0.39	9.5	10.2	(6)
211853		WN6 + O6I[a]	0.35	10.1	11.5	(7)

Notes to Table I.
[a] Classification from Kuhi and Smith Atlas.
[b] Classification from Conti and Smith (1972).
References to Table I:
(1) Ganesh, K. S. and Bappu, M. K. V.: 1967, *Kodaikanal Obs. Bull.*, No. 183.
(2) Bracher, K.: 1967, Thesis, Univ. of Indiana.
(3) Cowley, A. P., Hiltner, W. A., and Berry, C.: 1971, *Astron. Astrophys.* **11**, 407.
(4) Hiltner, W. A.: 1951, *Astrophys. J.* **113**, 317.
(5) Ganesh, K. S., Bappu, M. K. V., and Natarajan, V.: 1967, *Kodaikanal Obs. Bull.*, No. 184.
(6) Münch, G.: 1950, *Astrophys. J.* **112**, 266.
(7) Stepien, K.: 1970, *Acta Astron.* **20**, 117.

HD 190918). This feature will provide the basis for further speculation by Lindsey Smith and I will comment no further on it here.

2.3. LUMINOSITIES

The luminosities of WR stars were discussed thoroughly by Smith in 1968 and have been reviewed again by her at this symposium. As far as binaries are concerned the major new contribution comes from γ_2 Vel which has received considerable attention from various workers recently. At the Boulder conference Smith assumed that the WC 8 star was the more luminous and derived an absolute visual magnitude of $M_V = -6.2$. This was ~ 1.5 mag. brighter than the other WC stars which all had about the same value of M_V. Baschek (1970), however, in a reanalysis of γ_1 Vel, suggested that $M_V = -5.6 \pm 0.4$ for the combined system of γ_2 Vel and hence that the visual continuum of γ_2 Vel must come largely from the O-type companion. In order to verify this Baschek and Scholz (1971) compared the absorption-line spectrum of γ_2 Vel to that of standard O stars and classified the companion as O 8 on the basis of

TABLE II

Separations of Wolf-Rayet binaries

HD	Name	Sp. Type	$a_w \sin i$ $(R_\odot)$	$a_w \sin i$ $(R_\odot)$	$\sin i$	Base of $\sin i$	$a/R_\odot$
186943		WN4 + B	29.7 44.3	12.6	0.77	Assume $M_{WR} = 11$	64:
190918		WN4 + O9I	108	17.6	0.36	Assume $M_{WR} = 11$	350
MR 114	CX Cep	WN5	12.1		>0.94	Eclipsing	< 26[a]
193576	V444 Cyg	WN5 + O6	25.6	10.0	0.98	Eclipsing	36
211853		WN6 + O6I	27.5	13.1	0.96	Eclipsing	53
228766		WN7 + O6V	50	10.6:	≈1	Lge. mass function	61
214419	CQ Cep	WN7	9.6		≈1	Eclipsing	< 20[a]:
193793		WN8 + O5	5.0				
			Period 3 years (Conti-private comm.) large!				
152270		WC8 + O5–8	28.2	7.6	0.52	Assume $M_{WR} = 11$	69
68273	γ_2 Vel	WC8 + O9I	234	65.7	1	Lge. minimum masses	300
168206	CV Ser	WC + B0	86.5	26.4	1	Was an eclipser in the past	113

[a] Assuming $M_{WR}/M_{OB} <$

line ratios. They then also compared the equivalent widths in γ_2 Vel to those of a standard O8 star and deduced a ΔM_V of -2.3 ± 1.3 mag. with the O-star being the brighter. Combining their magnitude difference with that of Brown *et al.* (1970) (i.e. $\Delta M_V = 1.2 \pm 0.6$ mag.) they obtain $\Delta M_V = 1.8$ to 1.0 mag. This gives $M_V = (-3.6$ to $-4.2) \pm 0.4$ mag. for the WC 8 star. Conti and Smith (1971) also redetermined the luminosity ratio but by considering the emission-line to continuum ratios instead since this should lead to a much more accurate estimate if the WR star is the fainter star. Conti reclassified the spectrum of the companion as O9I on the basis of HeI 4471/HeII 4541 for spectral type and SiIV 4089/HeI 4143 for luminosity class. For the emission lines between $\lambda 3850$ and $\lambda 4700$ they found a range of 0.59 to 0.85 in the log of the ratio of equivalent widths in HD 192103 (WC 8) to those in γ_2 Vel. Since some lines were clearly more sensitive to slight differences in excitation than others they eliminated them by looking at HD 164270 (WC 9) to see which lines changed the most. Their final result was 0.68 ± 0.05 for the log of the ratio or a $\Delta M_V = 1.4 \pm 0.1$. The same ratio from the absorption line spectrum is considerably less certain i.e. $\Delta M_V = 0.6 \pm 0.6$, and consequently the value of -1.4 ± 0.1 was adopted. This is in good agreement with the previous redeterminations as well as that mentioned by Bappu at this symposium. However, Conti and Smith prefer to use a distance modulus of 8.3 mag. as derived by Graham (1965) and confirmed by the $H\beta$ work of Brandt *et al.* (1971) on the association to derive $M_V = -4.8 \pm 0.3$ for the WC 8 star. Since these estimates are all within the range of values for the other WC stars it seems safe to conclude that all WC stars in our galaxy have the same absolute visual magnitude. Conti and Smith's method assumes that the emission-line strengths of binary stars are the same as those of single stars and that those strengths do not vary appreciably. The observed variations in binaries are all less than 20% and mostly less than 10% so that the latter assumption seems all right. There is also no evidence to the contrary of the former assumption.

Other complications involving the interpretation of WR binaries such as γ_2 Vel will be discussed by Virpi Niemela who will comment on the observational changes with phase in this system as well as in others. These complications confuse the interpretation and do not help our understanding of the WR stars. Consequently I have chosen not to dwell on such details but to discuss instead only one such system, namely CV Ser, because of its role in the solution of the atmospheric stratification problem.

3. CV Serpentis

CV Serpentis was originally discovered to be an eclipsing system by Gaposchkin (1949) from a study of Harvard patrol plates. The photographic eclipses were quite shallow: 0.2 mag. at primary and 0.10 mag. at secondary. Hjellming and Hiltner (1963) obtained a photoelectric light-curve covering the deeper eclipse as well as a revised period of 29.640 days. They recorded a primary eclipse of $\sim$0.6 mag. but had no observations at secondary eclipse presumably because of poor weather conditions. However, the increased depth of primary minimum was striking. Because of the long period this system was observed by Kuhi and Schweizer (1970) in order to investigate the atmospheric stratification (see below). Various emission lines and continuum points longward of $\lambda5500$ were measured with a photoelectric spectrum scanner and no eclipses larger than the observational errors were detected. Figure 1 gives typica

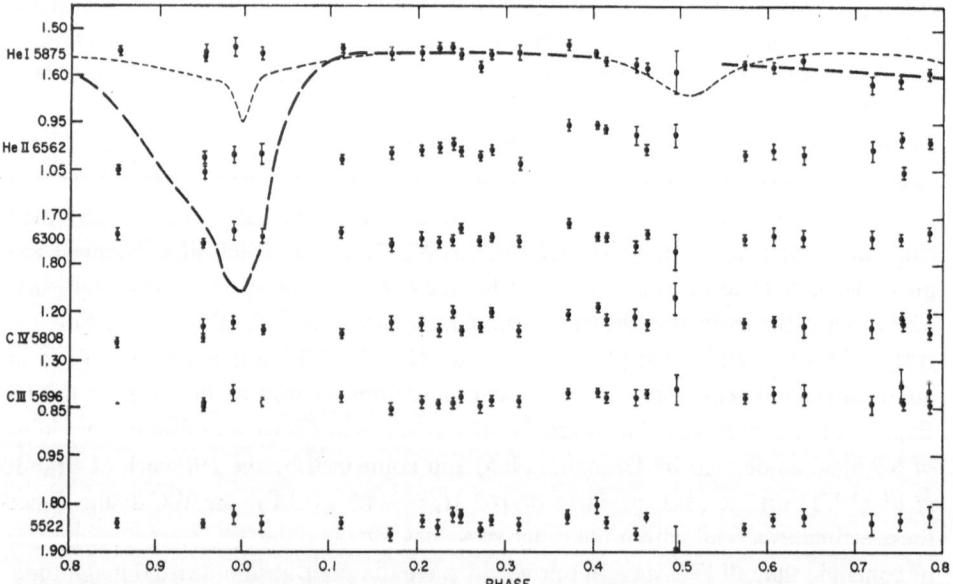

Fig. 1. Photoelectric spectrum scanner observations of CV Ser as a function of phase. The fine dashed curve is due to Gaposchkin; the thick dashed curve is due to Hjellming and Hiltner. Both have been normalized to the scanner data for 5875. No eclipses are evident.

lightcurves in He II, C IV as well as those obtained earlier by Gaposchkin and Hjellming and Hiltner. The earlier curves are normalized to the data for $\lambda 5875$ to indicate the depths of eclipse expected. Clearly none is visible. Stepien (1970) also observed the star in UBV and again failed to detect any significant eclipse at either time of minimum. The system had stopped eclipsing!

Since that time a number of developments have taken place. The most important is the complete redetermination of the orbital elements by Cowley *et al.* (1971) from spectroscopic data obtained in 1970. This work has led to two major changes: (1) a new period of 29.706 days and (2) the interpretation of the deep minimum (primary) as the B-star being in front of the WR star instead of the opposite which had been assumed by all previous investigators. The new period implies that the expected time of secondary eclipse used by Kuhi and Schweizer as well as by Stepien occurred four days too soon. However, using the revised period Cowley *et al.* showed that there was still no evidence of eclipses in the photoelectric data. On the other hand Tcherepaschuk (1969) observing with a 90 Å bandwidth filter at $\lambda 4652$ and the adjacent continuum still managed to see eclipses of ~ 0.15 mag. in the C III–C IV emission line and ~ 0.07 mag. in the continuum. In contrast observations longward of $\lambda 5500$ by Kuhi and Schweizer still failed to show any eclipse in 1971. A closer examination of the spectroscopic behaviour of the C III–IV 4650 blend by Kuhi and Schweizer reveals the appearance of a strong absorption feature to the violet during the time of secondary minimum. This absorption component is strong for 5 to 6 days and actually occurs when the WR star is in front of the B star. If one makes a crude determination of the expected blue profile of the emission line by reflecting its red profile about the zero-velocity point one can then estimate the strength of this absorption component and hence compute the change its presence would produce on the total intensity measured in the bandwidth of the filter. The result is an 8.6% decrease in the line and no change in the continuum. This compares favourably with the value of 60% of the eclipse observed by Tcherepaschuk computed by Cowley (private communication) considering the uncertainty in the determination to the profile. It is very likely that the above estimate is a lower limit since any occultation effects would tend to decrease the redward emission profile and hence lead to an underestimate of the equivalent width of the absorption component. Hence, it seems possible that one can account for a large fraction of Tcherepaschuk's emission-line 'eclipse' entirely by the appearance and disappearance of the absorption feature. However, there is no way of explaining the continuum observations in a similar way and one is left with a real puzzle: i.e.

> *what happened to make the large eclipses disappear and why is there still a small eclipse in the blue continuum and not in the red?*

Kuhi and Schweizer suggested that the eclipses had been caused primarily by the inner dense parts of the envelope and that for some unknown reason the envelope had increased considerably in size so that the optical depth was no longer large enough to cause an eclipse. However, this explanation cannot be correct since an optically thin envelope would be expected to produce flat-topped emission profiles and no gross changes in the profiles have been observed. Alternatively one might suggest a shrinking of the envelope which would produce no major change in the emission line profiles and hence would be consistent with the observations. We can offer no argument against such a hypothesis although it does not seem particularly attractive on theoretical grounds. Specifically it seems very difficult to get such an envelope to suddenly collapse upon the star without causing a noticeable brightening of the star itself. No such changes were observed.

However, bearing in mind the gas streams and common envelopes occurring in V444 Cygni (Kuhi, 1968a) it is quite likely that material between the two stars contributes significantly to the emission lines. In fact Cowley *et al.* have suggested that variations in the amount of this material could cause the changes observed and that the 'eclipses' were initially produced by occultation of this material by the two stars. They also mention that whereas this would be consistent with the present observations it would be somewhat difficult to reconcile the large eclipse observed by Hjellming and Hiltner. The suggested variations would also account for the shallow eclipses observed by Gaposchkin as being due to the averaging of deep and shallow eclipses with an incorrect period. Unfortunately Liller (private communication) has had the Harvard patrol plates reexamined for such changes and apparently has not found any. This seems to make the variable material hypothesis somewhat untenable since it now requires the amount of material to remain constant for ~ 70 yrs and then suddenly decrease by a large amount. Such constancy seems very unlikely if V444 Cygni is a representative example.

Yet another possibility, namely a change in the orbit, was also ruled out by Cowley *et al.*, since there is no spectroscopic evidence for changes in the radial velocity curves. Figure 2 illustrates the agreement between old and new observations for the C III–IV 4650 blend.

Consequently we are left with no viable hypothesis that is consistent with the present observations. There remains, of course, the possibility that Hjellming and Hiltner observed some other star or committed some other error but this seems quite unlikely considering the large number of observations involved. It is to be regretted that their observations did not cover a longer time period. Clearly CV Serpentis deserves close attention during the next observing season to unravel its mysterious behaviour.

4. Atmospheric Stratification

One of the basic correlations discovered by Beals many years ago was the fact that lines from ions of high ionization potential had much narrower widths than those of low. This relation is most readily seen in the WN stars for which the line widths are easily measured. For example, in HD 192163 these half-widths vary from 350 km s^{-1} for N v, 700 km s^{-1} for N IV, to 1150 km s^{-1} for N III and from ~ 1000 km s^{-1} for He II to 1350 km s^{-1} for He I. Figure 3 shows that the correlation is extremely well defined. The relation also exists for the WC stars but because of very severe blending problems the measured line widths show a much larger scatter and hence define the relationship

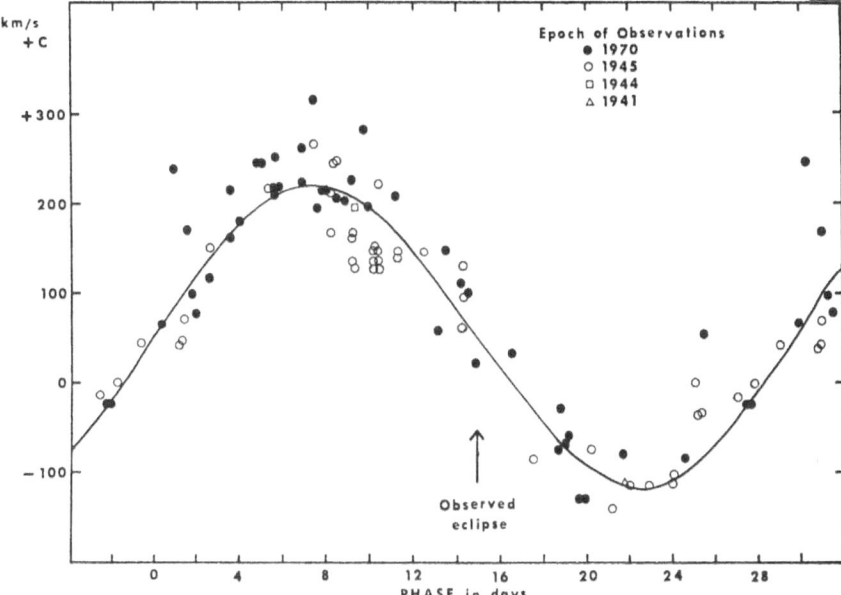

Fig. 2.   Radial velocity measurements for CV Ser from Cowley *et al.*

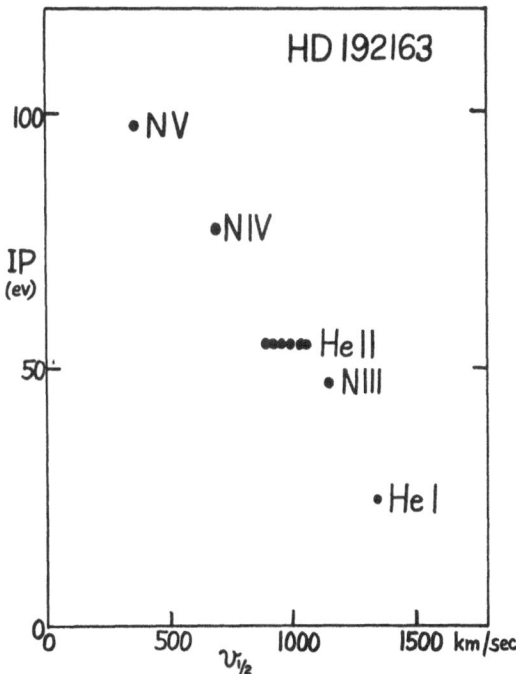

Fig. 3.   Correlation of emission line-width and ionization potential for HD 192163 (WN6). $v_{1/2}$ is the half-width at half-intensity in km s^{-1}.

much more poorly. The correlation is also upheld in the velocities of the violet-displaced absorption components when such lines occur e.g. He I, C IV 4650, N IV 3483, etc. This has already been pointed out by Bappu in his talk on spectral line identification and typical values are tabulated there.

The correlation immediately implies the presence of stratification in the WR envelope since such a large number of stages of ionization cannot exist in the same physical location. It also implies that the velocity field is either an accelerating or a decelerating one depending on the temperature structure. Beals assumed that the dominant mechanism of ionization was akin to that in planetary nebulae, i.e. ionization by far UV photons and the resultant recombination followed by cascading produced the emission-line spectrum observed. Because the higher energy photons would be absorbed closest to the star this picture led to the 'ionization temperature' decreasing outwards with increasing radius. This immediately implies an accelerating velocity field. Today we are much less certain about the ionization and excitation mechanisms and would hesitate to apply Beals' scheme to all the lines formed in the WR envelope. However, it seems likely that many lines are indeed recombination lines and hence that Beals' mechanism may still play an important role although not necessarily a unique one. Consequently a model with temperature decreasing outwards is a very viable one even though the exact mechanisms of excitation, ionization and acceleration are not known. The outward increasing velocities could be provided by radiation pressure for example if there were enough flux available to do the job.

A different suggestion was made by Thomas in 1949 who interpreted the velocities in terms of random motions or large scale turbulence. However, the very large temperatures ($\geqslant 10^7$ K) required did not prove acceptable to most investigators and consequently the model died quietly although it was briefly revived during the Boulder conference on WR stars.

A model with temperatures increasing outwards was put forth by Münch in 1950 in order to interpret the spectroscopic (Münch, 1950) and photometric (Kron and Gordon, 1950) observations of V444 Cygni. The small separation of the system ($\sim 35 R_\odot$) made it very difficult for Munch to see how the temperature could decrease outwards under the very strong heating influence of the O-type companion. In particular he found it impossible to reconcile the existence of He I in the outer parts of the WR envelope which he felt would be subjected to the ionization produced by the UV radiation from the O star. Therefore, the temperature must increase outwards and consequently the velocity must decrease with increasing radius. If the material were ejected ballistically then gravity could provide the decelerating mechanism.

The rotating model proposed by Limber (1964) in which material was ejected equatorially because of rotational instability would also suggest a structure with temperature increasing outwards with radius. Presumably the material at larger distances from the star would have slower rotational velocities than the material close to the surface because of the conservation of angular momentum. Consequently the emission lines should be narrowest at large radii and hence the linewidth-ionization potential correlation implies that the temperature must increase outwards.

An attempt was made by Kuhi (1968a) to resolve the question by obtaining eclipse curves in the light of individual emission lines for V444 Cygni (Period = 4.2 days). If the temperature decreased outwards then one might expect to find a shallower broader eclipse for lines from ions of low ionization potential and very deep narrow eclipses for lines of high ionization. Qualitatively the observations obtained at secondary minimum (i.e. O star in front) did suggest this but the emission lines decreased by even larger amounts at primary eclipse (WR star in front) leading to the conclusion that a significant fraction of the light was produced in the region between the two stars. Other peculiarities were also observed in the behaviour of individual lines and these were attributed to the interaction effects arising between two hot stars so close together. Typical light curves were given at the Boulder Symposium as well as in the original paper and need not be repeated here. Consequently it was impossible to disentangle the stratification problem from the interaction effects.

The other known eclipsing systems (all discovered by Gaposchkin) are CQ Cep (P = 1.6 days and hence even closer than V444 Cygni), CX Cep which is too faint, and CV Ser with a period of 29.7 days. CV Ser should, therefore, have a large enough separation so that interaction effects would be minimal. It had already been observed by Hjellming and Hiltner (1963) to have a primary eclipse of 0.5 to 0.6 mag. Thus it seemed like the best candidate on which to repeat the emission-line experiment but unfortunately it stopped eclipsing (Stepien, 1970; Kuhi and Schweizer, 1970) as discussed above. Consequently this simple method of attack was totally doomed to failure.

In 1970, Brown, Davis, Herbison-Evans and Allen used the intensity interferometer to study the multiple star γ_2 Vel. They made measurements of γ_2 Vel (WC8 + O9I) both in the continuum at 4430 and in the light of C III–IV 4650 to determine the angular size of the star and of the C III–IV emitting region. The star would have a radius of $17 \pm 3 R_\odot$ and the envelope $76 \pm 10 R_\odot$ or $(0.24 \pm 04)a$ where a = semi-major axis if the distance were 350 ± 50 pc. This result was obtained by assuming a uniform brightness distribution across the disk of the C III–IV emitting region. Interestingly enough the mean radius of the critical Roche lobe for stars in the mass ratio 0.28 is $(0.26 \pm 0.01)a$. This suggests that the C III–IV producing region fills the critical Roche lobe around the WR star. The measurements also provided the first incontrovertible evidence that the line emitting region was indeed much larger than that responsible for the continuum, in this case ~ 4.5 times larger. Unfortunately no other lines were measured so that no information was obtained that could settle the stratification problem.

It now seems, however, that the question can be answered quite definitely from the behaviour of the emission line profiles as a function of spectral subclass (e.i. temperature) at least for the WC stars. Numerous authors have pointed out the strikingly flat-topped appearance of C III 5696 in the hotter WC stars but no one has interpreted the change in its profile with temperature. Because the line is normally very strong and hence quite difficult to analyze photographically I have obtained profiles for each spectral subclass from WC9 to WC5 with the Wampler photoelectric spectrum scanner at Lick Observatory using an exit slit of ~ 4 Å. This provides adequate resolution to map out the entire profile with an error of only a few per cent. Figures 4 to 8 illustrate

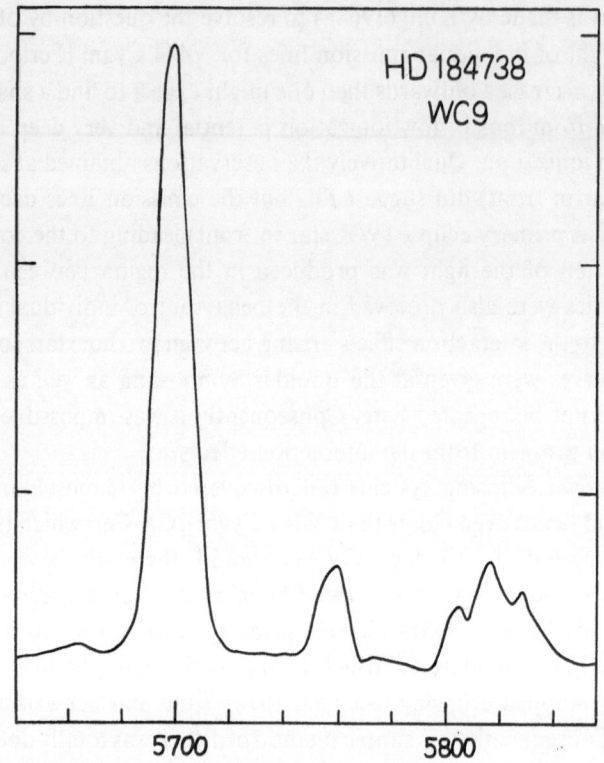

Fig. 4. Photoelectric spectrum scanner profiles of C III 5696 and C IV 5808 in HD 184738 (WC9). The vertical scale (arbitrary) is linear.

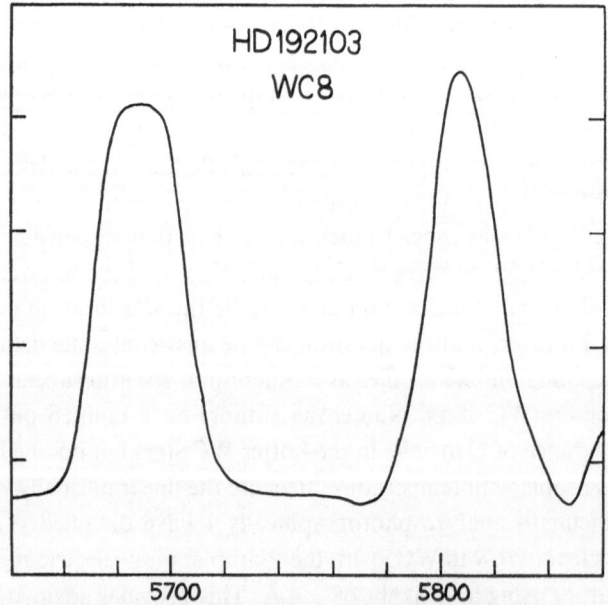

Fig. 5. Scanner profiles of C III 5696 and C IV 5808 in HD 192103 (WC8).

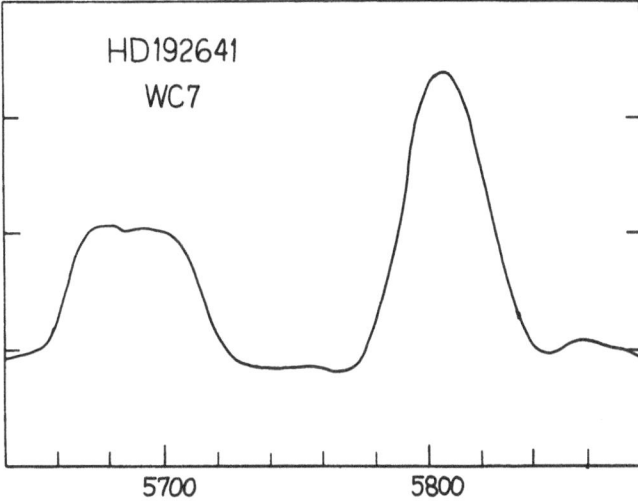

Fig. 6.   Scanner profiles of C III 5696 and C IV 5808 in HD 192641 (WC7 + Be).

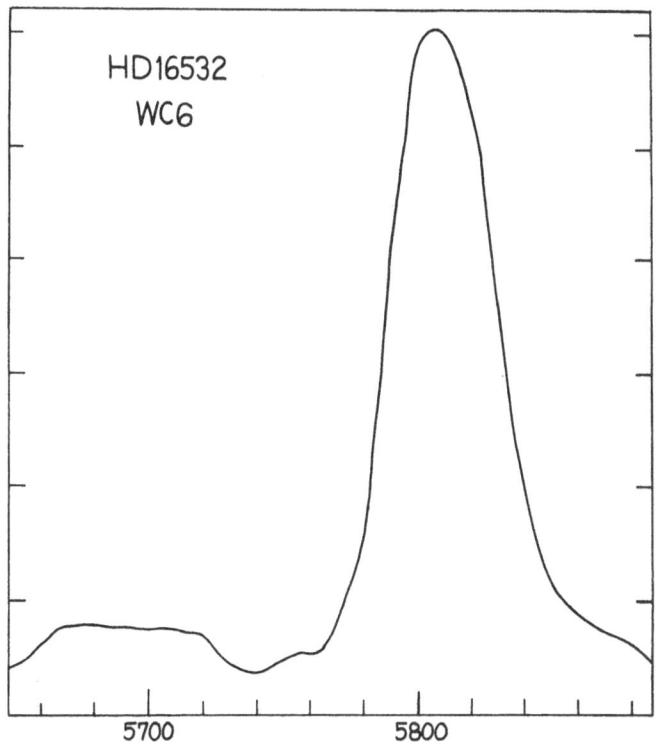

Fig. 7.   Scanner profiles of C III 5696 and C IV 5808 in HD 16532 (WC6).

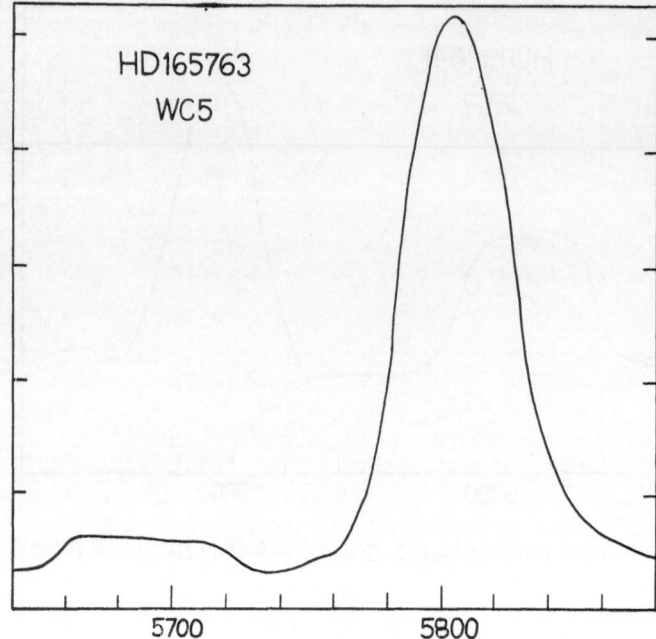

Fig. 8.   Scanner profiles of C III 5696 and C IV 5808 in HD 165763 (WC5). Note the dramatic flattening that has taken place in C III 5696.

the dramatic change from a distinctly round-topped profile in WC9 through progressively flatter profiles in the hotter types to a very flat-topped profiles in WC5. The increase in line width used to set up the the WC subclasses is also readily apparent.

It is my contention that this behaviour allows only one interpretation regarding the temperature stratification under reasonable assumptions concerning conditions in the envelope. We can apply Sobolev's theory for expanding envelopes as modified by Castor (1970) and Castor and Van Blerkom (1970) since their basic condition is definitely satisfied i.e. the velocity of expansion and hence the velocity gradient in the line of sight is very large, compared to the thermal motions. Hence the problem is dominated by the expansion of the envelope and emission and absorption can be assumed to take place only over the distance corresponding to two Doppler half-widths. I will not discuss the details of such calculations here since they are fully covered in the article by Van Blerkom in his discussion of the 'On-the-Spot-Approximation'.

However, the basic conclusion reached is that a flat-topped profile is produced in an optically thin region which is expanding with constant velocity. Applying this to C III 5696 means that in the hottest WC stars the line must be produced in some part of the envelope which meets these conditions. (Please note that optically thick or thin as used here refers to the optical depth in the distance in the envelope corresponding to two Doppler thermal half-widths according to the Sobolev approximation. Once outside this region the envelope is completely transparent). The most likely location is at some distance from the star so that the density and hence the optical depth has become sufficiently low. However, this also means that the C III in the interior regions must not be producing any λ 5696. Otherwise, since the density is high and consequently the optical depth large a round-topped profile would be produced instead. This can be accomplished (1) by increasing the degree of ionization so that C IV is the dominant ion (if T decreasing with radius) or (2) by cutting off the special mechanism that may be responsible for λ 5696. The constant velocity requirement must be met by the acceleration (or deceleration) mechanism which is still unknown.

Now if one accepts a decreasing temperature and an accelerating velocity field, then the above indicates that as one increases the stellar temperature as indicated by the spectral sub-class C IV becomes the dominant ion and produces a round-topped profile at all times which simply gets stronger as the type gets earlier. The temperatures reached are not high enough to produce sufficient C V so that it never becomes the dominant carbon ion. This C IV emission occurs throughout the envelope in high density regions as well as low and consequently in no case could it have a flat-top. The behaviour of C IV is not specified in the second case where the special excitation mechanism is destroyed. On the other hand with an increasing temperature and a decreasing velocity field then C IV must occur in a region exterior to C III 5696 i.e. in a region of even lower density and hence of even lower optical depth. Consequently the C IV lines must exhibit flat-topped profiles which would most likely become less flat-topped as one went to earlier spectral type because one would expect the C IV region to encompass more material at higher density with increasing temperature. This would apply in both cases since the region of higher temperature is always exterior to the C III region.

The observed behaviour of the C IV 5808 multiplet is also shown in Figures 4 to 8 and it is clearly seen that it is always round-topped and hence the increasing temperature alternative must be ruled out i.e. *the temperature must decrease radially outwards from the star*. This behaviour can also be demonstrated theoretically by computing the line profile expected from a radially expanding spherically symmetric envelope with some assumed velocity and temperature distributions using the 'On the-Spot Approximation'. We assume a temperature decreasing with increasing radius and a velocity law of the form

$$v^2 = v_0^2 + (v_\infty^2 \, v_0^2)\left(1 - \frac{R_0}{r}\right)$$

as might be expected if radiation pressure is responsible for the acceleration. v_∞ is the velocity at infinity, v_0 the initial velocity, r the radial coordinate and R_0 the stellar radius. This law guarantees the accelerating envelope where the density is large and the approximately constant velocity in the outer envelope where the density is low. We can then impose an arbitrary cut off to the C III 5696 emission without requiring a detailed knowledge of its excitation mechanism. That is to say, we can compute a series of profiles produced by the envelope with differing values of some critical inner radius inside of which no C III 5696 emission occurs. If the temperature profile were

even crudely correct then we should be able to mimic the qualitative behaviour discussed above. Figure 9 indicates the results: as the cut off radius increases the profile
becomes more and more flat-topped exactly in the manner observed. Similar calculations could be made for the temperature increasing with radius case but these

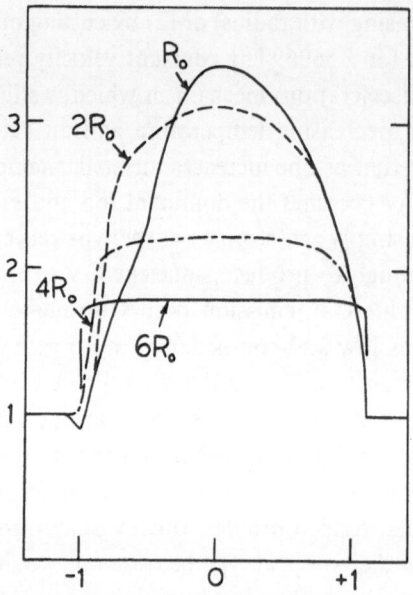

Fig. 9.   Theoretical profiles computed for a spherically symmetric radially expanding envelope. R_0
is the stellar radius, the continuum is normalized to 1.0, and the velocity scale to 1.0 for the maximum
velocity. The profiles show the expected change as the inner
cutoff radius becomes progressively larger.

were not performed since the observations already rule out the qualitative predictions.
Therefore, we can safely conclude that over most of the radius the temperature must
decrease outwards. This of course immediately rules out Limber's rotational model.

    The distinction between a simple increase in ionization or the destruction of a special
excitation mechanism is not so easily made. The difficulty is the severe blending that
is present in most emission lines in WC spectra and which makes precise profile determinations very hazardous. However, it does appear that other C III lines do exhibit
the same type of behaviour as $\lambda 5696$ (e.g. $\lambda 3608$, $\lambda 6720$) but not so well defined
because of blends. Also the flattening does not always start at the same spectral subclass: $\lambda 3608$ behaves similarly to $\lambda 5696$ but $\lambda 9710$ does not really look flat-topped
until WC6 or WC5. Consequently it seems likely that both possibilities for the inner
cutoff occur but that the special excitation mechanism starts getting destroyed before
the conversion of the C III region into predominantly C IV by increased ionization takes
place. The C II lines in stars hotter than WC9 are so weak and badly blended as to
render them useless in profile determination. They should of course become flat-topped
even more rapidly than the C III lines and remain so until there is virtually no C II left.

    In principle it also seems possible to set definite dimensions to the C III emitting region if one had a satisfactory velocity and temperature distribution. Further work along the lines of Castor and Van Blerkom's theoretical studies is clearly in order before such estimates can be made.

The final picture for the temperature profile requires one additional detail, namely reconciliation of the very high electron or kinetic temperatures required by the presence of lines from O VI and other highly ionized atoms and the considerably lower photospheric temperatures implied by the continuum (Kuhi, 1966) and other temperature determinations (Morton, 1970). Since the observations rule out any increase in temperature occurring over a large fraction of the radius the only acceptable solution is to have a very rapid temperature increase near the surface occurring over a very small (and hence unobservable) range in radius, probably less than 1% of the total radius. This would then correspond to the region in which the energy required to maintain the envelope is dumped by some mysterious as yet totally unknown mechanism. The final temperature profile then would have a very rapid initial rise to a few 100000 K followed by a gradual decrease with increasing radius.

Therefore, one might conclude that the temperature stratification problem is definitely solved for the WC stars. Unfortunately no such firm statement (except by analogy or by faith) can be made for the WN stars since no flat-topped lines are observed other than He I. Comparing He I profiles which are flat to He II profiles which are round-topped in all types would strongly suggest, however, that the same conclusion applies to the WN stars as well. Ionization equilibrium calculations would help considerably here since the conclusion implies that N III persists longer in large enough quantities in the WN stars than C III does in the WC stars. Otherwise the N III lines should be more flat-topped than they are observed to be in the hotter WN stars. Alternatively the temperature structure may remain the same among different subclasses while the extent of the envelope changes.

Acknowledgements

I would like to thank the National Science Foundation for its support of some of the work discussed here via grant GP-12824. I would also like to thank Mr. Pui Kuan for his assistance in the calculations and Dr. L. F. Smith and P. S. Conti for many valuable discussions and prepublication communications.

References

Baschek, B.: 1970, *Astron. Astrophys.* **7**, 318.
Baschek, B. and Scholz, M.: 1971, *Astron. Astrophys.* **11**, 83; **12**, 322.
Bracher, K.: 1967, Thesis, University of Indiana.
Brandt, J. C., Stecher, T. P., Crawford, D. L., and Maran, S. P.: 1971, *Astrophys. J.* **163**, L99.
Brown, H. R., Davis, J., Herbison-Evans, D., and Allen, L. R.: 1970, *Monthly Notices Roy. Astron. Soc.* **148**, 103.
Castor, J. I., Smith, L. F., and Van Blerkom, D.: 1970, *Astrophys. J.* **159**, 1119.
Castor, J. I.: 1970, *Monthly Notices Roy. Astron. Soc.* **149**, 111.
Castor, J. I. and Van Blerkom, D.: 1970, *Astrophys. J.* **161**, 485.

Conti, P. S. and Smith, L. F.: 1971, *Astrophys. J.*, in press.
Cowley, A. P., Hiltner, W. A., and Berry, C.: 1971, *Astron. Astrophys.* **11**, 407.
Gaposchkin, S.: 1949, *Perem. Zvezdy* 7, 36.
Graham, J. A.: 1965, *Observatory* **85**, 196.
Hiltner, W. A.: 1951, *Astrophys. J.* **113**, 317.
Hjel ming, R. M. and Hiltner, W. A.: 1963, *Astrophys. J.* **137**, 1080.
Kron, G. E. and Gordon, K.: 1950, *Astrophys. J.* **111**, 454.
Kuhi, L. V.: 1966, *Astrophys. J.* **143**, 753.
Kuhi, L. V.: 1968a, *Astrophys. J.* **152**, 89.
Kuhi, L. V.: 1968b, in K. B. Gebbie and R. N. Thomas (eds.), *Wolf-Rayet Stars*, N.B.S. Special Publ. No. 307, p. 103.
Kuhi, L. V. and Schweizer, F.: 1970, *Astrophys. J.* **160**, L185.
Limber, D. N.: 1964, *Astrophys. J.* **139**, 1251.
Morton, D. C.: 1970, *Astrophys. J.* **160**, 215.
Münch, G.: 1950, *Astrophys. J.* **112**, 266.
Smith, L. F.: 1968a, *Monthly Notices Roy. Astron. Soc.* **138**, 109.
Smith, L. F.: 1968b, in K. B. Gebbie and R. N. Thomas (eds.), *Wolf-Rayet Stars*, N.B.S. Special Publ. No. 307, p. 21.
Stepien, K.: 1970, *Acta Astron.* **20**, 13.
Tcherepaschuk, A. M.: 1969, *Astron. Cirk.*, 509.
Thomas, R. N.: 1949, *Astrophys. J.* **109**, 500.

DISCUSSION

Sahade: We started talking about percentage of binaries and the ways in which we can detect them. Perhaps Virpi Niemela could add a few words about this question. I am happy to see that the figure that represents the percentage of binaries among WR stars surpasses 70%; it is such a large percentage that there can be little doubt that they all must be binaries. I feel certain that the next meeting will find everybody agreeing on this.

Niemela: I should like to show some spectra of γ_2 Velorum taken by Perrine in 1919 (Figure 10), where you can see the violet-shifted, variable absorption edge of the He I 3888 line. The variations of

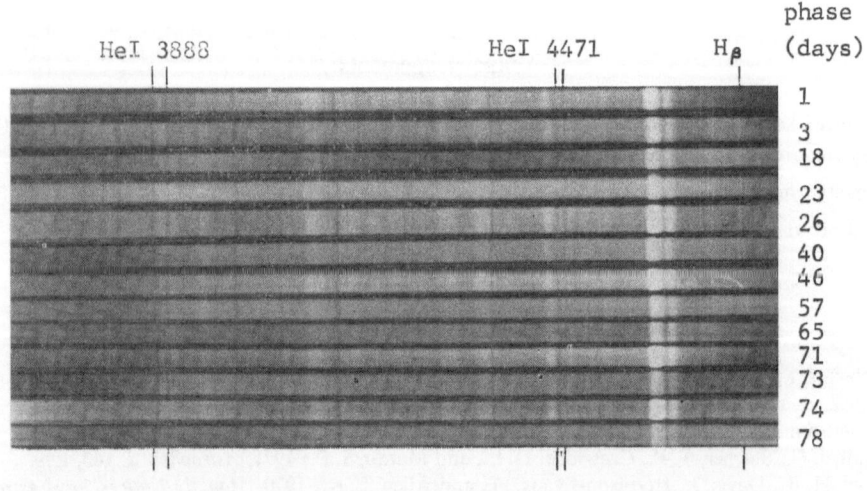

γ_2 Velorum 1919

Fig. 10.

the shape of this absorption line depend on the phase of the binary, and so do the V/R variations of the Balmer and the He I emission lines. The same thing is shown by the spectra taken by Dr. Sahade in Cordoba, between 1948 and 1962. When the O star is in front, the violet absorption edge of He I 3888 becomes very strong, and the emissions have their peak intensity on the violet side. When the Wolf-Rayet star is in front, the violet-displaced absorption edge of He I 3888 is very broad and very shallow, and the emission lines have their intensity peak shifted to the red. These V/R variations in the emission lines could perhaps be explained in terms of Doppler effect as mass flows from the WC to the O star. In microphotometric tracings this kind of variations can be better observed. Figure 11 shows the He I 3888 emission; at phase 20 days (according to Ganesh and Bappu), the violet part is much stronger and the violet absorption edge very intense, while at phase 60 days, the violet displaced absorption is wide and shallow, and the emission line intensity peaks on the red side. The same kind of behavior can be observed on the spectra taken by Perrine, so it seems that no changes have occurred in 50 years time. The microphotometric tracings of Figure 12 show similar V/R variations in Hβ on Perrine's plates.

Coming back to the violet-displaced absorption of He I 3888, at phases near 20 days, and perhaps also near 64 days, there appear two components, one of which changes velocity very fast, in 24 h, or so, and then disappears. At phase of about 17 days, there is a remarkable splitting into two very strong components. He I 4471 seems to exhibit the same kind of behaviour. In Figure 13 we can see the variations in radial velocity of the violet-displaced He I 3888 absorption. The circles are velocities relative to H8, since Perrine's plates have no comparison spectra. The velocities go from

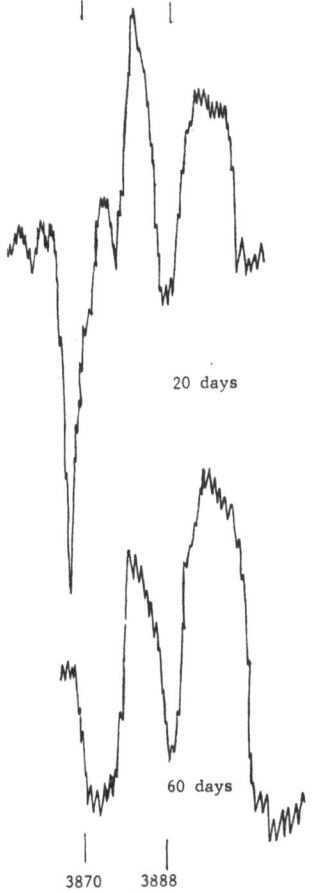

20 days

60 days

3870   3888

γ_2Velorum 1960

Fig. 11.

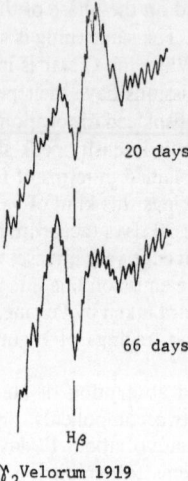

20 days

66 days

Hβ

γ₂Velorum 1919

Fig. 12.

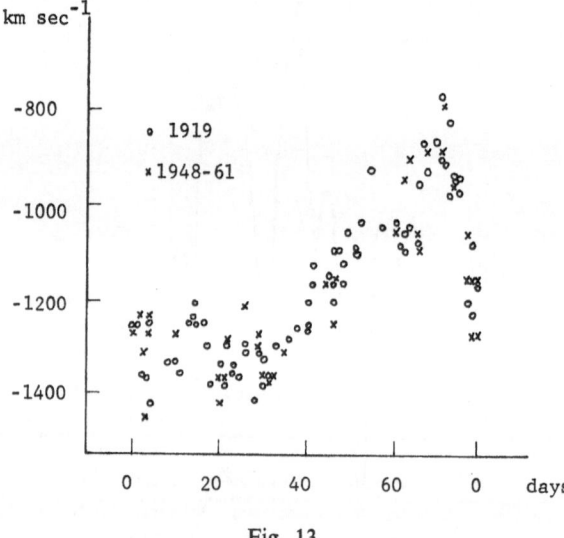

Fig. 13.

800 km s⁻¹ to − 1400 km s⁻¹, so the amplitude is twice as large as that derived from the velocities of the emission lines.

Bappu: And this reproduces every period?

Niemela: Yes, Perrine's plates cover 2½ periods. The same curve is derived from the plates Dr. Sahade took in Cordoba between 1948 and 1962. The superposition of the two curves has permitted me to improve Ganesh and Bappu's period to 78.5004 days.

Sahade: The behaviour of the violet-displaced He I 3888 with phase is similar to what we find in stars like β Lyrae, for instance. There is a certain variation in the shape, in the structure, and in the profile of He I 3888 which changes with phase, and this must be connected, in the case of β Lyrae, with the gas stream that comes from the B8 component towards the unseen companion, and the variations must be also connected, not only with the geometry of the problem, but also with the out-flow of matter. Allen Batten and I are studying new material on β Lyrae, and we hope to being able to throw more light on the problem of He I 3888.

Underhill: A part of the larger amplitude of λ3888, may be due chiefly to increased rates of ejection at those phases. Either you say, the largest amplitude is the amplitude of the stellar motion, or you say the smaller one is, and you are always getting more outflowing gas.

Sahade: I think that λ3888 is formed way out from the system, and, therefore, the amplitudes might be connected with the shape of the outer envelope, and perhaps with the way in which material is replenished to the envelope. We have to check this.

Conti: Even with such a sharp line, you think it is due to the system, rather than...?

Sahade: But sometimes it is not very sharp, and sometimes you see many components.

Underhill: The question I am trying to get at is, whether that displacement is the real amplitude, or whether it is just a false amplitude due to seeing it at moments of greater outward driving force.

Conti: I would think that that is the correct amplitude.

Sahade: We have to observe the star in many cycles and see whether there is any correlation between the intensities of the line, the behaviour velocity-wise, and so on.

Underhill: How big is the orbit? It is a long period, 78 to 79 days, but with respect to the size of the stars (you know the nominal size of an O9 and the nominal size of a Wolf-Rayet say 20 solar radii), what is the size of the orbit? How much space is there in between? Because, if you are going to see λ3888 in absorption over all that space, the absorbing gas must have something to be projected against, which makes you wonder if there is a lot of what I call 'false' light or extra light in the system due to a lot of electron-scattering.

Smith: The separation of the system is 300 solar radii, according to Bappu's figures.

Kuhi: Could I say something about λ3888 also? In V444 Cygni for example, which is a close binary, that line, does not seem to have the motion of the binary system, although at certain phases, it doubles by quite a large amount. So, you might say that in a binary system in which the stars are close together, the He I emitting region really extends past both stars and it is very large, surrounding the entire system. In the case of γ_2 Velorum, the period is so long and the separation quite large, that it may be that the He I emitting region again probably being comparable in size to what it was in V444 Cygni is sharing the binary motion because it is still far enough away from the O-type star. However, it is still close enough so that there is enough interaction so that you actually end up with streams of material perhaps flowing around the O-type star and seen in projection against it. I do not really know, I am just suggesting the possibility.

Conti: I will throw in something else. HD 193793 is the star in which this violet-shifted absorption line at λ3888, discovered by Anne Underhill seems to be a binary of a rather long period. I have plates extending over 3 or 4 yrs and I think the period is a year or a few years. I have measures of λ3888, which is quite broad, plus the violet-shifted absorption component of the C III triplet. Both show the same large violet shift, and they show a motion which is opposite to that of the absorption lines in the O star. So in HD 193793, which is apparently a very wide binary, the λ3888 appears to follow the star.

Sahade: There is a case, HD 211853, where the opposite was true i.e., in which the velocities from the violet-displaced line of He I 3888 followed the O star, rather than the WR star.

Bappu: Almost in every case that I can think of, starting from CQ Cephei, the behaviour of λ3888 is different from the rest of the lines.

Kuhi: It is certainly always different, but is it really out of phase with the other lines?

Bappu: I have never come across a situation where it was out of phase.

Underhill: That is very reasonable for the spectroscopic characteristics. If you are ever going to see any reasonably hot, thin gas, you are going to see it by means of that line. The only better line would be λ10830, but it is too difficult to observe.

If it goes with the speed of the star, it will not go any faster than the star around the orbit. However, if you are making an apparently larger K, it must be going faster. Then you have to say that those changes that give you an idea of an apparent extra K, are actually changes in outflow velocity. And that means that in the case of γ_2 Velorum we are observing changes of outflow velocity on the order of 1200 km s^{-1}, which is a considerable amount.

Sahade: Now, as regards the location of the source, I think that we have to remember that Stecher's UV observations showed us the intercombination line, which should arise from a less dense region, and the suggested expansion velocity is the same as the one you obtained from He I 3888. I remember Stecher reported these observations at Boulder.

Morton: Stecher's resolution and wavelength scale are not adequate for estimating a velocity shift of λ1909. However, one Princeton UV spectrum shows that the C III 1175 line does show an ejection velocity of a few hundred kilometers a second, but that could be part of the overall Wolf-Rayet ejection effect. I see no reason to suspect that it is connected with the kinetic problem.

Sahade: Summarizing, I think we should accept that the envelope where the violet-displaced He I 3888 is formed, surrounds the whole system, rather than only the W star.

Niemela: Then, why does He I 3888 show a radial velocity curve of such large amplitude?

Sahade: I still think that the reason must be connected with what I said a little bit earlier.

Voice: And perhaps the large distance between the components.

Morton: The distance between the components does not particularly worry me, because the interferometer data shows that the Wolf-Rayet shell is still in the Roche lobe, and with a situation like that, you can imagine some additional complications.

Underhill: What does worry me, is the apparent changes in the ejection velocity; these are considerable changes.

Bappu: This particular plot of yours of λ 3888 shows up the fact that you do find this amplitude of the velocity curve from cycle to cycle, and that the curve is well defined.

My question now is whether the apparent increases in the amplitude are confined to a very limited range of phase values around zero. It is around zero, that you get this sharp, violet-displaced feature which comes at about 1300 km s⁻¹, over and above the normal value of about 800 or so for the Wolf-Rayet star.

Now, if you get it sometimes, it is possible that at phase zero, your points go up to 1300 and then increase the overall velocity value. Could this the be reason, or is the cause something else which is systematic and smooth, right through the entire velocity curve?

Sahade: I suppose you have to evaluate the whole thing by considering the way the profile changes, before you try to draw any conclusions.

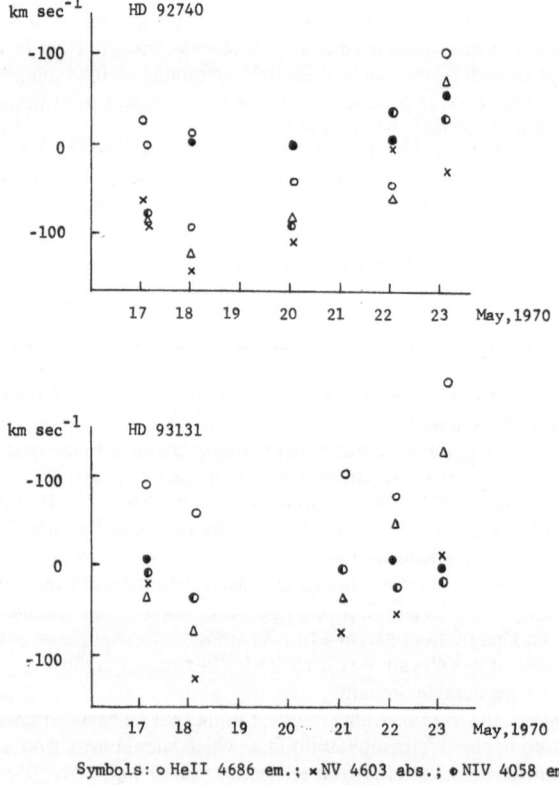

Symbols: o HeII 4686 em.; x NV 4603 abs.; ● NIV 4058 em.;

△ NIII(mean) em.; ● CaII int.abs.;

Fig. 14.

Bappu: No, I am not trying to draw any conclusions. I am only saying, that if you plot over the conventional λ3888 absorption line velocity curve, you will find that around phase zero, where you do get some of these sharp displaced features an additional 600 km beyond, then, apparently, the whole amplitude of the velocity curve would change by another 600 km.

Niemela: I never used the two components. I always used the most intense component, and when it is double it only lasts for some 24 h. And 24 h in almost 80 days is a very small fraction that could not affect the velocity curve. If I may, I would like to show you observations of another two WR stars, HD 92740 and HD 93131, both WN7 and in Carina, that show the Balmer lines in weak absorption. The observations were made at Tololo. Figure 14 show plots of the measured radial velocities; they look like parts of velocity curves. The two stars have the same apparent magnitude, the same spectral sub-type; have they about the same period?

De Groot: How are the curves for different elements, do they cross over or are they parallel?

Niemela: They are approximately parallel. This is what usually happens in WR stars, the different elements suggest different γ-velocities. There is another thing: N v in absorption is not violet-shifted, while the N iv 3482 absorption edge is violet-shifted some 500 km s^{-1}.

Kuhi: Should you not point out that radial velocity value of the interstellar line at about phase 23 days, which indicates that there may be something wrong with the spectrograph?

Niemela: Yes, in this case there may be something wrong. In the others cases, the velocities of the interstellar line were just the same all the time. In both stars the hydrogen absorption lines are seen up to H11. The radial velocity appears also to be variable, the difference between extreme values being larger than 100 km s^{-1}.

AN INTERPRETATION OF THE WC STARS

LINDSEY F. SMITH

*NASA Goddard Space Flight Centre, Greenbelt, Md., U.S.A.**

I wish to draw your attention to a (quite unexpected) property of the WC binaries. I believe that it gives us a clue to one of the parameters that controls the subclasses of the WR stars.

1. Observations

Table I collects the information available regarding the separations of all WR binaries for which there are adequate observations. The data are taken from the same references

TABLE I

Separations of Wolf-Rayet Binaries

HD	Name	Sp. Type	$a_w \sin i$ ($R_\odot$)	$a_0 \sin i$ ($R_\odot$)	$\sin i$	Basis of $\sin i$	$A/R_\odot$
186943		WN4 + B	⎰ 29.7 ⎱ 44.3	12.6	0.77	Assume $M_{WR} \approx 11 M_\odot$	64:
190918		WN + O9 I	108	17.6	0.36	Assume $M_{WR} \approx 11 M_\odot$	350
MR 114	CX Ceph	WN5	12.1		>0.94	Eclipsing	26 [a]
193576	V444 Cyg	WN5 + O6	25.6	10.0	0.98	Eclipsing	36
211853		WN6 + O6I	27.5	13.1	0.96	Eclipsing	53
228766		WN7 + O6V	50	10.6::	≈ 1	Large mass function	61
214419	CQ Ceph	WN7	⎰ 9.6 ⎱ 5.0		≈ 1	Eclipsing	20: [a]
193793		WC7 + O5	Period 3 years (Conti, private communication large!)				
152270		WC7 + O5–8	28.2	7.6	0.52	Assume $M_{WR} \approx 11 M_\odot$	69
68273	γ_2 Vel	WC8 + O9I	234	65.7	1	Large minimum masses	300
168206	CV Ser	WC8 + B0:	86.5	26.4	1	Was eclipsing in the past	113

[a] Assuming $M_{WR}/M_{OB} < 1$

used to compile Table VI of Smith (1968a) and Table I of Kuhi (this Symposium). In many cases the inclination of the system is unknown; I have estimated it by assuming a mass of 11 $M_\odot$ for the WR star. (Since the quantity derived from the observations is $M_w \sin^3 i$, the derived value of $\sin i$ is not sensitive to the assumed mass, and the values obtained for $\sin i$ should be reasonably accurate.) It is immediately obvious

* Present address: Max Planck Institut für Radioastronomie, 53 Bonn, Germany.

M. K. V. Bappu and J. Sahade (eds.), Wolf-Rayet and High-Temperature Stars, 228–234. All Rights Reserved.
Copyright © 1973 by the IAU.

that the separations of the binaries containing WN stars are all, with one exception, less than $65R_{\odot}$, while those of binaries containing WC stars are all greater than $65R_{\odot}$, often much greater. I think we may safely deduce that the wide separation is a universal property of WC binaries and must be a consequence of the evolution that creates WC stars.

The difference in separation characteristic of WN and WC binaries provides an explanation for the observation that spectra of WC stars are apparently unaffected by the presence of a companion (see, for example, the spectra of WC5 binaries in the Large Magellanic Cloud shown by Smith, 1968b); whereas the spectra of WN stars in binary systems appear to be significantly different from those of single stars with comparable degree of excitation (see Hiltner and Schild, 1966). The WC binaries are sufficiently widely separated that the WC atmosphere is unperturbed by the presence of the companion, while the companions of WN stars are closer and gravitational perturbation, reflection effects and limitation of the atmosphere's extent by the Roche lobe (see Limber, 1968) may each have an effect. This suggestion is confirmed by the observation that the spectrum of the WN4 star in the very wide binary, HD 190918, shows little sign of perturbation by the companion – in particular, the line widths are the same as in the spectrum of the 'single' star, HD 187282.

2. Interpretation

The binary WR stars have certainly all undergone mass exchange according to the original suggestion of Paczynski (1967). The separation of a binary changes during the mass exchange. The final separation is usually greater than the initial separation; however, its value depends on the initial separation, initial masses and also on the unknown amount of mass loss to the system as a whole (Paczyński and Ziolkowski, 1967). Thus, it is not obvious what values of the initial parameters will yield the observed final separations. However, let us consider the simplest possibility, that the large final separations of WC binaries result from large initial separations. (I suspect that mass loss is probably also vitally important but, in as much as its effect is not yet clear, I will not pursue that thought at this time.) Thus, I suggest that the initially closer binary systems produce WN stars, and the initially wider binary systems produce WC stars. That the separation should cause such a difference is not unreasonable, since the initial separation and mass ratio defines the stage in the evolution of the primary at which mass exchange commences. If the binary is wide, mass exchange will commence later. In the most extreme case (case C – Weigert, 1968) mass exchange does not begin until after ignition of helium burning in the core of the star; such a situation has the attractive consequence of creating plenty of carbon and oxygen for subsequent production of the characteristic spectrum of the WC stars.

It is of interest to ask if the initial separation might also be important in the definition of the subclasses of the WN stars. If it is so, I suggest that the separation increases from left to right in Table VIII of my review paper. Adding the WC stars and the initial separation to that table produces Table II.

TABLE II
The WC stars

$M/M_\odot$	M_{bol}								WC8
17	−9.5			⊤			⊤		⊤
14	−9.0		⊤	WN7		⊤	WN5	WN5–C6	WC6
11.5	−8.5	WN8		⊥	⊤	WN6	⊥	WN6–C7	WC7
8.5	−8.0	⊥			WN3	WN4			
7	−7.5				⊥	⊥			⊥

		(H/He)	≪ 2.3	< 1.0	0.8:	0.4	0.0		0?
		T_{eff} (K)	23 000°			{ 46 000 ⟩ 40 000 ⟨ 29 000 }	53 000		50 000?

Increasing Initial Separation ⟶——⟶——⟶——⟶————⟶————⟶

3. Discussion

I offer the following observations in support of the hypothesis that the order of the subclasses given in Table II represents a smooth progression of the basic stellar parameters and may be due to a progression of increasing initial separation.

(I) *The intermediate WN-WC types* always have WN subclasses found at the right of the diagram, e.g. HD 62910, WN6-C7; HD 90657, WN4-C; HD 117688, WN6-C; MR 76, WC7-N6; NGC 6751, WC6-N5.

(II) *The WN7 binary, CQ Cep* is the closest binary known, consistent with the position of the WN7 stars on the left side of the table. If a binary is initially very close, mass exchange can commence before exhaustion of hydrogen in the core of the star (Kippenhahn and Weigert, 1967; Case A); the comparatively high H/He ratio in WN7 and WN8 stars may result from such an evolution. (It should be noted that the WN5 binary, CX Cep, also appears to have a very small separation, indicating that the scheme is not yet perfect.)

(III) *The 'single' WN5 and WN6 stars* commonly found in ring nebulae may be interpreted as very widely separated or disrupted binaries (see Paczynski, 1967), consistent with their place on the right side of the diagram. Note that, for disruption of a binary to occur, significant mass loss from the system as a whole is probably required. This may account for the presence of the nebulae around these particular subclasses, and also for the efficient removal of nearly all of the hydrogen from the surface of the star.

(IV) *Hydrogen abundance* decreases to the right among the WN stars. Thus, the suggested placement of the WC stars implies that they also have no hydrogen in their atmospheres. Kuhi and I have not attempted a derivation of the H/He ratio of the WC stars because of the severe blending of carbon lines with the Pickering series. However, Paczynski, in his reveiw (this conference) gives us a choice between cosmic abundance and zero hydrogen for WC atmospheres. If the H/He ratio were 10/1, as for the cosmic abundance, then among the later lines of the Pickering series (which presumably eventually become optically thin) the even-n lines would have 10 times the strength of the odd-n lines. I think that, despite blending with carbon, such a contrast would be immediately noticeable. Thus, I anticipate that a H/He ratio of 10/1 will be ruled out by the observations.

It is worth noting that the strength of the helium lines is much less in the WC spectra than in the WN spectra; this is in accord with Paczyński's prediction that the helium abundance will begin to be significantly depleted in the extreme case where all the outer envelope of the star is lost, and the products of helium burning have been somehow mixed to the surface.

From the spectrum of γ_2 Velorum, Castor and Nussbaumer (1971) derive a lower limit to the number ratio of carbon to electrons of $2.5\ 10^{-3}$. That number is the order of magnitude expected if all hydrogen is gone and the helium is fully ionised. However, since it does not include C IV ions, Castor and Nussbaumer point out it may be an underestimate by a large factor.

(V) *Temperature* increases to the right among the WN stars. Thus, Table II implies that all the WC stars have very high photospheric temperatures. This may seem surprising considering the quite low excitation of spectra such as that of the WC 9 star in γ_2 Velorum. However, consider the fact that that star has a minimum mass of 15 $M_\odot$ (Ganesh and Bappu, 1968). If it is a pure helium star, its bolometric magnitude should be about -9.2. Thus, an observed visual absolute magnitude of -4.8 implies a bolometric correction of -4.4 and a temperature of 50 000 K.

The planetary nebulae associate with WC8 and WC9 nuclei are usually of quite low excitation. Johnson's suggestion (this Symposium) that the atmosphere of the WR star itself may be capable of 'smothering' the UV radiation is of interest in this regard. If the stars are as hot as I suggest, they must be quite small; $\simeq 15 R_\odot$ in the case of γ_2 Velorum. The atmosphere of γ_2 Velorum is, however, about 100 $R_\odot$ (Hanbury Brown *et al.*, 1970). (Note a distance of 460 pc is used to obtain this value; Hanbury Brown *et al.* use 350 pc.) The large size of the atmosphere relative to the core together with the profusion of broad lines in all regions of WC spectra may mean that line opacity is dominant at all wavelengths and the amount of radiation escaping from the photosphere may be negligible. I think this is probably an equivalent statement to Johnson's 'smothering'.

It is of interest that, while the planetary nebulae around WC8 and WC9 nuclei are of low excitation, NGC 6751, which has a WC6 nucleus, is of moderately high excitation; thus, a temperature sequence like that among the WN stars may be indicated.

4. Conclusions

The observations are in reasonable accord with the suggestion that Table II represents a smooth progression of physical properties of the WR subclasses.

Ultimately, we would like to be able to arrange the classes according to the 'causal' parameters. To this end it has been suggested that the initial separation may be one of the more important such parameters and that its relationships to the subclasses and other final properties of the stars may be approximately as indicated. However, it seems certain, from both observational and theoretical considerations that a one-parameter family, as suggested by Table II, is an over-simplification. The evolution of the stars is probably sensitive not only to the initial separation, but also to the initial masses, angular momenta and chemical composition of the stars. The relationship of these initial parameters to the final subclasses, temperature, H/He ratio, or separation may not be simple, or even single valued. Thus, the absence of an obvious and simple relationship between the final properties and the inferred initial properties is not surprising.

It would be an interesting experiment to deduce the initial configurations from the observed final configurations. The unknown amount of mass loss from the system represents a major uncertainty; however, calculations ignoring that problem would be a first step.

Acknowl^{e}dgements

The manuscript was prepared while the author was at the Institut d'Astrophysique, Liège, and was supported by an ESRO Fellowship.

References

Castor, J. I. and Nussbaumer, H.: 1971, *Monthly Notices Roy. Astron. Soc.* **155**, 293.
Ganesh, K. S. and Bappu, M. K. V.: 1968, *Kodaikanal Obs. Bull.* **183**, A77.
Hanbury Brown, R., Davies, J., Herbison-Evans, D., and Allen, L. R.: 1970, *Monthly Notices Roy. Astron. Soc.* **148**, 103.
Hiltner, W. A. and Schild, R. E.: 1966, *Astrophys. J.* **143**, 770.
Kippenhahn, R. and Weigert, A.: 1967, *Z. Astrophys.* **65**, 251.
Paczyński, B. E.: 1967, *Acta Astron.* **17**, 355.
Paczyński, B. E. and Ziolkowski, J.: 1967, *Acta Astron.* **17**, 7.
Limber, N. E.: 1968, in K. B. Gebbie and R. N. Thomas (eds.), *Wolf-Rayet Stars*, U.S. Government Printing Office, Washington, D.C., p. 233.
Smith, L. F.: 1968a, in K. B. Gebbie and R. N. Thomas (eds.), *Wolf-Rayet Stars*, U.S. Government Printing Office, Washington, D.C., p. 44.
Smith, L. F.: 1968b, *Monthly Notices Roy. Astron. Soc.* **140**, 409.
Weigert, A.: 1968, *Mitt. Astron. Ges.* **25**, 19.

DISCUSSION

Underhill: There is one problem. In some of the WC's, I think HD 192103 is one of them, Hβ, with the blends and all, is very considerably stronger than λ4541.

Smith: Ten times?

Underhill: What has the factor of 10 to do with anything? If all the things that are making that line are pretty optically thick, it doesn't matter if you add one hundred times; you do not increase the intensity that much.

Smith: Yes, that is correct, and Hβ is still thick, but by the time you get down to the thin lines...

Underhill: I am not talking about any optically thin lines, I am talking about Hβ and λ4541, both of which, I believe, are probably rather optically thick.

Smith: Yes, but there are high lines which are eventually thin.

Underhill: Try to remember that there is a very definite intensity difference between those two lines. And you are saying there should be practically none. I cannot interpret that intensity difference without going to the theory, because I am pretty certain that it is not an optically thin situation.

Smith: I agree. From the thick lines you cannot say anything because differences of line strengths of a factor 2 are more likely to be due to blending with C than with H. A definite answer is only possible when the lines are thin so that the full factor 10 will show up if H is present in its cosmic abundance.

Conti: I have a comment here. First of all, if one wants to put an element like hydrogen in there, then, you must see it. Also another interpretation one could have instead of the initial separation arguments which goes exactly the same way, is that the change in the period is a clue to the change in the mass ratio.

Smith: Yes, I am aware of that. That is the other possibility.

Conti: This argument says that a WC star would have peeled off more of its envelope, and become further separated.

Underhill: Were you first trying to infer that it is not true that the intensity of Hβ is not strong, considering the blends?

Conti: I was just saying you can play games if you see something or if you do not see something, even *before* you have a theory. But you cannot say "just because you do not see it, it is there, although I do not have a theory to explain it".

Underhill: You cannot say it is not there, if you consider that things are pretty well optically thick.

Conti: No, but you cannot say that it is there, if you do not see it.

Thomas: But do you believe that there is hydrogen in these stars?

Underhill: It depends on what 'seeing' is. What do you mean by 'seeing' something?

Conti: Hydrogen is seen in every star, except in these kinds of stars. That means there is definite observational evidence.

Thomas: Let me remind you the remark that Shapley once made, in discussing solar physics. He said, "Once you have seen one of Menzel's invisible prominences, you have seen them all". I think that is a very appropriate thing here.

Niemela: There are four stars, in Lindsey Smith's Catalogue, classified as intermediate objects. I took spectra of two of them, HD 117688 and HD 90657, and I could not confirm that they really are intermediate objects, they are just normal WN stars. HD 117688 was classified as an intermediate object, because of the emission line at λ4325 which was assigned to C, but as you can see in Figure 1, the line is present in other normal WN stars: it is actually a N III line, not a C line. Another reason for the classification as an intermediate object, was that the N III emission at $\lambda\lambda$4634-40-41 had a contribution from C. On Figure 2 you can see that the line is centered at λ4638, so it should be N III.

Smith: I can always be wrong. HD 117688 is one of the milder cases where the C contribution appeared to me much weaker than the N contribution. However, in cases such as HD 62910 and MR 76 I think there is little doubt, the spectra look like double exposures between WN and WC stars. I am very glad that you have continued the investigation, because it has obvious relevance to this somewhat revolutionary proposal I just made.

Niemela: Another thing I need to point out are those V/R variations I was talking about in the first part of the discussion. They are seen in the Balmer lines, and in the He I lines, but not in He II. Could it mean that there is hydrogen in γ_2 Velorum?

Smith: Yes, that is the alternative explanation, of course. A naive interpretation of the mass exchange theory might suggest that if the separation is wider you may get more hydrogen left on the surface. My feeling was that the scheme presented fits the observations more consistently. Hopefully the theoreticians can solve this one, if the observations cannot.

Underhill: The answer to this question of leaving hydrogen on the surface, I think is pointed well towards the direction of hydrogen being on the surface, when you look at the spectra of two X-ray sources, that have optical spectra. There you see very strong hydrogen lines in emission as well as typical C, He II, O V and so on. There is no doubt about it: nobody has ever questioned that the hydrogen is there. Perhaps you would say, an X-ray source is a very old star, a helium star, very far long in evolution. However, it managed to keep some hydrogen which shows in the spectrum. That qualifies it, by the definition of the Wolf-Rayet spectrum, as a Wolf-Rayet object.

Conti: There is a theory I heard very recently about X-ray sources, which suggests that an X-ray source is a binary system in which material is falling in to one star, and, of course, that would all be hydrogen. So, the argument can go either way.

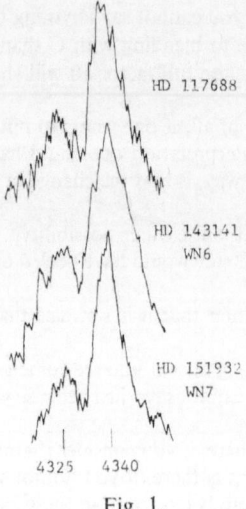

HD 117688

HD 143141
WN6

HD 151932
WN7

4325 4340

Fig. 1.

HD 117688   4340   4638

Fig. 2.

Van Blerkom: Can I ask about the temperature in the envelope of the X-ray source. It must be very hot.

Underhill: The region that is creating the X-rays, according to one calculation I read, is between 10^7 and 10^8 degrees and N_e is 10^{16}. The spectrum that you see, the Wolf-Rayet like spectrum looks like a very cool WC, except that it has very strong hydrogen lines.

Van Blerkom: The fact that you see hydrogen strongly in emission means that it probably has not been ionized away, even in these very hot envelopes, and it certainly would not be ionized away in the cooler Wolf-Rayet star.

Underhill: It would be an interesting question; exactly what happens to it and how it emits so strongly. So, we have now a question. For every Wolf-Rayet star, and there are some where there is very definitely Hβ stronger than $\lambda 4541$, can we always throw in a third star to provide us with the hydrogen?

SECTION VIII

CHAIRMAN: R. N. THOMAS

SUMMARY OF PROBLEMS AND CONCLUSIONS ON THE NATURE AND PHYSICAL STRUCTURE OF WOLF-RAYET STARS

ANNE B. UNDERHILL

Goddard Space Flight Center, NASA, Greenbelt, Md., U.S.A.

1. Introduction

The Wolf-Rayet stars were the subject of a symposium held at the Joint Institute for Laboratory Astrophysics in Boulder, Colorado in June 1968. Here there was considerable discussion about whether one should talk about a class of objects called 'Wolf-Rayet stars' or whether it was more appropriate to talk about something called the 'Wolf-Rayet phenomenon' (Gebbie and Thomas, 1968). The point of view taken in this review is that it is more advantageous for obtaining an understanding of Wolf-Rayet objects to collate the available material under the broad category *Wolf-Rayet phenomenon* than to attempt to demonstrate that there is a homogeneous set of stars which can be called Wolf-Rayet stars and which occupy a significant stage in the evolution of stars, this stage being traversed routinely by all stars of a certain range of mass.

On the average, galactic Wolf-Rayet stars in binary systems have masses in the range 6 to 15 solar masses. Normally, hydrogen-burning stars with such masses would have a B-type spectrum or, if evolved somewhat, the spectrum of a late-type giant. Wolf-Rayet objects do not have such spectra, by definition. Why is this so? What special factor produces the Wolf-Rayet spectrum? These are questions to which we are seeking answers.

The words 'Wolf-Rayet' refer to a particular type of spectrum which Wolf and Rayet (1867) were the first to recognize as they surveyed the spectra of stars in Cygnus with a visual spectroscope. Since their original discovery, about 127 Wolf-Rayet stars have been recognized in our galaxy (see Smith, 1968a, for a catalogue). Fifty-eight are known in the Large Magellanic Cloud and two in the Small Magellanic Cloud (Smith, 1968b). Twenty-five objects with Wolf-Rayet-like spectra are now known in M 33 (Wray and Corso, 1971).

At the 1968 conference, a Wolf-Rayet spectrum was defined to have the following characteristics (Thomas, 1968):

(1) The spectrum consists of emission lines on a continuous spectrum which has an energy distribution rather like that of an O or a B star.

(2) A few absorption lines may occur as shortward displaced satellites on the edges of some of the emission lines. There is no general absorption-line spectrum as known for normal spectral types. If such a spectrum is seen, it is attributed to a companion star.

(3) The emission lines seen in any one object represent a wide range of excitation

M. K. V. Bappu and J. Sahade (eds.), Wolf-Rayet and High-Temperature Stars, 237–263. All Rights Reserved.

and ionization, the excitation of the line spectrum generally being much higher than that estimated from the shape of the continuous spectrum.

(4) Most of the emission lines are broad with widths corresponding to hundreds to thousand km s^{-1}. The widths are not the same for all lines in any one stellar spectrum.

(5) Most of the spectra fall into two groups: (1) the WC stars in which the lines from ions of C and O dominate, and (2) the WN stars in which the lines from ions of N dominate. Both groups show strong lines of He II.

The galactic Wolf-Rayet stars are the prototypes of the class. Some central stars of planetary nebulae have a spectrum which satisfies the five criteria given above. The optical spectrum of the X-ray source Sco X-1 (Sandage *et al.*, 1966) and of WX Cen (Eggen *et al.*, 1968), which may be the X-ray source Cen X-2, have similar characteristics. In fact the spectrum of WX Cen looks quite like that of a WN 7 star except that He I is weak and H is strong. All objects with spectra satisfying criteria 1 to 5 may be classed as Wolf-Rayet objects. This review will attempt to indicate the nature of these objects. Criterion 2 excludes Of stars, Be stars and stars like P Cygni from the class 'Wolf-Rayet object'. The behavior of the spectra of Of, Be and P Cygni stars, however, has much in common with that of Wolf-Rayet objects as defined above and illuminates the physical theory required.

2. Wolf-Rayet Objects and the HR Diagram

It is true for stars with predominantly absorption-line spectra that the spectral type, when it includes an indication of luminosity, serves as a shorthand notation for the position of the star in the HR diagram. The spectral type can be correlated uniquely with effective temperature and the visual absolute magnitude may be used to indicate the radius of the star. If the stellar mass is known and the composition, normal or otherwise, is specified, one can transform the position in the HR diagram into a stage of evolution. The transformation from the M_V and spectral type diagram to the luminosity-effective temperature diagram to stage of evolution is done by means of model atmospheres and synthetic spectra and by comparing with computed evolutionary tracks which result from models of stellar structure and energy generation. The Wolf-Rayet objects are puzzling because no unambiguous set of rules has yet been established about how to do the transformations starting from a spectral type, the visual absolute magnitude and the mass of the star. One can argue from the visual absolute magnitude and the masses of stars in binary systems that the interiors of these stars can be represented by nominally pure helium stars burning helium.

The Wolf-Rayet stars can be put fairly consistently into a few spectral types which are defined by explicit stipulations concerning the relative intensities of some emission lines and the widths of the emission lines. The visual absolute magnitudes of the stars can be found from consideration of membership in groups of stars which contain stars of known luminosity and from consideration of the H II regions which surround some 56 Wolf-Rayet stars. A recent study of this subject by Crampton (1971) gives the following visual absolute magnitudes (see Table I).

TABLE I

Visual absolute magnitudes of Wolf-Rayet stars

Spectral Type	M_v	No. Stars	Spectral Type	M_v	No. Stars
WN3, 4, 5	− 3.7	7	WC5, 6	− 3.6	3
WN6	− 4.8	7	WC7	− 4.4	3
WN7	− 6.5	4	WC8	− 5.0	1
WN8	− 5.0	2			

It is interesting that in each sequence there is a trend of visual luminosity with spectral type. One of the problems concerning the nature of Wolf-Rayet stars is to decipher what this trend means. Is it a reflection of differing bolometric corrections? Where do the central stars of planetary nebulae with Wolf-Rayet-like spectra fall on this sequence?

A further point that requires clarification, in view of the fact that the emission lines seen in Wolf-Rayet spectra seem to be the optical result of a hot plasma generated by mechanical effects, is how the bolometric correction that should be defined. The bolometric correction classically is a measure of that part of the radiation field from a star which does not fall into the visual band. It is an expression of the law of conservation of energy when the simplifying assumption is made that energy appears only as radiation. In objects like Wolf-Rayet stars, however, part of the observed radiation field is a consequence of mechanical effects.

Radiative and mechanical energy may be conserved within a system of two or more objects, for instance the two stars of a binary system or one star which is shedding material into the circumstellar medium and thus interacting with the circumstellar medium. In such cases a careful look must be taken at the meaning assigned to a bolometric correction and subsequently to the visual absolute magnitude. Certainly the meaning is not the same as for normal absorption-line stars. One wonders whether an effective temperature estimated from the shape of the continuous spectrum over a short region in the visible spectrum represents accurately enough the total rate of energy generation and emission by the star. The transformations which are used to develop a picture of stellar evolution from the positions of normal stars in the HR diagram would seem to be of doubtful validity for Wolf-Rayet objects.

The magnitude of the problem can be seen by the following simple calculation. If the bolometric correction of a Wolf-Rayet star is − 3.0 and the visual absolute magnitude is − 4.4, the radiative luminosity of the star is $6.3 \times 10^4 \, L_\odot$. Such a star radiates at a rate of about 2.4×10^{38} erg s^{-1}. Suppose that the particle density in the expanding atmosphere is 2×10^{12}, which is a value like that in an early type supergiant atmosphere, and that the average molecular weight is 0.5. Then the density is about 1.6×10^{-12} g cm^{-3}. If each cubic centimeter moves at 1000 km s^{-1}, the kinetic energy of each cubic centimeter is 0.8×10^4 erg. Suppose that the effective outer radius of the dense expanding atmosphere is 40 $R_\odot$. The kinetic energy transported each second across this boundary is then 8.5×10^{37} erg s^{-1}. The ratio of the kinetic energy lost per second from the star to the radiative energy emitted per second is then about 0.35. These numbers are rough estimates and they do not take into account the excitation and ionization energy contained in the ejected matter.

This example makes it clear that galactic Wolf-Rayet stars may radiate mechanical energy at a rate comparable to the rate at which they lose energy in the form of radiation. In the case of a central star of a planetary nebula, if the density of the expanding atmosphere is an order of magnitude smaller than that of a galactic Wolf-Rayet star, about the same result may be attained because the luminosity of such a star is at least a factor ten smaller than that of a galactic Wolf-Rayet star. For precise consideration of the *total* energy radiated by a Wolf-Rayet star, the amount of energy leaving the Wolf-Rayet star as kinetic and excitation energy should be included in the sum. The change in the bolometric correction to take account of a factor x in the total energy emitted in all forms is $2.5\log x$. This factor can easily become significant in comparison to the bolometric correction estimated from purely radiative model atmospheres.

Those central stars of planetary nebulae which have Wolf-Rayet-like spectra have visual absolute magnitudes fainter than about -3; see Webster (1969) and Smith and Aller (1971). Their masses are considered to be of the order of one solar mass. These stars are smaller and less massive than normal Wolf-Rayet stars. According to Gatewood and Sofia (1968), the X-ray star Sco X-1 is an evolved compact object of the nature of a white dwarf. The visual absolute magnitude lies in the range $+6$ to $+7$ and there is no evidence of a companion star. Mook and Hiltner (1970) have reviewed all the available information about the distance of Sco X-1. If it is more distant than Gatewood and Sofia estimate, the visual absolute magnitude is brighter than given here. Gatewood and Sofia suggest that the star once had a mass in excess of 10 solar masses and that it has shed sufficient mass to become an evolved compact object. Nevertheless, the spectrum of the star has characteristics similar to those of Wolf-Rayet objects and to old novae. The emission-line spectrum varies in a matter of days (Sandage *et al.*, 1966) and there is some suggestion of a shortward displaced absorption component associated with CIII 4650.

These observations are cited to demonstrate that the factors causing a Wolf-Rayet-like spectrum to appear are not limited to one area in the HR diagram. In fact the position of an object in the HR diagram does not seem to be a constraining factor on causing a Wolf-Rayet spectrum. It is, however, clear that Wolf-Rayet-like emission lines are seen only in objects which have blue continua rather like the continua of B and O stars. This fact together with the fact that lines from the second, third and fourth ions of carbon, nitrogen and oxygen are seen indicates that one is observing radiation from a hot plasma with an electron temperature greater than 3×10^4 K.

The mechanism for heating the line emitting regions of a Wolf-Rayet atmosphere is unknown, although it seems inevitable that the deduced high electron temperatures come from a conversion of mechanical energy to heat. In addition, the matter in the atmosphere must be accelerated to the outward velocities which are observed. No satisfactory mechanism for achieving the needed acceleration has been proposed.

The following remarks may help to define the problem. (1) A range in outward directed velocities is observed, those ions with the lowest ionization potential having the highest outward directed velocity component. If the acceleration mechanism provides kinetic energy equally to each particle, the resulting velocities should vary inversely as the square root of the mass of the ion. The observed trend is compatible with this suggestion. If the mechanism gives a constant velocity to all particles (such as might occur with the acceleration of a condensation in the atmosphere), some differential accelerating process is required which operates differently in the regions where each of the observed ions occurs predominantly. (The outward directed motions are determined best from shortward displaced absorption components, none of which is narrow enough to define sharply a locally changing acceleration.) (2) It is considered that the presence of magnetic fields is essential for the efficient heating of the solar chromosphere and corona by the deposition of mechanical energy. There is no observation which excludes the presence of a magnetic field in the atmosphere of a Wolf-Rayet star, nor, on the other hand, is there any spectroscopic observation demanding the presence of a magnetic field. Magnetic fields have been measured in some B type stars and these fields are believed to be remnants of the interstellar magnetic field which was compressed as the star formed. A magnetic field might be attached by the same process to the proginator of a Wolf-Rayet star as it formed. Since stellar magnetic fields are believed to decay with time, an old Wolf-Rayet star (one showing the Wolf-Rayet phenomenon at a late evolutionary stage after a fairly lengthy lifetime) might have a negligible magnetic field. The presence or absence of a magnetic field might affect significantly the spectroscopic behavior of the line-emitting regions of the stellar atmosphere.

Whatever the evolutionary stage of Wolf-Rayet objects, the physical conditions in their outer atmospheres must be such as to generate the observed spectrum. Because Wolf-Rayet objects are rather few in number, it seems that the appropriate physical conditions do not occur with large probability. Because Wolf-Rayet objects, except objects like Sco X-1, seem to be brighter than $M_V = 0$, one would not expect to see a companion star of type later than A0 unless it were a bright giant or supergiant. We literally do not know whether Wolf-Rayet objects can occur in close association with late-type stars because the range in visual absolute magnitude between Wolf-Rayet stars and late-type stars is too large to permit late-type companions to be observed.

3. Gas Distributions Around Wolf-Rayet Stars

The first systematic study of the association between Wolf-Rayet stars and H_{II} regions was made by Smith (1966) who demonstrated that a number of WN stars were surrounded by a ring-like nebula as well as being associated with H_{II} regions. The ring nebula can be understood as a sweeping up of the interstellar material by the material ejected from the star. These observations have been reviewed and extended by Crampton (1971) who also lists the radio sources known to be associated with some of these H_{II} regions and with Wolf-Rayet stars. It is common for Wolf-Rayet stars to be immersed in interstellar gas. The radio radiation seems to be thermal in origin.

Spectroscopic observations of binary stars [for instance Kuhi (1968a) on V 444 Cygni and Cowley and Hiltner (1971) on CV Serpentis] demonstrate that the amount of material which emits line radiation in the vicinity of Wolf-Rayet stars varies in an erratic manner. The spectral variations occur at irregular intervals and they involve enough material to cause quite conspicuous changes in the line and continuum intensities. Whatever the cause is of the ejection of matter from Wolf-Rayet stars and the excitation of this matter to radiate the observed emission-line spectrum, this cause is not a fully regular, spherically symmetric process.

Any effects of the tidal force of a companion star on the gas in a binary system should be periodic. The observed changes in spectrum have not been shown to be strictly periodic in any Wolf-Rayet star.

The detailed spectrophotometric observations of Kuhi (1968a) on V 444 Cygni confirm the following characteristics of Wolf-Rayet atmospheres.

(i) The Wolf-Rayet envelope is optically thick in many lines.

(ii) The emission lines behave in a completely individualistic manner, without regard to ionization and excitation potential. This means that a model consisting of stratified layers with ionization potential decreasing or increasing outward through a circumstellar shell is inappropriate for Wolf-Rayet stars.

(iii) The effective size of the emitting region is enlarged by electron scattering.

(iv) Simple models of uniform radial expansion are inadequate.

The suspected X-ray star WX Cen (Eggen *et al.*, 1968) shows irregular short-term optical fluctuations, a pronounced UV excess and night to night optical brightness variations with a total range of 0.4 mag. In these respects it is like some of the more irregular Wolf-Rayet stars. In the spectrum of Sco X-1 the complex emission band at 4630 to 4655 Å and the hydrogen lines undergo striking variations (Sandage *et al.*, 1966; Westphal *et al.*, 1968). The X-ray sources with Wolf-Rayet-like spectra appear to show more conspicuous changes in their optical spectrum than do most galactic Wolf-Rayet stars. This may be because the optical thickness of the part of the atmosphere emitting the optical spectrum is less for an X-ray source than it is for galactic Wolf-Rayet stars, with the result that the emission lines from an X-ray source lie predominantly on the linear part of the curve of growth. Such lines are more responsive to changes in the physical state of the atmosphere than are optically thick lines.

Insufficient information exists about the spectra of central stars of planetary nebulae with Wolf-Rayet characteristics to place these stars in the picture as regards spectral variations. One might expect central stars to have an emitting atmosphere of intermediate optical thickness in the lines. Such an atmosphere would tend to reveal smaller fluctuations in the density and exciting conditions of the line-emitting region than are detected for galactic Wolf-Rayet stars. An emitting region that is optically thin in all lines of optical wavelength is the most sensitive indicator of changes in the physical conditions in the atmosphere. Such a region is only readily detected for galactic Wolf-Rayet stars when the more dense regions are occulted by a companion because the sensing systems used are biased to embrace the dynamic range about the intensity of the most intense, thus optically thick, lines. The lowest line intensity that can be observed is closely limited by the brightness of the continuous spectrum in the neighborhood of the line.

4. The Spectroscopic Problems Posed by Wolf-Rayet Stars

The information required for making identifications in Wolf-Rayet spectra is now in a pretty good state with tables of the spectra of the carbon ions (Moore, 1970), of Nv and Niv (Hallin, 1966a, b) and of Oiv (Bromander, 1969) and Ov (Bockasten and Johansson, 1968) having appeared. Two of the greatest difficulties remaining are recognizing emission lines which have low intensity (10 to 20% of the continuum) and resolving blends, particularly when the Wolf-Rayet spectrum is blended with the spectrum of an O or a B star. These problems are aggravated by the width of each line and the variation of the intensities of the emission lines. Since weak lines are on the Doppler part of the curve of growth, they will change more in apparent intensity than will moderately strong conspicuous lines for a given change in the abundance of the carrier ion.

In WC spectra the following identifications have been made:

definitely present – He ı, He ıı, C ıı, C ııı, C ıv, O ııı, O ıv, O v, O vı, N ııı, N ıv, S ııı,

probably present – H, N v, A1 ııı, Si ıv, Ti ıv,

possibly present – Mg ıı

In WN spectra we have:

definitely present – He ı, He ıı, C ıv, N ııı, N ıv, N v, O v, Si ıv,

probably present – H, O ıv,

possibly present – N ıı, O ııı, O vı, S ııı.

There is a considerable difference between the subclasses of any one sequence. Which lines are conspicuous depends upon which spectral range is observed and how the strong lines of the carbon, nitrogen and oxygen ions are distributed over wavelength. The line widths have a distinctive trend in WC stars, in fact this trend is one of the classification criteria. Among the WN stars, however, the case is different. Some WN stars have both sharp and broad lines in their spectra, others have only broad lines. These differences in line width indicate that the various emission lines may be formed under quite different physical conditions.

Ultraviolet observations of γ_2 Velorum (Stecher, 1968) show that a low density shell [C ııı] 1909 surrounds this WC 8 star. Observations (Bless, 1971) of HD 50896, WN5, with the OAO-II spectrum scanner show the [N ıv] 1486 line which implies a low density shell containing N^{+3} ions around this WN star.

In addition to the emission line spectrum, which appears to be due chiefly to recombination and cascade, with some superposed selectively excited emission lines, a few lines appear as shortward-displaced absorption cores. These are invariably lines which arise from levels which will be strongly populated in a low density gas through which a radiation field not in equilibrium with the kinetic temperature of the gas is flowing. The superposed, selectively excited emission lines are the lines which appear in emission in Of stars.

The designation of a WC sequence and a WN sequence for Wolf-Rayet stars has often been considered to imply gross abundance differences in carbon and nitrogen between the two types of star. This idea can be supported no longer. The ultraviolet observations of γ_2 Velorum by Stecher (1968; 1970) clearly show a strong emission at about 1720 Å. There are no lines of C III and C IV in the neighborhood of this wavelength; the only probable identification is as N IV 1718.55. The N III spectrum can be detected in WC8 and WC9 stars at several places in the spectrum, see Underhill (1959; 1962), thus the conclusion of Kuhi (1968b) that the only plausible explanation of the differences between WN and WC spectra is gross differences in the carbon and nitrogen content of these stars is unjustified. It is widely recognized that the C IV spectrum appears in WN stars. Ultraviolet spectral scans from OAO-II show C IV lines in the spectrum of the WN star HD 50896.

The reason why WC spectra look so very different from WN spectra is that the WC spectra are dominated by lines of C III which has a spectrum full of quite strong lines in the spectral ranges observed with ground-based instruments. The spectra which are prominent in WN spectra are C IV, N IV and N V in addition to He I and He II. All of these contain only a few conspicuous lines in the usually observed spectral range.

Kuhi (1968b) has mentioned weak unidentified lines in WC stars at 7426, 6503 and 5700 Å. The first two may be blended multiplets of O V, see the line list of Bockasten and Johansson (1968). A line at 5700 Å would be masked by broad C III 5696 in most WC stars. It is not clear to what Kuhi is referring in this case. The unidentified lines Kuhi mentions in WN stars at 10430 and 5200 Å are due to He II (6–13) and to the N IV 2^3P^0–3^3D multiplet, respectively.

Before returning to the problem of understanding the implications of the groups of ions represented in Wolf-Rayet spectra and of the relationship of Wolf-Rayet spectra to the optical spectra of X-ray sources and of central stars of planetary nebulae, let us summarize briefly the theoretical studies that have been made to understand the meaning of the shapes of the emission lines in Wolf-Rayet spectra.

Most of the emission lines have a rounded shape and only a few lines are flat-topped. Any adequate theory of the formation of the Wolf-Rayet spectrum should be able to explain this characteristic. A number of attempts have been made to develop a detailed theory for the formation of the emission lines seen in Wolf-Rayet spectra. In a recent study, Castor (1970) has used the escape probability method for treating the transfer of radiation in a stellar envelope which is in rapid radial expansion. Castor has applied the method to a two-level atom and has calculated representative line profiles for different distributions of the density of the absorbing atoms in the case of a fixed velocity distribution. He finds that flat-topped profiles, with or without violet absorption, depending on details of the model, are produced by an optically thin hollow sphere of emitting material, which, indeed, is confirmation of an old result. An optically thick expanding envelope produces a rounded line profile. This work demonstrates that although sufficient, it is not necessary to postulate large (of the order of 700 km s^{-1}) chaotic velocities to account for the rounded lines which are commonly observed in Wolf-Rayet spectra.

Two possible simple models for Wolf-Rayet atmospheres can be considered. (1) A radially expanding atmosphere which is optically thick in most lines. Such an atmosphere may be like that used as an example in Section II for estimating the loss of kinetic energy from a Wolf-Rayet star. That model implies a mass loss of 2.5×10^{-4} solar masses per year. The outer part will become optically thin in a few lines which are not excited in the inner dense part of the atmosphere. These lines will have a flat-topped profile. One may account for the observed flat-topped lines by a suitable adjustment of the ionization level as one moves outward. (2) An inner, dense atmosphere in which large, randomly directed motions have been generated by some unspecified process, is surrounded by an exosphere which is visible optically in a few lines, all optically thin. This model is rather artificial; it can, however, be adjusted to imply a rather small rate of mass loss if that is considered desirable. Both models ignore the irregularities known to exist in the radial flow from a Wolf-Rayet star.

Castor and Van Blerkom (1970) have applied the theory developed by Castor to the He II spectrum of two WN6 stars. They find that the observed intensities of He II lines and the rounded shapes of the lines can be represented by plausible combinations of temperature, density and radial flow. This work indicates that models of Type 1 are relevant for Wolf-Rayet stars. Castor and Van Blerkom estimate suitable, representative values of N_e and T_e in a WN6 atmosphere to be $(2.5 \times 10^{11}, 5 \times 10^4 \text{ K})$, $(2.5 \times 10^{11}, 10^5 \text{ K})$ or $(5 \times 10^{11}, 2 \times 10^5 \text{ K})$. The suggested electron temperatures are higher than have sometimes been suggested for WN stars while the electron densities are fairly low. The adopted radiation temperature of the photosphere which provides the radiation illuminating the expanding atmosphere is set to be 40000 K. It is assumed that typically the radius of the line-emitting atmosphere is 3 times the radius of the photosphere.

The calculations of Castor and Van Blerkom indicate that the populations of the higher levels of the He^+ ions in a Wolf-Rayet atmosphere relative to the number of He^{+2} ions have the distribution expected according to the Boltzmann law at the electron temperature. That the high levels will be in thermal equilibrium with respect to the number density of the next higher ion at temperatures and pressures of astrophysical interest was first shown for hydrogen-like ions by McWhirter and Hearn (1963). We shall assume that this is true for the upper levels of all the lines one observes in the normally observed spectral region of Wolf-Rayet stars. For example, to understand the strengths of the lines in the C IV spectrum, one requires the populations of the upper levels of these lines in the C^{+3} ion. We assume that these populations can be found from the abundance of the C^{+4} ions and use of the Boltzmann law.

The fraction of the light elements in each stage of ionization at a given electron temperature has been calculated by Jordan (1969) for a thin plasma in which the radiation field is negligible taking account of collisional ionization, collisional excitation followed by autoionization, direct radiative recombination, radiative recombination via bound levels and dielectronic recombination. The fraction of any element in each stage of ionization in the case of thermal equilibrium at a temperature T and electron pressure, $\log P_e$, can be calculated by successive applications of Saha's law.

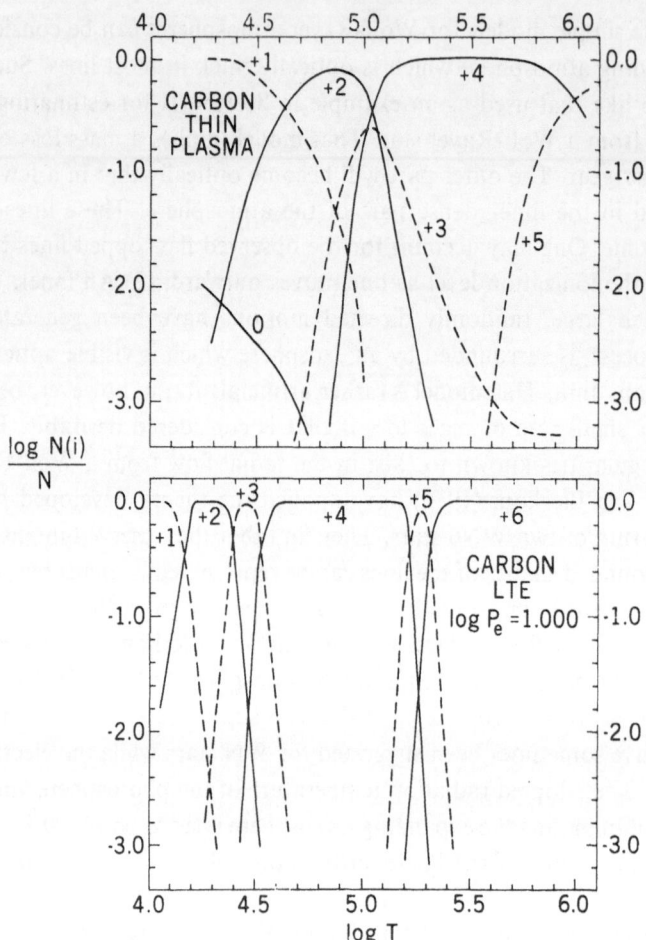

Fig. 1. The fractional ionization of the carbon ions as a function of temperature. Upper panel: in a thin plasma (Jordan, 1969); lower panel: in a plasma in LTE with $\log P_e = 1.000$.

These two cases bracket what may be expected to occur in a Wolf Rayet atmosphere. The results for carbon, nitrogen and oxygen for some temperatures of interest for Wolf-Rayet stars are shown in Figures 1, 2 and 3. The LTE results have been taken from Sparks and Fischel (1971), see also Fischel and Sparks (1971). They are displayed for the case $\log P_e = 1000$, which implies an electron density of 3.6×10^{12} when the electron temperature is 20000 K and 3.6×10^{11} when the electron temperature is 20000 K.

Thermal equilibrium ion ratios can only be maintained in an optically thick plasma. The true case for a Wolf-Rayet atmosphere lies between the two extremes presented here. Note that the electron temperature required to produce the maximum ionization fraction for any ion is greater in a thin plasma than in a plasma in LTE. Also the temperature domain in which the ionization fraction of any ion is greater than a set level, say 10 percent, is considerably wider in a thin plasma than in an LTE plasma.

Consider a WC7 spectrum. Here the C III and C IV lines are strong. N III and N IV

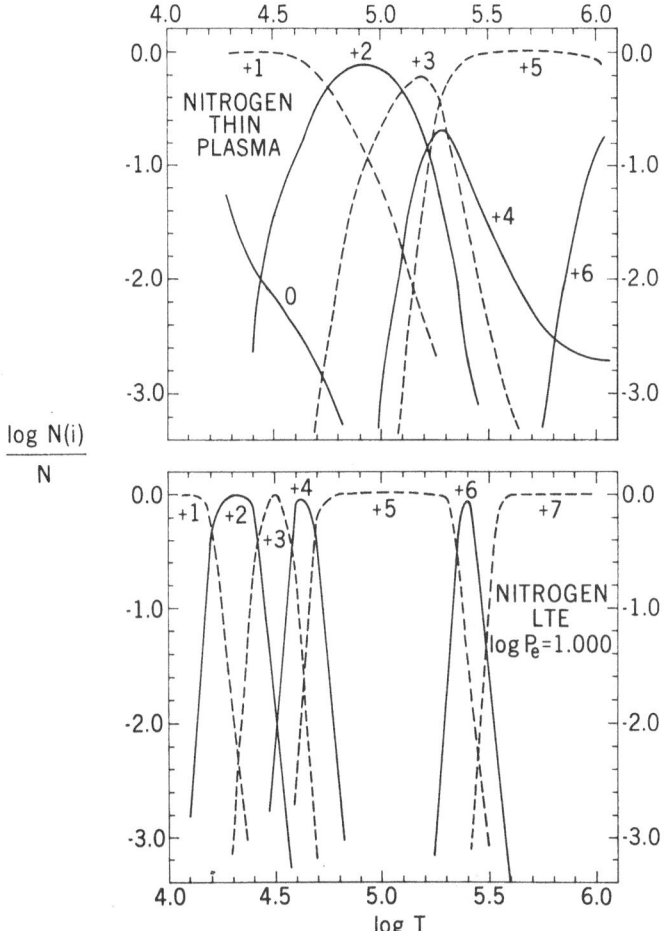

Fig. 2. The fractional ionization of the nitrogen ions as a function of temperature. Upper panel: in a thin plasma (Jordan, 1969); lower panel: in a plasma in LTE with $\log P_e = 1.000$.

are weak and N v is not seen. Lines of O III and O IV are seen, but neither O v nor O VI is conspicuous. To obtain these spectra in emission one would like to find a temperature where the C^{+3} and C^{+4} ions are abundant, the N^{+3} and N^{+4} ions are moderately abundant but N^{+5} is not very abundant, and finally that O^{+3} and O^{+4} are moderately abundant, but O^{+5} is not very abundant.

If the plasma is thin the following temperatures are needed:

Carbon $4.95 < \log T < 5.10$
Nitrogen $5.10 < \log T < 5.25$
Oxygen $5.20 < \log T < 5.40$.

If the plasma is in LTE and $\log P_e = 1.000$:

Carbon $4.45 < \log T < 4.60$
Nitrogen $4.52 < \log T < 4.62$
Oxygen $4.68 < \log T < 4.78$.

In the case of WN stars one requires that the carbon appears predominantly as

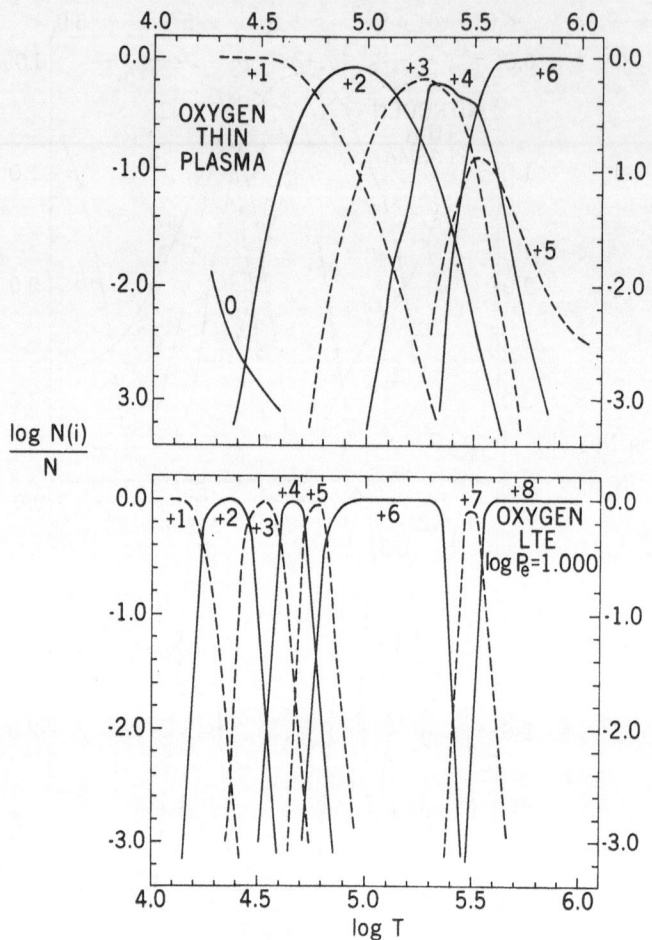

Fig. 3. The fractional ionization of the oxygen ions as a function of temperature. Upper panel: in a thin plasma (Jordan, 1969); lower panel: in a plasma in LTE with $\log P_e = 1.000$.

C^{+4}, nitrogen as N^{+4} and N^{+5} with some N^{+3}, and oxygen as O^{+4} and O^{+5}. In the case of a thin plasma:

Carbon $5.20 < \log T < 5.90$
Nitrogen $5.30 < \log T < 5.90$
Oxygen $5.50 < \log T < 5.70$.

In the case of a plasma in LTE with $\log P_e = 1.000$:

Carbon $4.60 < \log T < 5.10$
Nitrogen $4.65 < \log T < 5.20$
Oxygen $4.65 < \log T < 4.85$.

Clearly the electron temperature in the extended atmosphere of WC stars is lower than that in the atmosphere of WN stars. Thus, the WC and the WN stars do not form parallel sequences. Very roughly, if the thin plasma calculations are appropriate, the electron temperature is approximately 1.6×10^5 K in WC atmospheres and 5×10^5 K in WN atmospheres. The lower limits given by the hypothesis of LTE are approximately 4×10^4 K in WC stars and 6.5×10^4 K in WN stars. These temperatures are higher than may be inferred from the shape of the continuous spectrum. They indicate that the outer atmospheres of Wolf-Rayet stars are being heated by mechanical energy.

If an object shows the O VI lines in emission, as some WC stars do, the temperature must lie in the range $5.4 < \log T < 6.0$ if the plasma is thin or $4.8 < \log T < 5.4$ if the plasma is in LTE. Temperatures such as these are sufficient for the star to generate soft X-rays by bremsstrahlung. However, since most galactic Wolf-Rayet stars are in the galactic plane, this X-ray emission would never reach the earth in observable quantity. The peculiar object HD 6327 which only shows emission lines of He II (Hiltner and Schild, 1966) according to these ideas has an atmosphere in which the electron temperature is so high that all the carbon, nitrogen and oxygen atoms are in stages of ionization too high to show lines in the visible spectrum. A temperature of more than 10^6 K would be required.

According to Kestenbaum et al. (1971), the X-ray continuum intensity distribution of Sco X-1 corresponds to a plasma at 8×10^7 K and an electron density near 10^{16}, yet the emission-line optical spectrum (Sandage et al., 1966; Johnson et al., 1967; and Westphal et al., 1968) corresponds to a fairly low level of excitation with strong H and He I and no C IV. The optical thickness of the line-emitting region of the spectrum seems to be sufficiently small that very little of the X-ray energy is degraded into optical radiation. The optical observations of Mook, Hiltner and Lynds (1971) give evidence of a sporadic mass flow at 900 km s^{-1}. Since there is no evidence for fluctuations in the X-ray intensity of Sco X-1, Boldt et al. (1971) have concluded that the energy input into the plasma must be essentially continuous compared with the time scale of the optical variations. The irregular variations in optical wavelengths would seem to be generated by density fluctuations in the ejected gas and these density fluctuations seem to occur independently of the energy input causing the X-ray emitting plasma.

Optical observations of WX Cen (Eggen et al., 1968) indicate a somewhat similar pattern. Here the optical plasma corresponds to a level of excitation like that of a WN7 star, except that He I is surprisingly weak and H is strong. The faint lines shown shortward of He II 4686 correspond to N V (effective wavelength near 4605 Å) and N III and C IV (effective wavelength near 4642 Å). The line at 4931 Å unidentified by Eggen, Freeman and Sandage, is surely a blend of O V and N V lines. Eggen, Freeman and Sandage argue that there are no good reasons to couple a variation of the X-radiation with the optical fluctuations. One may infer that the optical variations are largely caused by irregularities in the flow of matter through a thin outer shell.

The optical spectra of X-ray sources indicate that a major difference between X-ray sources and Wolf-Rayet sources is the optical thickness of the outer atmosphere which generates the optical spectrum. A radial flow of matter which contains density irregularities is seen for X-ray sources and for Wolf-Rayet stars. The ultimate source of energy for creation of a high temperature plasma may be the same in both cases. In X-ray sources the high-temperature plasma is visible owing to the thinness of the fringe atmosphere; in Wolf-Rayet stars the high-temperature plasma is barely visible, if at all, and the line emission of the fringing atmosphere dominates the optical observations. The irregularities in the density structure of the expanding atmosphere remind one of the coronal condensations and coronal plumes of the Sun. Possibly they are related to a magnetic field structure in the stellar atmosphere. The presence of a magnetic field seems to be necessary for deposition of mechanical energy in the solar chromosphere and corona.

The spectra of two Wolf-Rayet-like objects in M33 (Wray and Corso, 1971) seem to be WC-like. The relative weakness of the He I and He II lines in comparison to the C IV lines is a bit strange. It is premature to discuss these spectra now; their existence demonstrates one more place where the Wolf-Rayet phenomenon occurs, but the relative line intensities of galactic Wolf-Rayet stars are not seen.

5. Summary

Any acceptable model of a Wolf-Rayet-like object must take account of the following characteristics of Wolf-Rayet-like objects:

(1) Galactic Wolf-Rayet stars have masses in the range 6 to 15 solar masses; Wolf-Rayet-like central stars of planetary nebulae presumably have masses an order of magnitude smaller as do the Wolf-Rayet-like stars which emit X-rays. We have no direct evidence on this last point.

(2) The optical light from Wolf-Rayet-like objects may fluctuate in intensity and the optical spectrum may change in an irregular manner. For galactic Wolf-Rayet stars the changes are small. For stellar X-ray sources the optical changes are fairly conspicuous but they are not accompanied by detectable changes in the X-ray spectrum. No changes have been noted for the Wolf-Rayet-like central stars of planetary nebulae, but the observations are not sufficient to rule out the occurrence of small changes.

(3) There is a low density outer atmosphere around the inner atmosphere. In the case of galactic Wolf-Rayet stars the inner atmosphere is optically thick in most lines.

(4) There is a radial outflow of matter from the star; irregularities in the density of the outflowing gas occur.

(5) In WC stars the electron temperature may lie in the range 4.0×10^4 to 1.6×10^5; in WN stars the range is 6.5×10^4 to 5.0×10^5 K. If a model corresponding to a thin plasma in which radiative excitations are not important is appropriate, temperatures near the top of each range may be expected.

(6) The electron temperature in the line-emitting regions seems to be higher than that in the regions which give the continuous spectrum at optical wavelengths. In the case of X-ray sources, the temperature in the regions where the continuous X-ray spectrum is formed is of the order of 10^7 K.

(7) The electron density in the optically thick line-emitting regions lies in the range 10^{11} to 10^{12} particles per cm^3.

There is no reason not to assume a normal stellar composition for the atmosphere of a Wolf-Rayet-like object. The theory of spectrum formation in a Wolf-Rayet atmosphere is not yet sufficiently well developed to permit accurate deduction of abundances, but qualitatively the observed line intensities follow the theory of spectrum formation in a hot plasma of normal stellar composition at the temperatures and densities indicated above.

The problems which require further study (in addition to the problems of the theory of line formation) are related to the fundamental question still requiring an answer, namely, what special factor produces a Wolf-Rayet spectrum for stars in the range 6 to 15 solar masses or 0.6 to 1.5 solar masses? The action of this factor is not limited to one region in the HR diagram. This factor is not active for every star in the specified mass ranges. In fact, only a few stars of the specified masses show a Wolf-Rayet spectrum. Either being a Wolf-Rayet object is a relatively short-lived stage of evolution traversed by all stars of the correct mass, or some special combination of physical factors produces the hot plasma which is required to generate a Wolf-Rayet-like spectrum.

One suspects that the high electron temperatures required to produce a Wolf-Rayet spectrum are the result of converting mechanical energy to optical radiation by high speed particles impinging on a low density gas. If this is so, then it is not obvious what meaning is to be attached to the trend of visual absolute magnitude with spectral type found in the WC and in the WN sequence. This problem raises the question of how the bolometric correction should be defined when the star may be ejecting as much energy in the form of kinetic and excitation energy of particles as it is loosing in the form of radiation. What spectral features, if any, are true indications of effective temperature when effective temperature is defined to represent the total rate of energy loss from the star?

The Wolf-Rayet-like objects seem to be somewhat similar to each other in nature, their spectroscopic differences reflecting merely a somewhat different visibility of the three major parts of their atmosphere. These three parts are (1) a dense hot plasma, (2) a moderately dense expanding atmosphere and (3) an exosphere in which radiative excitation is not important.

5.1. STELLAR X-RAY SOURCES

(1) A high-temperature plasma ($T \approx 10^7$ K) which radiates X-rays is seen. X-rays escape through the thin overlying atmosphere largely unimpeded.

(2) The expanding atmosphere is a thin fringe which radiates an optical spectrum corresponding to an electron temperature in the range 3×10^4 to 8×10^4 K. The density in this expanding atmosphere varies irregularly with time.

(3) An exosphere has been observed for the X-ray source WX Cen. The exosphere is seen most clearly by the forbidden C III and N IV lines in the ultraviolet. Observations of adequate spectral resolution to resolve these lines have not been made for stellar X-ray sources so far.

5.2. WOLF-RAYET-LIKE CENTRAL STARS OF PLANETARY NEBULAE

(1) The dense high temperature plasma is partially obscured by an overlying expanding atmosphere. Hard X-rays may be degraded to soft X-rays which escape and excite the surrounding nebula.

(2) The moderate density expanding atmosphere may have an electron temperature near 10^6 K in some regions. This very hot plasma will radiate O VI lines; other Wolf-Rayet lines would be radiated from cooler parts of the plasma. In order to have O VI lines and lines of N V, C IV and lower ions present, it is necessary to have plasma at 10^6 K and at 10^5 K or cooler.

(3) Ultraviolet observations have not yet been made at adequate resolution to confirm the presence of an exosphere. The planetary nebula can be considered to be an extreme extension of the exosphere.

5.3. GALACTIC WOLF-RAYET STARS

(1) A high density, high temperature plasma, barely visible through the overlying expanding atmosphere, may be postulated. This may be the source of the UV excess of WN stars. It is doubtful if a sufficient X-ray flux would escape and be observable at the earth.

(2) The optical spectrum is given by an extensive, fairly dense expanding atmosphere with electron temperatures of the order of 10^5 K, the temperature being higher in WN atmospheres than in WC atmospheres.

(3) An extensive exosphere has been observed in C III] 1909 for γ_2 Velorum, WC 8, and N IV] 1486 for HD 50896 as well as by means of flat-topped lines and shortward displaced absorption cores. The exosphere eventually meets with the surrounding interstellar medium, creating in the case of some WN stars a visible ring-nebula around the Wolf-Rayet star. Because ring nebulae are not seen around WC stars, one must conclude that the energy transferred from the gas of the exosphere to the interstellar medium is insufficient in the case of WC stars to create a nebula observable in the hydrogen Balmer lines or that the filamentary material has dispersed. There does not seem to be any significant difference between the distribution of WC and of WN stars in the interstellar medium. Thus, the absence of detectable ring nebulae around WC stars should be construed to mean a difference in the energy carried by the exosphere of WC stars from that carried in the exosphere of WN stars or a difference in stage of evolution.

A three-part model of the general type given above will account for the observed spectrum of all Wolf-Rayet-like objects. The hot, dense plasma is the origin of the energy which causes the overlying layers to radiate a Wolf-Rayet type spectrum. One may speculate that this hot plasma is the outer part of a shell in which hydrogen is burning. Why such a shell should occur so close to the stellar surface that it is visible or nearly so is a problem in the theory of stellar structure. It may be because of mass exchange in a close binary (Paczynski, 1967). However, the stars which show ring nebulae or planetary nebulae seem to be single stars. Studies which take into account hydrodynamical flow in the outer layers of stars will be required to provide a proper understanding of the mass flow from Wolf-Rayet-like objects and of the cause of the density irregularities which seem to appear in the flow. The Wolf-Rayet stars are still peculiar objects, but some pattern for understanding them is beginning to appear.

References

Bless, R.: 1971, unpublished OAO-II spectrum scans.

Bockasten, K. and Johansson, K. B.: 1968, *Arkiv Fysik* **38**, 563.

Boldt, E. A., Holt, S. S., and Serlemitsos, P. J.: 1971, *Astrophys. J. Letters* **164**, L9.

Bromander, J.: 1969, *Arkiv Fysik* **40**, 257.

Castor, J. I.: 1970, *Monthly Notices Roy. Astron. Soc.* **149**, 111.

Castor, J. I. and Van Blerkom, D.: 1970, *Astrophys. J.* **161**, 485.

Cowley, A. P. and Hiltner, W. A.: 1971, *Astron. Astrophys.* **11**, 407.

Crampton, D.: 1971, *Monthly Notices Roy. Astron. Soc.*, in press, **154**.

Eggen, O. J., Freeman, K. C., and Sandage, A. R.: 1968, *Astrophys. J. Letters* **154**, L27.

Fischel, D. and Sparks, W. M.: 1971, *Astrophys. J.* **164**, 359.

Gatewood, G. and Sofia, S.: 1968, *Astrophys. J. Letters* **154**, L69.

Gebbie, K. B. and Thomas, R. N.: 1968, *Wolf-Rayet Stars*, Natl. Bur. Std. Special Publ. 307.

Hallin, R.: 1966a, *Arkiv Fysik* **31**, 511.

Hallin, R.: 1966b, *Arkiv Fysik* **32**, 11.

Hiltner, W. A. and Mook, D. E.: 1970, *Ann. Rev. Astron. Astrophys.* **8**, 139.

Hiltner, W. A. and Schild, R. E.: 1966, *Astrophys. J.* **143**, 770.

Johnson, H. M., Spinrad, H., Taylor, B. J., and Peimbert, M.: 1967, *Astrophys. J. Letters* **149**, L45.

Jordan, C.: 1969, *Monthly Notices Roy. Astron. Soc.* **142**, 501.

Kestenbaum, H., Angel, J. R. P., and Novick, R.: 1971, *Astrophys. J. Letters* **164**, L87.

Kuhi, L. V.: 1968a, *Astrophys. J.* **152**, 89.

Kuhi, L. V.: 1968b, in K. B. Gebbie and R. N. Thomas (eds.), *Wolf-Rayet Stars*, Natl. Bur. Std. Special Publ. 307, p. 108.

McWhirter, R. W. P. and Hearn, A. G.: 1963, *Proc. Phys. Soc.* **82**, 641.

Mook, D., Hiltner, W. A., and Lynds, R.: 1971, *Astrophys. J. Letters* **163**, L69.

Moore, C. E.: 1970, *Selected Tables of Atomic Spectra* (C I, C II, C III, C IV, C V, C VI), NSRD-S-NBS Section 3.

Paczynski, B.: 1967, *Acta Astron.* **17**, 355.

Sandage, A. R., Osmer, P., Giaconni, R., Gorenstein, P., Gursky, H., Waters, J., Bradt, H., Garmire, G., Sreekantan, B. V., Oda, M., Osawa, K., and Jugaku, J.: 1966, *Astrophys. J.* **146**, 316.

Smith, L. F.: 1966, Thesis, Australian National University.

Smith, L. F.: 1968a, *Monthly Notices Roy. Astron. Soc.* **138**, 109.

Smith, L. F.: 1968b, *Monthly Notices Roy. Astron. Soc.* **140**, 409.

Smith, L. F. and Aller, L. H.: 1971, *Astrophys. J.* **164**, 275.

Sparks, W. M. and Fischel, D.: 1971, *Partition Functions and Equations of State in Plasmas*, NASA SP-3066.

Stecher, T. P.: 1968 in K. B. Gebbie and R. N. Thomas (eds.), *Wolf-Rayet Stars*, Natl. Bur. Std. Special Publ. 307, p. 65.

Stecher, T. P.: 1970, *Astrophys. J.* **159**, 543.

Thomas, R. N.: 1968, in K. B. Gebbie and R. N. Thomas (eds.), *Wolf-Rayet Stars*, Natl. Bur. Std. Special Publ. 307, p. 2.

Underhill, A. B.: 1959, *Publ. Dominion Astrophys. Obs.* **11**, 209.

Underhill, A. B.: 1962, *Astrophys. J.* **136**, 14.

Webster, B. L.: 1969, *Monthly Notices Roy. Astron. Soc.* **143**, 113.

Westphal, J. A., Sandage, A. R., and Kristian, J.: 1968, *Astrophys. J.* **154**, 139.

Wolf, C. J. E. and Rayet, G.: 1867, *Compt. Rend. Acad. Sci. Paris* **65**, 292.

Wray, J. D. and Corso, G. J.: 1971, *Astrophys. J.* in press.

DISCUSSION

Kuhi: It is very dangerous to extrapolate the conditions that we have in a binary system, such as V444 Cygni, γ_2 Velorum, and so on, to the single stars. I think that it is the fact that the Wolf-Rayet star is in a binary system that causes all or most of the peculiar structure, intensity fluctuations, etc., that we see. This is not to deny that there could be iregularities as well in the single Wolf-Rayet stars, but observations on that point are sadly lacking. The reason I feel this is that the O star in the system is a very hot object that is very close by, and clearly must influence the ionization structure, etc. of the Wolf-Rayet star. The observations also suggest flows of material from the Wolf-Rayet star to the O star, and perhaps streams in an outward direction as well. I just wanted to make a little bit clearer that those particular observations refer to the binaries and that I am not sure that we can really safely extrapolate them. Then, just one question: the suggestion of the rings with the WN stars and the nuclei of planetary nebulae being somehow related in an evolutionary sense, raises an interesting point. The nuclei of planetary nebulae that are Wolf-Rayet stars, seem to be primarily WC stars; I think without exception, is that right?

Smith: There is just one exception. M1–67 has a WN8 nucleus.

Kuhi: We know that the ring objects are entirely Wolf-Rayet stars, so that if we are going to put planetary nebulae into the same category as the ring objects, are you implying some sort of evolution from WC to WN, or not?

Underhill: That follows as a deduction, yes.

Kuhi: I wonder what Paczynski might say about that.

Underhill: What Paczynski made clear to us is that the changes in the chemical composition in the inner part of the stars during their evolution, make it possible that you can have WC in two different stages of the star's evolution, as you uncover deeper and deeper parts of the star.

Paczyński: It is quite possible that there is an evolutionary sequence. I am not prepared to discuss this as the problem appeared just at this Symposium and I would rather spend some time looking at those objects, before I comment upon them.

Thomas: It seems to me that what you were willing to discuss is that point that Lindsey Smith and I had some interchange about: the fundamental difference between the Wolf-Rayet star in the planetary nebulae and that one which was not in the planetary nebulae; that is, the small mass and the big mass. The biggest item in your mind is, maybe there is not a difference in mass; at least, if I understand well, it is open to question whether there is really a difference in mass. Is that right?

Paczyński: What I really had in mind, when I spoke a few days ago about these problems, was that we know fairly well masses and luminosities of those Wolf-Rayet stars which are the components of binaries. The evolutionary status of those stars and the population assignment does not lead to much doubt. However, we have two other classes of objects in which the Wolf-Rayet type spectra are found. These are: nuclei of planetary nebulae and central stars of ring nebulae. If we assume that these objects are related to each other rather than related to either normal planetary nebulae or to Wolf-Rayet binaries, then we have to start from the very beginning. Their luminosities, temperatures, masses, their evolutionary status was assumed to be known because those objects were believed to be related to planetary nebulae and binaries respectively. And now, if we assume that they are related to themselves we just cannot say anything about them, at the moment.

Thomas: Can I summarize what you are not willing to say about them? One, you are willing to say, for those masses like 5 plus or minus something, where the Wolf-Rayet 'class' lies on the evolutionary scale, you are not really willing to try to make a fine distinction between different WN types or between the WN and WC types, other than a very qualitative speculation. Is that correct?

Paczyński: Sorry for making so much confusion! I believe that as a result of the discussion we had here a few days ago, our knowledge about the nature of central stars of planetary nebulae and central stars of ring nebulae which show Wolf-Rayet type spectra is smaller than it had been one week earlier. And, therefore, I would not like to comment about the evolutionary status of those stars, their masses, or their population assignment. You have to distinguish the stars which do show nebulae around them from those stars which appear in binaries. As far as I know, there is no star in common to the two categories. And, therefore, I would not like to discuss the evolutionary nature of the stars that are in nebulae. Previously, I thought that some of them are like nuclei of planetary nebulae, and then their evolutionary status would be very clear. And I thought that the others are related to massive Wolf-Rayet binaries, and their evolutionary status would also be more or less clear. Now, if we are willing to put the two kinds of stars which show nebulae around them together, you have to break their relation to the Wolf-Rayet binaries and 'normal' planetary nebulae, because Wolf-Rayet binaries and 'normal' planetary nebulae have nothing in common with each other. Therefore it is very difficult for me to say anything about the evolutionary status of those Wolf-Rayet stars which show nebulae around them, until I convince myself that these are either similar objects or dissimilar objects.

Thomas: You talked perpendicular to what my question was! If I decide to divide WR stars into 3 classes, those whose mass is around 10 (which we only know from the binaries); those single stars which have some nebulae around them, and those single stars, central stars of the planetary nebulae, then what I asked was: looking only at these, if I understood what you said the other day, you were willing to discuss where, on the evolutionary scale of a massive star, the Wolf-Rayet spectrum phase is liable to occur, namely when it becomes roughly a helium star. That is what you associated this spectrum with.

Paczyński: I would like to stress again that no stellar interior calculations can provide you with spectra.

Thomas: All right. Let us associate you with what Lindsey Smith said. In some way, by some train of logic, without worrying about right or wrong, she suggested that possibly the Wolf-Rayet atmospheric phenomenon is associated with the central star, which is a helium star. Now, once you adopted that, then you were willing to discuss the evolutionary configuration which contributed the helium star. That is all I have said.

Paczyński: Yes.

Thomas: Now, if I go off again by some chain of logic, from Lindsey Smith or anyone else, and say, of these Wolf-Rayet stars, there is a set of WN's and there are some WC's, then I can make some conjecture about what this may or may not mean. You are willing to again engage in some kind of conjecture; you say, all right, there is a helium star, maybe there is a phase of C burning also, and that would make some kind of a behaviour on the surface. I understand; you do not want to get involved with the atmosphere, you only get involved with the interior. So, you are willing to discuss where the helium configuration occurs, and something before that and something after that, in terms of the change in effective temperature or luminosity or something like that. Is that a good way to put it?

Paczyński: I am not sure.

Thomas: You are awfully cautious on the last day, as compared to the earlier days!

Paczyński: No, I believe all of my discussion referred only to the Wolf-Rayet binaries.

Thomas: Let us distinguish, then, two things about a binary: First, for binaries, I know what the masses are, and second, there is a mass exchange problem in them. You have two distinct things. If I had a separate star, whose mass is like 10, then you are willing to say when the helium phase might come.

Paczyński: Yes. There can be little doubt that Wolf-Rayet stars in close binaries are burning helium and are largely helium stars. And this is based on the fact that they have masses around 10 solar masses, and the luminosities above 10^5 solar luminosities, and fairly high temperatures.

Thomas: And a way of getting rid of hydrogen from the outer part.

Paczyński: Right. And I am really in a very different position when we go to single stars.

Thomas: We are just asking conclusively, what you are willing to do for binaries in terms of mass exchange, and speculatively, in terms of single stars.

Paczyński: Yes, as far as binaries are concerned, mass exchange does only one thing, it removes the massive hydrogen envelope from the star. Therefore, the star becomes overluminous.

Thomas: Right. Do we understand that that is your main point?

Paczyński: Yes.

Thomas: Now, if you let me put that aside for one minute, and say – because I keep trying to come back to Lindsey Smith's speculation if you now go to the single stars, then maybe I have two kinds of single stars that I can talk about, and maybe only one. I have two alternatives, either there is a class of single stars with masses like 10 $M_\odot$, that are like the Wolf-Rayet components of binaries, except that they remove the possiblity of mass exchange. Then, maybe I have stars with one solar mass, which will somehow exhibit essentially a Wolf-Rayet spectrum. Or, maybe, these two classes of single stars are the same, only I just do not know the mass well enough; they may all be mass 10, 1 or 5 $M_\odot$. Now, given those possibilities for the single star, if I have no way of removing the excess hydrogen from the envelope, then you are worried and you do not really want to talk. If you could find some way of removing that excess hydrogen from the envelope, then you would be willing to say that the 10 $M_\odot$ ones, at least, might be like the stars in the binaries.

Paczyński: I will put it in a different way. We do observe that at least some hydrogen was removed from those stars, as we do observe nebulae expanding away from those stars.

I am willing to speculate that those stars lost most of their hydrogen envelope. However, as long as we are not sure whether we can relate those stars to other kinds of stars we know, we are not sure about their masses. As long as we do not know their masses we cannot say anything definite about their internal structure.

A few days ago I thought that some of those stars are burning helium in the core, and others are burning helium in the shell. Today I do not know, because I do not know the masses.

Thomas: That is right. But if you were told by someone that they really had a mass one, then, maybe you would go off into your discussions about the C burning as well as the He.

Paczyński: It would be helium shell burning.

As I mentioned previously, if we believe that there is neutrino emission due to the universal Fermi interactions, then carbon burning is out of question. It lasts for too short a time.

Thomas: That is all for the stellar interior standpoint. Now, when you say you are sure that these things have ejected the hydrogen shell, then there comes the question: I have those Wolf-Rayet stars which do not give us evidence of shells or nebulae, for whatever reason, but there is no real difference in the spectrum, so far as the presence or absence of hydrogen, between these stars and those stars with nebulae. Is that a correct observational statement? All I am really doing is being the devil's advocate, in trying to follow these lines of thought. Paczynski concludes that these things have a mechanism ejecting the outer part of hydrogen shells, because they have nebulae around them, so that the star itself does not have any hydrogen in the surface, and hence can be overluminous. My comment is, these stars, spectroscopically do not really differ that much from those with no shells, with regard to those observed facts that led Lindsey Smith to say that there is no hydrogen in the big Wolf-Rayet stars – not the planetary nebulae ones, but the big Wolf-Rayet stars. So that the common characteristic of those stars, in your mind, is no hydrogen, is that it?

Smith: My hypothesis is that lack of H is a necessary condition for a WR spectrum in both kinds of star. The facts are: (1) the single, Population I WR stars with nebulae are of classes WN5, 6 and 8; I have reported their H/He ratios and they are all very low. (2) The Population I binaries appear to have the same H/He ratios as the single stars of the same subclass. (3) The planetary nuclei are all WC's except one. The spectrum of the only WN nucleus is WN8 and appears to visual inspection, to be identical to the spectra of the Population I WN8 stars, so I presume its H/He ratio will also be low. The WC planetary nuclei have the same problem as the Population I WC stars, regarding H/He ratio determination – I have given my indirect reasons for believing that they have no hydrogen.

Underhill: I think it is essential that a spectroscopic method of estimating the hydrogen to helium abundance with some precision be devised. At the moment we have only guesses. I will say no more about that. The second point I brought up, which I think is important to do, is to determine if those central stars of planetary nebulae with Wolf-Rayet-like spectra, do indeed belong kinematically to population II. This is a straightforward, observing program, though a difficult one, but I think it is essential, because we do not know whether stars of Population I can have such dense envelopes around them that they look like planetary nebulae.

Thomas: Can you really do a kinematic study from five or six stars?

Underhill: For those stars for which you have accurate measured proper motions and radial velocities directly.

Thomas: Would you believe it if you had five or six stars?

Underhill: For each group you could tell whether they are high velocity objects or not. These stars are presumably rather faint intrinsically. Whether they are at the limit of the instrumental capabilities for such measurement, it is very difficult to say. I think it is of sufficient interest to make it a worthwhile problem to look into. It is not easy to get the proper motion; it may be completely impossible, but if it is all possible, I think it would be very worthwhile doing because it is a big generalization that you have been making all the way through, when you say of the central stars of the planetary nebulae that they may be all belong to one polulation. The very few stars we have talked about actually are not spread very much in galactic latitude. So, they may be Population I objects. It is just something to be investigated.

Conti: Yes, I would like to come back to the point Thomas raised. You were worried that these single stars without nebulae do not show any evidence of having ejected any material and, therefore, you say since the spectra looked like those which have ejected material you do not know whether the hydrogen has come off or not. Well, it seems in astronomy that one often takes another field of astronomy for granted. For example, if the radio atronomers tell us something, we say that is all right. It may well be that the reason that they do not have any nebulae, is not so much that they have not ejected or lost all the material, but rather that the conditions around the star, in the interstellar medium, are such that you just do not see the nebulae.

Thomas: Yes, that is the point we raised earlier. That is indeed a point that I really feel one should look very carefully at, both that one and the non-spherical nature of the ejected shells, and I hope that you or somebody would push that point further.

Conti: That is a nice problem for somebody who knows something about the interstellar medium.

Thomas: No, it is a problem for somebody who wants to learn something about the interstellar medium. It seems to me, this is the kind of thing I would pose to the people at Berkeley. Johnson, you are used to this material. Could you sit down right now and measure the ellipticity of the nebulae, and then come up with a conclusion on the differential distribution of the interstellar medium and throw it up to the Berkeley people and say, does this make possible sense?

Morton: One always wonders when you see a non-spherical nebula, how much of it is due to the interstellar medium, and how much of it is due to some effect like the rotation of the star or the ejection mechanism, particularly if the nebula is elliptical.

Smith: (added after the Symposium) Losinskaya has shown that the major axis of NGC 6888 is parallel to the direction of elongation of the Cygnus arm. This implies that magnetic fields restrain expansion along the minor axis – a very simple explanation in this case.

Thomas: All you can do, is examine the possibilities. And I agree, is there some way that one can make some investigation of rotation possibilities here? With these kind of lines?

Underhill: That is difficult, but the idea of uneven ejection, in different rates, of course, is very well documented. That it occurs, was known actually in 1905 or so. Very definitely there were bursts of gas coming out of Nova Aquilae, which were followed for many years. They were not spherically symmetrically distributed about that star.

Kuhi: That happens for most novae, and, in general, one can say that the ejection is not spherically symmetric. And certainly Nova Delphini 1967 is another good case. But, I think that with novae, one is dealing with a much more catastrophic event which takes place on a short time scale compared to what we are talking about here.

Thomas: One should just look at all of these possibilities. Possibly, sitting down and listing these possibilities and investigating them, one by one by one.

Conti: I would like to bring up a slightly different topic now, which concerns the binary frequency for Wolf-Rayet sstars. What I want to make is a statistical argument and only a statistical argument, that it could be that all the Wolf-Rayet stars are binaries, and the arguments that are important are the following: first, if you suppose that you have a Wolf-Rayet system which had equal masses, then you can realize immediately, that you would not observe the lines of the companion, because the Wolf-Rayet star is some 2 or 3 mag. overluminous for its mass. Normally an absorption line star has to be within 1.5 mag. of a brighter companion, otherwise you do not see the absorption lines. So, you see that there is some room for a more massive companion than the Wolf-Rayet star, but let us just suppose they are equal masses. Then, you do not see the companion as far as any absorption lines are concerned. So then, the only other way to detect whether the Wolf-Rayet is a binary, is in variations in the emission-line velocities. And now, of all the binary systems that we know about (some of which have been measured much more carefully than Wolf-Rayet stars that appear to be single, because we already see the absorption lines), the smallest velocity amplitude (2K), in any known Wolf-Rayet system for the emission lines, is about a hundred kilometers per second. Now, it is also true that of all the known Wolf-Rayet binary systems, I do not think any of them have the inclination less than 30°. If you think of random orientations of axes and inclinations, then sin i is also proportional to the fractions which have those inclinations. Therefore, since sin 30° = 0.5 you could also easily imagine that there is another 50% of the Wolf-Rayet stars, that have sin i less than 30°, and, therefore, the maximum velocity range may well be less than 100 km s^{-1}. We have not detected these systems. And Dr. Kuhi has already said something like 60% or 70% Wolf-Rayet stars are already detected as binaries. So that, I would just want to point out that you can make this statistical argument that says they could all be binaries.

Thomas: Except, let us ask Paczynski one question. To be overluminous, a star has to shed a lot of mass. Let us talk about the possibility of any configuration you want to start with, then what can I say in terms of this mass shedding, as to the mass of the secondary, when I reach the Wolf-Rayet stage of the primary? Can they be as small or equal?

Paczyński: If you assume that there is not much mass loss from the system as a whole, then the mass of the original primary, after the mass transfer, is smaller than the mass of the other star, the non-Wolf-Rayet star.

Thomas: What does it do to your theory, Conti? Is that not embarrassing?

Conti: I am not so sure, because I could imagine the situation where you start with a 20 $M_\odot$

star and a 5 $M_\odot$ star, and then the Wolf-Rayet star loses half of its mass. Is that sufficient or not?

Paczyński: Oh, sure, it can lose ....

Conti: And then they each have about 12 $M_\odot$. I mean, that is an example, or start with 25 and 5.

Morton: I wonder if we can turn Conti's argument around. He paints such a pessimistic picture of finding binaries among Wolf-Rayet stars, and we find so many. Is the fact that we are so successful in finding binaries really trying to tell us something about the nature of the systems?

Conti: Yes, I agree that is an alternative explanation. What I think you are saying is, does the fact that we find them so easily, mean that really there has always been such a large mass exchange, that the new primary ends up massive enough so that we almost always see it. Am I right?

Thomas: Right! So, if it is not seen as a binary, it probably is not a binary.

Underhill: No! Suppose your original companion, is a low-mass A5 star and you have this great big mass of a Wolf-Rayet that generates a Wolf-Rayet spectrum at a certain stage as it starts to get rid of matter in a hurry. Now if you only have a little star there to catch the matter, I wonder if the loss from the system might not be considerable. So, when it has caught some mass it still is not big enough to be seen. It is a point I skipped over in my summary. To shorten it, I did not mention the fact that the only companions we possibly do detect are very bright ones.

Thomas: Until one solves the problem of how much mass is captured. Has that been done?

Paczyński: It will not be done!

Thomas: It will not be done within the range of methods that you see at the minute, but always somebody will. Anything which can be done, will be done, it is just a matter of when.

Paczyński: As far I remember, there are single line Wolf-Rayet stars, which do not show O-type spectra, but which do show orbital motion. Is that right?

Smith: That is right.

Paczyński: So, we have observational evidence that there are binaries, whose O-type component is invisible. No matter what is the theory of mass transfer and mass loss.

Sahade: The companion star does not have to be O.

Paczyński: We just do not see it.

Thomas: So, in the argument now, which way are you going?

Paczyński: From purely empirical point of view, there are binaries with only the Wolf-Rayet spectrum visible. Therefore, if such a binary has an inclination of the orbit so low, or the period so long that you cannot detect variations in radial velocities then you have no way to say that this is a binary. So this is in favour of Conti's suggestion.

Smith: Yes, if you had a significant amount of mass loss from the system, which is what you suggested for formation of the nebulae, then the companion would be low mass and very difficult to observe. So, that there still is a very strong possibility that all WR stars are, or were, binaries.

Paczyński: Well, we do observe outflow velocities from planetary nebulae to be as low as 30 km s⁻¹. It implies that if there is a binary, its velocity of orbital motion must be less than 30 km s⁻¹. Otherwise all the matter would be captured by the second component. And that implies that if it is a binary then it is very wide as not to interfere with the mass loss.

Thomas: Except, if it were elliptical, rather than circular.

Smith: Yes, I would not argue particularly strongly that the nuclei have to be binaries. They are quite distinct objects from the ring nebulae and have obviously a different evolutionary history. Stars that become planetaries obviously have their own mechanism for lifting off the H atmosphere; one does not need the binary mechanism.

Kuhi: I would just like to say something about the undetected binaries by making use of the correlation between single WN stars and those found in binaries by Hiltner and Schild, viz., that the line widths in the binaries were narrower than those in the supposedly single stars. In the case of HD 190918, the orbital separation was, I think, the largest of the WN binaries and it has the broadest lines among the binaries. Therefore, I would suggest that the 'undetected binaries' must be very widely spaced binaries indeed, according to this correlation, because the presently single stars, do indeed have much broader lines. Therefore, that would make them very wide binaries, with very low orbital velocities, and consequently we have not found them.

Sahade: I just wanted to say that the criteria that Virpi Niemela mentioned yesterday are powerful means to detect the binaries, in those cases in which it is very difficult to detect them by radial velocity variation.

Van Blerkom: One thing that has not been mentioned is that the mass exchange mechanism should never leave you with two Wolf-Rayet stars. Is it true Virpi Niemela found a binary system in which one star is WC and one star is WN?

Niemela: No, I just suspected it.

Smith: You know, there are two stars which are classified as WN6-C7 and WC7-N6, respectively. I remember that I said flippantly that it looked like a double exposure on the plate, and Virpi Niemela said, I do not know how equally flippantly, that may be it was a binary with a WN6 and a WC7 star.

Thomas: It would be very interesting if it turned out that way.

Smith: It would be rather surprising.

Underhill: There is one star, HD 45166, we have not really discussed; I think that we have mentioned it once. It is a Wolf-Rayet star, that was and was not, and now appears to be some sort of very hot Of type object, plus an A star i.e., a binary. I bring it up because on the low dispersion spectra, which Carol Anger had at Harvard, there were very definitely broad N and C complexes in the $\lambda 4600$ region, and then it changed. The fact that it changed and how the spectrum varied, is being investigated by Sara Heap on some more modern spectra. I will not go into the details, but it is particularly interesting in understanding the significance of the material in the Wolf-Rayet atmosphere, that we did get these changes. It looks much more like an Of star at the moment than a Wolf-Rayet star, casually said. But I think, we are fundamentally dealing with a spectroscopic phenomenon, and a fairly small, maybe, change in density and extension of the velocity field is going to give us quite a different spectrum.

Thomas: I agree, it seems to be one of the most interesting points to push things in the direction which Lindsey Smith summarized. If I remember well, she said it shook her a little bit, to see the tendencies for some of the Of and other central stars of the planetary nebulae, not to differ much from their counterparts in the non-nebular cases. So I would personally be very much interested, as always looking for something which shakes what you are happy with, in this present sort of euphoric state where we think we understand things. If we can, through the planetary nebulae, introduce discord, either by having shown the ordinary stars really have small masses, or that the Of stars blend into the Wolf-Rayet stars in the planetary nebulae, that would be very nice.

Conti: I would like to make a comment on my paper, where I distinguished between a Wolf-Rayet star and an Of star by the fact that the latter had absorption lines which were not violet-displaced. And so, I would ask Anne Underhill, if, by her comment on HD 45166, she means that now you see absorption lines or whether it had to do something about the emission lines widths or strengths?

Underhill: I think the original classification was on strengths and widths of the emission lines. They were objective prism spectra, and my suspicion is that Miss Anger really did not see the absorption lines. The hydrogen absorption lines that are present, are indeed A-type hydrogen lines, like A0, i.e., broad and diffuse. They may very well not have appeared on the objective prism plates. Now things look sharper; when you say the spectrum is an Of on the whole you mean you observe broad hydrogen lines and fairly sharp emission for all of the characteristic lines that you mentioned and no others. It is an interesting point, you see, that the star appears to be a binary, and as far as I remember, Sara Heap was telling me, you can make quite a good argument that one component is an A0 dwarf or just about on A main sequence. That gives you a visual absolute magnitude of 0 to -1, at the most. So, the companion that is producing the emission lines, if this is indeed the proper interpretation of the combined phenomenon we see, must have somewhat the same visual absolute magnitude, which pushes it down into the region of some central stars of the planetary nebulae. As far as I know, there is no particular sign of this star.

Van Blerkom: I would just like to make one request to people who are going to publish observational data, and that is: if usually equivalent widths and line widths, let us say half-maximum, are given, one very useful piece of data that could be given with no more effort, is the central intensity above the continuum of the lines that are observed. It would help a great deal, I think, in the theoretical interpretation.

Sahade: Anne Underhill was talking about ejection velocities when referring to the velocities derived from different atoms or ions. I think the expression is misleading; one should talk about the velocity of a certain layer, rather than the ejection velocity.

Underhill: No, I think one is justified in speaking of ejection velocities, in the sense that those velocities we measure are all greater than any estimated velocity of escape from the star.

Thomas: No, the point that Sahade is making is this. I have the photosphere layer out here, somewhere away from the star. That I observe C IV, with a velocity being produced here, C III with a velocity here and helium out here, does not mean this is an ejection velocity. This means that that is the velocity of the material at this point, and I may have no helium here, because it is all ionized, I can just have C IV. So, the final ejection velocity out here, is maybe 1000 kilometers a second. But the fluid motion down here is maybe only 200 kilometers a second for everything.

Underhill: Well, it is an outward-directed motion. Perhaps, I misled people; instead of saying the outward-directed component of motion, I just used one word, ejection. It appears to have led to a misunderstanding.

Sahade: They may not be increasing all the way out, you see.

Underhill: They may not be, no, but I really meant the outward-directed component of velocity.

Thomas: Bappu, do you want to make the remarks that you made to me in private about this point now?

Bappu: I thought Anne Underhill had covered it fairly sufficiently, but I think is is rather important to realize that if these violet-absorption edges can be measured for velocities, there are two possibilities that come through: it is obvious that the O VI ions have much lower violet-edge velocities than the He I atoms, and one infers from it, that perhaps it is indicative of a variation of velocity with the excitation conditions. On the other hand, parallel to this variation of excitation possibilities, there is the possibility of a variation in atomic mass, and you have the heavier ions, slower-moving and the lighter ions, faster-moving, a result of mass transfer, kinetic energy being converted into velocities. But I do think that such a possibility has to be kept in mind, rather than to make a rush for the ionization potential-velocity correlations.

Thomas: It is just interesting that the ionization and mass go the same way in this sense.

Kuhi: If indeed, that is what you are going to do, how do you explain the behaviour of the C III 5696 line? That was the one that I talked a little bit about with regard to the way it changes from a round-topped line in WC9 to a completely flat top line in WC5. How do you account for that?

Bappu: My feeling is that, depending upon the spectral sub-class, you have certain characteristic overall widths. Now, if you assume an extremely simple procedure, which I shall develop from core-expanding shells, that yield a set of individual profiles (Bappu, M. K. V. and Menzel, D. H., *Astrophys. J.* **119**, 503, 1954), the net profile which is the summation of the individual contributions gives you the observed profile. In other words, I visualize that you have a series of concentric shells of different velocities. The picture that you get finally is defined by the extent of the limits of the final velocity and the initial velocity. Once you bring this initial velocity close enough to zero your extent of the flat-top gets close enough to zero, and, therefore, you can build up almost any profile which looks like a Gaussian by having a lower limit of velocity which is close to zero. And in a higher excitation star like HD 165763, your velocities are much larger.

Kuhi: I do not understand that.

Underhill: I think that part of the problem is C III 5696. In the case of the WC8 Campbell's hydrogen envelope star, it is sort of at maximum excitation. We get from that star what we think is a low excitation spectrum on the whole and, therefore, the major part, which I call the dense part, probably is forming most of that C III. There would be some drifting out. That can exist over certain temperatures; it depends on the rate of cooling outwards. So, whether it is quite flat-topped or not, I am not sure. I have a feeling that it has a small flat-top. What is the difference between that and a slightly round one? Because in the early WC's, you get in the very broad lines the flat-top, but then the only time that you have C ions around, so that you can get C III 5696, would be in the outer regions, where the temperature-drop inside would be such that the ionization ratio did not favour production of C III lines. So, although your line-of-sight, in principle, integrates through the whole atmosphere, it would pick up nothing from the inner part. The only part it picks up is from the exosphere. The exosphere, by definition of being an exosphere, has this thin shell of outward-directed velocity, which is the geometrical construction one requires to get a flat-top line. I think what you saw, what you demonstrated, was a nice demonstration of facts, and perfectly straight forward.

Paczyński: It was mentioned here a number of times that perhaps the observed different velocity of outflow of different elements might be due to different flow velocities. One can verify such a hypothesis by means of very simple calculations. If we are going to have the different velocities of outflow for different elements then the density of the medium must be small, so that the flow with different velocities is possible. We have to produce the emission lines in the same region. As the density is low and the observed emission lines are strong we need a very large volume and mass of the line-emitting region. It is very difficult to understand how, under so low density conditions the absorption edges can be formed. I am sure that we have a continuous flow, and the different emission lines show different expansion velocities because they are formed in different regions.

Underhill: That is very likely, so that they are formed in points above where the absorption lines are, and you can only see a few and those few are the sort that come from levels that would be strongly populated, in a low density, low radiation, standard dilution effective. You do not see all of the lines, even if they have large f-values.

Van Blerkom: If the molecular weight determines the velocity, then I do not see how you can get rapidly moving He I and slowly moving He II.

Underhill: I do not think you do. I think it is a question you can only observe He II when the temperature and density are such that you have enough atoms around sitting in the level with $n = 4$ to give you the absorption line. If you do not have enough, you do not see it.

Kuhi: A point that I think is really important, concerning this last discussion, namely Anne Underhill's and Bappu's approximate correlation with the square root of one over the molecular weight, and that is this: there is no doubt that there is a good correlation between the line widths, say, of N V, N IV and N III, in the sense that N V is narrower than N III. Now, the velocity that you assign also includes a temperature term; it goes as the square root of the temperature over the mean molecular weight. Therefore, if you are looking at a region of high temperature, the velocity should be high; therefore, the velocity that you observe for N V should be large and the velocity for N III should be small, i.e. the correlation goes the wrong way from what is observed. Therefore, the hypothesis warrants no further discussion!

Smith: And the absorption lines go the same way as the line widths, so, the N V absorption lines have lower velocities than the N IV, and this is the best and longest observed correlation in the history of Wolf-Rayet stars.

Thomas: I am glad for the details, but the main thing is somehow we must make sense out of this point Paczynski raised. How do you get it out?

Underhill: That point may come under discussion at the IAU Colloquium which is going to be held at Goddard in February. Stellar Chromospheres is the name, purposely chosen to cover all sorts of highly excited outer atmospheres.

Thomas: I, personally, have gotten an enormous amount out of this Symposium; I have learned very much from it. The biggest thing I have learned is that when I came here I was very worried that this was only a couple of years since we had the Wolf-Rayet Symposium in Boulder, and I worried that if we concentrated mainly on the Wolf-Rayet stars, with only side aspects of other things, it would be a re-hash and repetition. I had thought that maybe what one should do is extend it to the hot stars with just casual, passing references to the Wolf-Rayet stars. I am delighted to have been so wrong, both in terms of lots of the things that I thought I knew about what the Wolf-Rayet stars were, and in terms of the organization of the Symposium. It is very clear that what we got from Boulder, was a look at a subject which had been lying dormant, but was slightly stirring in recent years. Problems were posed, uncertainties were raised, and lots of interesting questions were thrown open. Then, in the years since then, people have worked actively on these problems, and have come up with lots of ideas which they brought here. In the course of this week's discussion, a lot of these ideas which have jelled to a very considerable extent since Boulder, jelled even more. I keep hearing from all of you as you stood up: "Well, I had a thought and now, in the last few days, it has changed", even Anne Underhill, who told me when she came in, she had to work very hard to write her summary before the meeting started, in terms of what people should better say, or, if they had not, she was going to fix it. But, what she said today, had no semblance with what she told me she was going to say.

So, it seems to me that this Symposium has taught us two things: one, how to think in terms of what you ought to have, in terms of what the results were, and, second, how to organize the thing. One, the specific developments on the Wolf-Rayet and the hot stars, and second, the lesson on how to organize such a symposium as this. On the Wolf-Rayet atmosphere, I break it into three parts, in my usual attempt to over-simplify, the bottom, the middle, the top. And you will forgive me if this is highly abstracted. At the bottom of that atmosphere, we worry about T_e, and that is why I was trying to push Paczynski and Lindsey Smith so much in the first few days, because it seems to me, that what you have done here is to talk in terms of stellar models and stellar mass ejection and the like, as a basis for specifying what the effective temperature is. And then, in the middle, what we really had from all the things that Kuhi, and the very good summary that Van Blerkom gave about profiles, is a real resolution of this stratification program. I am delighted to have been so wrong in what I had thought before, not only in details, but in terms of general philosophy. We have the ionization direction and the velocity direction all fixed now, finally and unambiguously. If you remember I keep saying one should use the Sun and the Wolf-Rayet stars as extreme examples of stellar chromospheres; now, of course, the point from all of this is that you should use the Wolf-Rayet stars as an extreme example of what I call corona and exosphere, rather than a chromosphere. So, for how you use the Wolf-Rayet stars, in addition to what they are, that is very interesting.

And then the lower middle atmosphere is something which we have not debated, a region where we must have a temperature rise from the photosphere, if any of these conclusions are correct, from something like $T_e = 10^4$ K in the photosphere to $T_e = 10^5$ K in the corona. So, here, we have, literally just zero empirical data on the details of the temperature rise. It is interesting here, if one does indeed have such a rapid rise it may well be that the reason that I do not see it, is that I have such a small atmospheric extent for the region of the rise. So again one stresses the rocket ultraviolet where maybe I have enough optical depths to probe this region. Now, then, at the top of the atmosphere occurs the mass-loss problem, and we have just had a debate on that, so there is no point in going into it again. The big point here is: can I produce it by radiation pressure? Can I produce it by the boiling off processes. What is it? And I just think that Paczyński summarized it so well the other day that there is no point in belaboring it: namely, we do not know.

And so, it is a problem to do much thinking on. I do think that my old idea about the Sun and the Wolf-Rayet stars representing the extreme range in the non-classical atmospheres, is reinforced. So that one again looks at the Wolf-Rayet stars and the Sun as extreme examples, aiming always at getting a model atmosphere which applies to all stars. I make you a wager, now, that that kind of a model behaves like the following. I have a photosphere, then I have a temperature drop, the decrease in the temperature coming in the standard LTE model, then there is the rise, coming not from the mechanical energy we put in, but the change in what fixes the temperature from a collision-dominated quantitative radiation field to a photo-ionization dominated qualitative radiation field. And again I emphasize, go look at what John Liebecker, Bob Stein and others are doing now, because they conclude you do not really get mechanical heating, you do not get the filter action coming, until you get the temperature upturn. I am personally, trying to establish the thesis, that this upturn comes independently of any heating, the heating only comes higher after the upturn, and you then have a very rapid rise of heating, as is empirically shown in the Sun and seems to be empirically shown, by default, here in the Wolf-Rayet stars.

Smith: If you have any temperature drop above the photosphere, would you predict that we would see some absorption lines?

Thomas: What you see depends very much on the optical depth. And that is why I like to use the Sun, the Wolf-Rayet stars, and all things in between, to map a range of the kind of things you ought to see, but not really need an empirical guide to develop the theory.

Smith: I understood that the only way you could get the observed spectra, with no absorption lines, is to have a situation where, in the continuum you see only to the temperature minimum, not beyond, being different from the situation in the Sun, where we do see beyond the temperature minimum and, therefore, we do see absorption lines.

Thomas: I can get absorption lines in an atmosphere, but in a standard non-LTE way; it depends on the kind of line. So, the presence or absence of absorption lines, by itself, does not prove anything. Your point is correct in principle, even if I quibble about the details a little bit. What one has to do, is to have indicators which map all this behaviour out. For example, you can show, by computing on a pure hydrogen atmosphere, that you can get a steeper initial temperature fall than in the LTE case before the other non-LTE things take over; then you get the T_e rise. But your point is absolutely right in principle, you just have to map it. That is why I like the idea of using all the ranges of conditions given by considering all stars in order to guide the theory. Well, I do not want to dwell on it; these are just pure guesses. If you miss on one thing, start guessing again, that is the important thing, And so, I am delighted to have been wrong in the direction of the stratification, but this leads to the suggestion that most of the observed atmosphere is a corona rather than a chromosphere. I would like to push the corona-exosphere as focussing on the thinking that Paczyński and Lindsey Smith did. If we can resolve this mass loss, it gives us again a good picture of the stellar atmosphere. Now, so much for the specific details.

Let me come to this organization business, on which I think I learned much from. We had a couple of Symposia on stellar atmospheres in a row, Varena and, then, Nice. The idea was that this was the time to go into aerodynamic phenomena in the stellar atmosphere, after having considered aerodynamic phenomena in the interstellar medium, because atmospheres exhibited the aerodynamic effects at least as well as did the interstellar medium, and at least as mysteriously. But the way we did it was to say, let us look at the whole variety of aspects in two succeeding symposia; at the first one, look at three or four aspects, on the second one, look at another three or four aspects. You have a little bit of overlap, but not the whole two symposia being devoted to the same things. What you taught me here was, let us have two colloquia in succession, and reexamine the same problems.

Pick a subject, get some people together, summarize it, and look at it from all aspects. Go home and work, two or three years, or, better, do not fix it in terms of time scale, but in terms of productivity. Come back, do exactly the same thing as you did before, invite as many of the same people as you can get plus those other people whom you have drawn in by your frenzied activity in the field. Have exactly the same symposium again, and ask your progress. But look at the possibilities of getting other people who do not have resources, of whom nobody has heard of yet; those are the people you want to attract to these meetings. And, then, third, the idea of having large-scale summer institutes (or winter, I do not know wihch is winter or summer these days) but where you have lots of people from within your own locality, and 20 % from outside. We ran a symposium in Yalta a couple of years ago. We had 125 people there, bigger than normal, of them, about 35 were from the outside, 100 were from within Russia, who had never been any place, for one reason or another. It was very useful for them, simply from the standpoint of finding people whom they had seen before to argue with, to discuss concepts and argue.

I am sorry to hammer on these points, I realize it is a cracked phonograph record, but I do think it pays off, as is exemplified by the meeting here. It is small and it would have been nice to have had some more money to bring some more people in from outside, to talk especially, to the people of Buenos Aires. So, Sahade and Bappu, I am delighted to have been wrong on so many things, and so grateful to you for having had this thing for us here.